Mathematics for geographers and plan

Mathematics for geographers and planners

A.G. WILSON
AND
M.J. KIRKBY

Contemporary Problems in Geography

CLARENDON PRESS · OXFORD
1975

Oxford University Press, Ely House, London W. 1
GLASGOW NEW YORK TORONTO MELBOURNE WELLINGTON
CAPE TOWN IBADAN NAIROBI DAR ES SALAAM LUSAKA ADDIS ABABA
DELHI BOMBAY CALCUTTA MADRAS KARACHI LAHORE DACCA
KUALA LUMPUR SINGAPORE HONG KONG TOKYO

CASEBOUND ISBN 0 19 8740220
PAPERBACK ISBN 0 19 8740239

Set by E.W.C. Wilkins Ltd., London and Northampton
Printed in Great Britain by J.W. Arrowsmith Ltd., Bristol

Editorial Preface

Developments in geography during the past twenty years have brought with them a clearer identification of a related group of problems and approaches which distinguish research and teaching in the subject. These developments have practical as well as theoretical significance. There has inevitably been a lag in the emergence of books for undergraduate courses which examine carefully these conceptual and analytical advances and provide a critical review of the relevant literature.

Contemporary Problems in Geography seeks to fill the gap. The series is designed as a whole and its structure derives from that of an introductory course developed by the series editor at the University of Leeds. There will be five sections: Introduction to geographical problems and approaches; Types of analysis in geography; Systems of primary interest to geographers; Processes of adjustment in geographical systems; Planning approaches to geographical systems.

This, the first volume to appear in the series, fits naturally into the second section. Its concern with mathematical analysis identifies the most recent major development in geography (and not only geography). The general relevance of this development for the subject as a whole makes the volume an appropriate one with which to start the series.

University of Leeds William Birch
February 1975

The general editor of *Contemporary Problems in Geography* is William Birch, who moves from being Professor of Geography at the University of Leeds to become Director of the Bristol Polytechnic in September 1975. He was formerly on the staff of the University of Bristol, the Graduate School of Geography at Clark University in the U.S.A., and Chairman of the Department of Geography in the University of Toronto. Dr. Birch is President elect (1976) of the Institute of British Geographers.

Preface

THIS book aims to provide an introduction to mathematics for students of spatial analysis, in geography, planning and other related disciplines. Subjects such as geography have changed rapidly in recent years. One of the most striking changes is the degree of use of mathematical tools in new approaches to theory. The incoming University student often has a major problem in that his grounding in mathematics is inadequate. Thus, we mainly address the student whose mathematical education stopped at age 15 or 16 (with O-level, say), though the student with a higher mathematics background (say A-level) may find the book useful both as a refresher course, and to help to relate his mathematics to his geography. The time may also be upon us when at least parts of such a book will be useful to A-level geography students in schools.

Our aim is to provide a sufficient basis in mathematics so that the student can read much of the new geographical and planning literature (in journals, research monographs and text books) with some understanding. We will have really succeeded if he is led to 'proper' mathematics text books to develop further his skills and to deepen his understanding of the related subject-matter.

The mathematical needs of the spatial analyst are of a special kind. There are many technical details within school mathematics syllabuses (at both O and A-levels) which he *does not* need to know. There are other technically detailed pieces of mathematics not currently in University mathematics courses which he *does* need to know. Perhaps the most important need, though, is to develop an ability to relate real-world systems-of-interest, whether these are cities or land forms or whatever, to tools of mathematical description and analysis.

Given these mixed needs, we have proceeded as follows. We begin with a general account of mathematics in geography, and proceed to a discussion of algebraic variables and system description. We emphasize such connections throughout the book by giving examples wherever possible which relate mathematics to real-world systems. (And in this sense, if the reader follows up the references thus cited, he is provided with an introduction to mathematical geography and planning). However, we have arranged the book as a *mathematics* book! We try to progress in an orderly way from elementary topics to less elementary topics, the latter sometimes relying on the former, but, we hope, not vice-versa.

We have pared down the amount of detail needed to an absolute minimum – as must be obvious since we cover so much ground in one short book – but we do try to provide the basis of understanding of current geographic and planning literature. We must re-emphasize, however, that the reader who wishes to deepen his understanding, or to be an active research worker in this field, must be prepared to take his mathematical education much beyond this book, at least on selected fronts.

Since we are limited by availability of space (and since we have tried to eliminate 'unnecessary detail'), we rarely give the most rigorous possible proofs. However, we do not believe in the presentation of mathematical techniques on a purely 'cook book' basis, and so we always try to give some form of proof. This is essential to adequate understanding. (On the rare occasions when even this is not possible without intolerably delaying a line of argument, we say so.)

We have used the material of this book for teaching second-year geography students on a course called *The mathematics of theoretical geography.* We feel that it could equally well be used as a first year book if overall course structure permitted this. It would be possible to use only the first three, four or five chapters (Part 1) for an elementary course. Part 2 may be particularly appropriate for third-year courses. Exercises have been provided for most chapters and the teacher can provide additional ones along the same lines where necessary. Our own experience in teaching has suggested that student execution of such exercises is essential: there is a world of difference between apparent understanding of an easy text, and actual full understanding as demonstrated by successful completion of exercises.

Our individual interests and skills will be evident in particular chapters. We subdivided the bulk of the writing into Chapters 1 – 5, 9, 10 (A.G.W.) and Chapters 6 – 8 (M.J.K.). We would like to thank colleagues in the Department of Geography in Leeds for help and comments, and the series Editor, Professor J.W. Birch, and the staff of the Clarendon Press for encouraging us to write the book. Our thanks are also due to Catherine Boyce, Ann Errington, and Marje Salisbury for typing various parts of the manuscript and to Gordon Bryant for drawing the figures.

Leeds, January 1974

A.G. WILSON
M.J. KIRKBY

Contents

List of Tables

Part 1

1. Mathematical analysis in geography and planning

1.1. Why mathematical analysis is important

1.1.1. The context

GEOGRAPHY is the study of the spatial structures and processes associated with man, organizations and the physical environment. Planning is a generic term which describes the activity of making plans which, when implemented, will solve some problems, or achieve some goals, or merely, 'improve' some situations. In this book we are concerned with those parts of planning related to systems of interest to geographers – broadly speaking, then, with town planning, regional planning, resource management and environmental planning. It is convenient to begin with a review of the geographers' systems of interest on the basis that such systems must be 'understood' if any planning is to be effective. Thus, geographers and planners have a common interest in good analysis. First, however, we make a preliminary comment on the appropriateness of mathematics to geographical analysis and planning at the present time.

Most disciplines which are concerned with the understanding or analysis of one or more systems of interest make identifiable progress over time towards deeper understanding – to deeper levels of explanation. This process usually begins with the orderly description of the system of interest (and perhaps a concern with classification problems). Deeper understanding is reflected in theories about why the structure of the system of interest is as it is, how it 'works', how it changes, and so on. Acceptable theories are those which have withstood tests – the comparison of theoretical predictions with real-world observations – over a period of time (though we should bear in mind that most 'acceptable' theories are susceptible to refinement and modification in the light of more research). With certain kinds of systems of interest, those which have components or associated processes which lend themselves to quantitative description, deepening understanding is often associated with statistical or mathematical analysis. Indeed, it can be argued (cf. Wilson, 1973–A) that such deepening understanding is associated *first* with statistical analysis, and *second* with mathematical analysis as yet more progress is made.

When a well-developed theory can be represented as a set of mathematical equations, this set of equations in turn represents (albeit in simplified form) some real-world system of interest. This is then known as a *mathematical model* – a term which we will use frequently throughout the book.

It can be argued that geography (and hence associated fields of planning) has moved beyond the period of orderly description, *via* a so-called quantitative revolution into statistical analysis, and into a period when mathematical analysis is becoming both commonplace and fruitful. This argument will be illustrated by example in this book, though in relation to its more general aims, a broader investigation of the geographical literature is needed. One of the main purposes of this book is to provide the key to such literature for the reader whose mathematical equipment is relatively limited.

The systems of interest to geographers can be divided almost unambiguously into those of human geography and those of physical geography, but further sub-division involves alternative overlapping concepts. Human geography is concerned with people, their activities, and their spatial distribution (population geography, social geography), with organizations formed to produce a variety of goods and services (economic geography), and with the use of resources (resource geography, economic geography again, and specialized subjects such as agricultural geography). Spatial analysis is related to various forms of regions, and particularly important units are cities, or urban regions (urban and regional geography). There is also an older (but still important) use of the concept of regional geography as the synthesis of all aspects of the geographical study of a region. Human geographers are concerned with geographical processes in time, and in relation to long periods, such work may be referred to as historical geography.

Physical geographers study landforms and the associated processes of sediment transport (geomorphology); the distribution of soils and processes of soil formation (pedology); and the spatial relationship of plants and animals and their interaction with soil (biogeography). They are also concerned with the pattern of climate on a world and local scale, and the processes which influence climate, especially in relation to the soil and to the hydrological cycle (climatology, micrometeorology). Water is an important component of most physiographic systems and the patterns and processes governing its movement are therefore also of critical interest (hydrology).

There are potentially strong interactions between these two major fields of geography. Broadly speaking, we can say that the structure of human activity in a region is in part determined by the physical environment in that region. The economic structure is partly determined by the distribution of natural resources. Water resource management is a function of climate and the structure of the local river network. The interdependence is probably strongest in less developed societies. In countries with advanced technologies, man's control over major parts of his environment in cities to a major extent decouples (at least for analytical purposes) the human and physical systems. It is now increasingly argued, however, that in advanced technological societies, there are strong interconnections, the study and planning of which have been neglected, and that this neglect will lead to serious repercussions – at the extreme, the so-called eco-catastrophes. Geographers, and others, therefore, are increasingly involved in the study of this kind of ecology.

It should now be clear how different kinds of planning relate to this great variety of systems of interest. Town, or city, planning is an art of long standing; much of its scientific basis depends on the geographers ability to achieve an adequate analytical understanding of urban structure and development processes. Regional planning – where 'region' is usually considered to apply to an area considerably larger than that of a single city, though still a subdivision of a country – is a more recent activity. Connections between regional planning and geography are similar to those between town planning and geography (though involving a different spatial scale). Resource planning (or management) is usually related to the particular resource, such as water. The various forms of public planning are associated with organizations of government, either as a subdivision of a local authority with wide interests (as with town planning) or a body whose prime concern is that aspect of planning as with a Water Authority. The particular forms of institution are not our main concern here: we concentrate on the analytical basis of planning, and some planning methods.

Before we go on to discuss mathematical analysis in these various fields of geography and planning, it is useful to refer to some other interests of the geographer or planner. First, all geographers have certain common interests: most use the map as a descriptive tool, for example, and, increasingly, many use aerial photographs. Thus, certain mathematical methods are necessary to the geographers which are associated with such fields as cartography, surveying and photogrammetry. Second, it will be clear from the descriptions above of the geographer's wide ranging systems of interest that he is forced to take a deep interest in a number of associated disciplines. Cities and regions are studied by economists, for example, landforms are studied by geologists, soils and vegetation by biologists. Thus, the geographer not only needs the mathematical equipment necessary to the study of his own subject, but also that which is necessary for the study of 'adjacent' disciplines.

1.1.2. Levels of resolution

Two further preliminaries remain for discussion before we can proceed to describe the range of application of mathematical analysis in geography. First, we discuss the *level of resolution* at which the various systems of interest can be viewed. Second, in subsection 1.1.3 we go on to discuss various general aspects of geographical *methodology* which can be related to several systems of interest.

Consider the study of residential location within urban geography. What levels of resolution could be adopted? At the finest level, we could focus on an individual household and, in principle, analyse the residential location of each household in the city in turn. At this scale, the level of *spatial* resolution would be the household address. At the most coarse level of resolution, we could simply note the total number of households in the city and the total number of houses. The whole city is being taken as a single spatial unit. Clearly, as we move from the finest level to the coarsest level, there is a tremendous loss of information. However, we also have to recognize that in practical terms, we may not be able to cope with the volume of information available at the finest level, much as we might like to. Thus, the appropriate level of resolution for any piece of geographical analysis will be determined partly by information needs (not to be finer than necessary) and partly by practical considerations (the analysis must be feasible). Sometimes, a compromise between these two kinds of objective is needed.

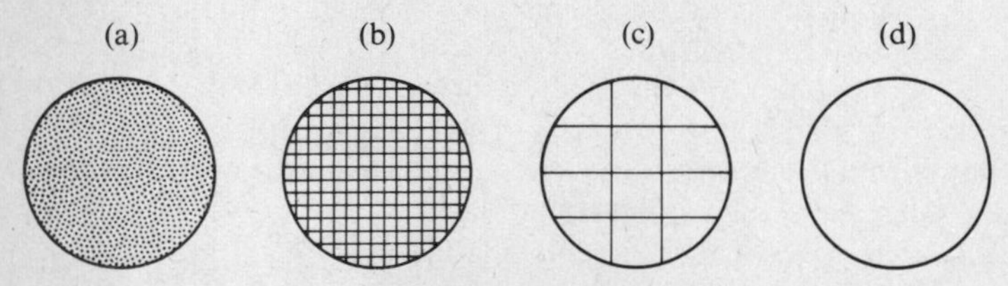

FIG. 1.1. Spatial levels of resolution

Then, there is likely to be a whole spectrum of levels of resolution for any particular piece of analysis. Examples in relation to spatial resolution and residential location are shown in Figure 1.1. Figure 1.1 (a) shows individual households located at their actual addresses; 1.1 (b) shows a zone system, involving a fairly large number of zones, and we may record numbers of households per zone; 1.1 (c) shows a coarser zoning system; and 1.1 (d) shows the coarsest level of spatial resolution where the whole city is shown as one unit – a single zone. When the individual component of the system of interest is identified, this will be referred to as a *micro*-scale level of resolution. The coarsest level of resolution will be referred to as a *macro*-scale. The intermediate levels will be referred to as a *meso*-scale (using a term first introduced in a geographical context by Haggett, 1965). Perhaps not surprisingly, much geographical analysis is conducted at a meso-scale.

The reader will find it easy to list his own examples of different levels of resolution (and particularly spatial resolution) in other fields of geography. In economic geography, the study of the firm is an example of the micro-scale; the distribution of some economic activity across the zones of a region, the meso-scale; the study of the total production of some good in a region, the macro-scale. In fluvial geomorphology, the study of water flowing across one cross-section of one river is a micro-scale study, the study of the structure of a river network is at a meso-scale; the study of total water production, say for possible extraction, by the network in a whole system is at a macro-scale. In this discussion, we have emphasized level of *spatial* resolution. There are associated concepts of *sectoral* resolution (related to the level of detail at which system components, their activities or their characteristics are described) and temporal resolution. However, we pursue such further discussion in section 2.2 of Chapter 2, (on methods of system description) in relation to examples used throughout the book. We leave it to the reader to list further examples of levels of resolution for geographical systems, and to give himself a good sense of what the concept of 'level of resolution' means. This is an important exercise, because the mathematical techniques and methods of analysis often differ considerably: the same formal problems can look quite different in mathematical terms at different scales.

1.1.3. General aspects of geographical methodology

There are various general aspects of geographical methodology, many of which are common to the study of a number of different geographical systems of interest. Since the geographer is primarily interested in spatial analysis, there are three main features of any geographical system of interest: *location* (of system components and their characteristics, and of volumes of 'activity'), *interaction* (between points, or zones, located in space), and *network flow* (since most interactions are 'carried' along the channels of a network rather than directly from place to place). To illustrate, we mention a number of examples for each of the three fields. The urban geographer is interested in the spatial distribution of residences and the characteristics of the people who occupy them; the location of natural resources is important; the physical geographer studies the location of types of landforms, soils or vegetation. Interaction phenomena are illustrated by the flow of people making the journey to work from one area to another; or by the flow of water from a reservoir to a city. The routes used by the workers, and corresponding link loadings, or the river system which carries the water, illustrate network concepts.

Theories of these spatial patterns have developed as location theory, spatial interaction theory and network theory. (The three fields turn out to be strongly related, of course). These theories often turn out to have considerable generality and to be applicable in a number of fields. They can be more or less *static* theories, offered as an explanation of pattern (or structure, or system 'state') at some point in time. Usually, they are then connected to some *equilibrium* assumption. It is possible also, in each field, to focus on the process of change, or time development, explicitly.

A variety of approaches can be used to study the development of a system of interest over time. At best, the mechanisms of change can be understood in detail and represented as a set of mathematical equations which form a good model of the system. Lack of such detailed knowledge will often force us to approximate and to build a more crude model. Whatever level of detail is achieved, a dynamic system model will have many features unique to the particular system. However, there are certain types of model which recur relatively frequently, and it is worth mentioning some of these and associated examples.

One particularly common situation occurs when the time development of a system is controlled by 'forces' which, following any 'external' disturbance generate a return to some *equilibrium* position. For example, if for the system of people making their journeys to work, a new road is suddenly made available, the journey to work pattern responds to this and, quite quickly, a new equilibrium position is reached. Similarly, a river system would quickly 'adjust' to utilize, say, a new flood-control channel, and achieve a new equilibrium position. There are mathematical techniques for representing this kind of process. One common situation of this type involves a *maximization* (or *minimization*) process: the traveller responds to the new road by maximizing his utility in the new situation; a river flow adapts itself to minimum work.

Some time-development processes involve the spatial redistribution of system components, and this is often usefully characterized as a *diffusion* process. Examples are population or animal migration. Some diffusion processes may also involve growth as well as redistribution: the spread of technological innovation, or plant species, from the point of introduction.

In other cases, we are mainly interested in the growth or decline of a population (system components) in an area or set of areas. Thus, economic change or a Pleistocene ice-cap could be studied in this way.

In some geographical systems of interest, it is useful to focus on *system inputs and outputs.* Another interesting set of time-development processes can then be characterized as having equal input and output: a manufacturing process, the flow of traffic across a section of a road, or water across a section of a river. These are sometimes called *steady state processes.*

We will see with the range of examples used throughout the book, that this range of types of theory, whether static or dynamic, can be represented mathematically in a variety of ways, according to level of resolution adopted, and according to geographical system of interest. We should perhaps also mention that although we have perhaps implicitly argued that we are dealing with deterministic processes above – and most of our examples will fall in this category – we will also introduce *probabilistic* concepts and *stochastic* models.

Despite the broad similarities of interest and types of process, there are human/physical differences which lead to different emphases in styles of approach. Theory in physical geography is often concerned with processes at a more micro scale. It is at this scale that the laws of natural science must be 'obeyed', and hence this scale may be the appropriate one for much theoretical development.

We can now return to the question of the relevance of mathematics to geography. First, consider location theory in relation to urban geography: various locational quantities, numbers of components (people, houses, jobs and so on) or volumes of activity (retail sales, goods produced), can be easily measured. In a non-mathematical (meaning, in effect, less advanced – theoretical) approach to the analysis of location, such quantities would be mapped, or listed in tables, and verbal explanations developed of what quantities were associated with what others. Although insights can be gained in this way, only a low level of explanation can be achieved. It is rather like understanding that the sun rises every 24 hours, but having no detailed mathematically expressed theories of planetary motion around the sun. Deeper explanation can only be achieved by an exploration of the mathematical relationships between the measured quantities, and the building of theories based on these. This comment applies at two levels of resolution at least: mathematically stated theories of location are available at the micro level – say for the household or the firm – or the meso level – in relation to the distribution of people, firms or activities across some zone system. A similar argument could be applied in relation to spatial-interaction theory or network theory. In the case of location theory and spatial-interaction theory, the mathematical techniques have been developed largely within geography and related spatial-analytical disciplines. In the case of network theory, appropriate tools have been developed in the field of operational research and have found applications in a very wide range of disciplines.

A similar range of comments can be made about the various process-oriented studies. In all cases, appropriate mathematical techniques have been developed, sometimes within geography, sometimes originating in operational research or applied mathematics, and finding application in geography.

There are few aspects of geography, therefore, which it is possible to study in any depth without having the appropriate mathematical tools available. It is hoped that enough has been said here to give some indication of the truth of such a proposition, though the detailed argument rests on the examples of the book, and the geographical and planning literature at large. The reader has to take something on trust in order to be able to acquire the tools with which he can satisfy himself about the overall argument!

1.2. The structure of the mathematical argument

Mathematics is a subject which has to be taught in order of increasing difficulty: later parts depend on earlier parts. The *structure* of our argument is determined by this condition. The *content* of our presentation is determined, as far as possible, by the relevance of each piece of mathematics to geographical application. The result may appear unsatisfactory to a mathematician, but it should enable the geographer to read much of his own literature with some understanding (both existing and to some extent, we hope, future). It should also point the geographer to appropriate mathematics texts for deeper treatments when required, armed with enough preliminary understanding to enable him to tackle such books.

We begin, in Chapter 2, with the concepts of elementary algebra. This is of crucial importance in that algebraic manipulation plays a role in almost all applications of mathematics. It is also crucial in the sense that it provides the basic tools – algebraic variables – which the geographer can use to describe his system of interest, and an important part of Chapter 2 illustrates various ways of doing this for a wide range of examples.

Chapter 3 is also concerned with basic and elementary concepts: functions, graphs and co-ordinate geometry and more algebra. As a step beyond the simple algebraic equations of Chapter 2, this provides the tools for expressing a relationship between two or more variables in a variety of ways, and so permits other kinds of algebraic manipulation. These two chapters complete what is essentially a refresher course in elementary mathematics, with some new concepts being introduced and, most importantly, an attempt being made to provide the reader with the skills with which the concepts of algebraic variables and functional relationships can be used in the description of geographical systems of interest, and for mathematical model building. Indeed, a wide range of important and useful models can be built with these tools alone.

In Chapter 4, there is a slight increase in order of complexity when the main notions of *matrix* algebra are introduced. The associated model building techniques are vital for some examples to be considered, and offer useful forms of shorthand in others.

The basic concepts of both differential and integral *calculus* are introduced in Chapter 5. Although many models can be understood by using the tools described in Chapters 2 and 3 only, there are many examples where the model, or the methods used in the construction of the model, can only be understood using calculus. This chapter should be considered to be part of a basic mathematical education!

There is then a further increase in order of complexity, and perhaps a more substantial one, from the first five chapters to the remaining chapters of the book. Chapter 6 is mainly concerned with more advanced aspects of calculus – differential and difference equations. In Chapter 7, various concepts associated with probability are introduced, and some associated models are presented in Chapter 8 on stochastic processes. We also take the opportunity in Chapter 7, to link the basic concepts of statistics with those of probability to provide the reader with a *mathematical* introduction to other statistics texts.

In Chapter 9, we discuss methods of maximization and minimization, using both the more traditional methods of the calculus, and the newer 'algorithmic mathematics', using such techniques as linear programming. Finally, we recognize that many kinds of geographical model building demand the synthesis of a number of the techniques which have been introduced. We discuss a variety of problem-solving methods on this basis.

The method we use throughout is to present the mathematical concepts and techniques in as straightforward and direct a way as possible, and then to illustrate these with geographical examples. The examples are usually interspersed throughout a chapter, though occasionally they are collected together at the end. In the following section (1.3), we list and describe qualitatively a number of examples (which relate to the broad subdivisions of geography and geographical theory introduced in section 1.1 above). These will be then used as a pool of examples to illustrate mathematical concepts in the rest of the book. In the following description, we note in each case the relevant subsequent chapters of the book, and so the reader can easily check how much mathematical progress he needs to make before he can fully understand various aspects of theoretical geography.

1.3. Some examples to be considered in more detail

1.3.1. Introduction

In this section, we list a number of fields of geographical analysis which mostly relate to the general headings defined in section 1.1, and which we will use as a pool of examples for the rest of the book. Here, of course, we give only a qualitative description. All the examples will be picked up again in section 2.2 of the next chapter, however, when, with the minimum algebraic tools, we will illustrate the principles of mathematical description of geographical systems.

There is inevitably some overlap between the fields covered by the headings used below, and so these should not be considered either exhaustive or mutually exclusive: they simply represent an attempt to identify certain main features for the purposes of this book. The existence of such a wide range of examples does, however, help to substantiate the general argument that mathematical analysis is essential and will be increasingly important in the future.

1.3.2. Micro-location theory 1: land use and rent

It is perhaps appropriate that land use and rent location models should form our first example, as they have their origins in the early nineteenth century in the work of von Thünen (1826). He was particularly concerned with agricultural land use. His work can be understood with a knowledge of basic algebra, used to calculate rents for different agricultural uses of land at different locations. Von Thünen's system of rings of land put to different uses around a market centre is well known. His work in the field of agricultural location was further developed by such workers as Dunn (1954) and Isard (1956), and the same methods were applied to industrial and residential land use – indeed to all land use – by Alonso (1964). The work of these later writers uses the economic theory of consumers' behaviour and can only be fully understood if corresponding constrained maximization methods involving differential calculus (Chapters 5 and 9) are used. It can also be stated in some circumstances as a linear-programming problem (Chapter 9). Thus the basis of this aspect of location theory can be understood using algebraic concepts only, but a deeper understanding needs further mathematical tools.

1.3.3. Micro-location theory 2: plant location and resource use

A second type of location theoretic problem involves the location of a plant and its utilization of resources. This problem also has a long history and was first studied comprehensively by Weber (1909). The situation is: given the locations (assumed to be points) of different kinds of resources, capacities at their locations and the costs of transport, some plant with a given resource-using input is to be optimally located. This is now sometimes known as a location–allocation problem (Scott, 1971): where to locate the plant, and how to allocate to it quantities of resources from various points. This theoretical problem has many guises in different situations, but is mainly studied as an industrial location problem. Weber showed how to obtain a *mechanical* solution, using string and weights! A mathematical solution involves minimization using methods of the calculus (Chapter 5).

1.3.4. Location theory via quadrat analysis

Another approach to location theory involves dividing the study area into quadrats, and theorizing about the probability of finding various quantities of some component or activity in a particular quadrat. This can be illustrated by reference to retailing in human geography (Rogers, 1969) or Karst phenomena in physical geography (McConnell and Horn, 1972). The assumptions involve algebra, as always, and a knowledge of the possible functions which can turn up as probability distributions (Chapters 3 and 7). At a deeper level, the types of distribution which can occur can be related to the underlying stochastic processes which bring them about (Chapter 8).

1.3.5. Meso-spatial interaction and location theory

Our first two location examples were at a micro scale in the sense that they were concerned with individual locators. In this case, we are concerned with, for example, predicting bundles of trips from any one area (or zone) within a study area to any other, and so this is the meso scale of analysis. These kinds of spatial interaction models also have nineteenth century origins in the work of Carey (1859) and Ravenstein (1885) among others. Spatial interaction models fit a wide range of phenomena, but are perhaps used most commonly for person trips for different purposes, and for migration studies. They might also be used for modelling physical flows over space measured at a meso scale, such as water flow (Wilson, 1973–B).

The main types of spatial interaction model are the gravity model (reviewed by Carruthers, 1956, Olsson, 1965, Wilson, 1971) and the intervening opportunities model (Stouffer, 1940, 1960, Schneider, Chicago Area Transportation Study, 1960). It has recently been realized that there is a whole family of such models (Wilson, 1970–A) each useful in different circumstances. The models can be understood with a knowledge of basic algebra and some key functions, such as the exponential and power functions (Chapters 2 and 3), though it should be emphasized that considerable manipulative algebraic skill is required. Understanding at a deeper level requires constrained maximization methods from calculus, either in relation to entropy maximizing approaches (Wilson, 1967, 1970–B) or utility maximizing approaches (Neidercorn and Bechdolt, 1969, Beckmann and Golob, 1971).

We should also note that, in many circumstances, such spatial interaction models function as meso-scale location models – when a locating activity, say residential location is related by some interaction with other locating activities such as workplace location. (cf. Wilson, 1969, for a review of elementary residential location models which have this character.)

1.3.6. Interaction and nodal structure

There is a substantial history in theoretical geography of the study of nodal structure of settlement patterns, particularly using the concepts of central place theory (Christaller, 1933, Lösch, 1940). However, much of this does not lend itself to mathematical analysis directly and we do not take it as one of our main examples. Rather, we consider in Chapter 4 Nystuen's and Dacey's (1961) analysis of an interaction matrix which aims to elicit the nodal structure of a region.

1.3.7. Network theory 1: descriptive concepts

The study of networks, and their mathematical representation, usually involves rather different methods to the ones mentioned hitherto. Classification schemes and length-order relationships for river networks (Horton, 1945, Strahler, 1952) can be handled, in the main, using the algebraic concepts of Chapters 2 and 3. It is also sometimes convenient to represent network structures in matrix form (Kansky, 1963), as we note below in Chapter 2. If a deeper understanding is needed of Horton's laws, using the concept of T.D.C.N.'s (*t*opologically *d*istrict *c*hannel *n*etworks, Shreve, 1966 reviewed in Werrity, 1972), then knowledge of stochastic methods is needed (Chapter 8) based on the techniques of combinatorial analysis (Chapter 7).

1.3.8. Network theory 2: shortest paths and flow loading

One of the most important problems in network theory is that of finding the shortest path from one node to any other node in a network. Algorithms have been developed which solve this problem (Dantzig, 1960, Moore, 1959, reviewed in Scott, 1971), and these have a wide range of application in geography. An obvious example is the set of shortest paths in transport networks. Bundles of trips can then be loaded on to shortest paths and network link loadings calculated. The various methods are outlined in Chapter 9. There are other kinds of network minimization problems: for example, the routing of water through a hydrological network (Calver, Kirkby and Weyman, 1972) which involves differential equations using the methods of section 6.7.13.

1.3.9. A note on the concept of 'equilibrium' and 'maximization' in time development of geographical systems

Most of the examples cited so far are static, in the sense that time development is not explicitly modelled. There is often an equilibrium assumption implicit in such models: that if the system remains undisturbed, then the variables will continue to be related as in the equations of the static model. If the system is disturbed – say a new motorway is opened in a transport system – then some parameter of the model can usually be changed and the new equilibrium can be calculated. This tells us nothing about the path taken by the system to *achieve* this new equilibrium. If the time taken to reach the new equilibrium is relatively short, then such 'transient' effects can be neglected, and we can proceed with 'comparative static' analysis. If not, then a fully dynamic model is required. Thus, some of the models in the example fields mentioned above can be used to represent time development under suitable conditions. Further examples of explicitly dynamic models will be mentioned below.

It is also convenient to comment on 'maximization processes' at this point. Many of the static models result from an assumption that some quantity is maximized (or minimized): utility, profit, rent, work done, and so on. If the system is disturbed, the new equilibrium is calculated so that the quantity is maximized in the new situation. This mechanism provides the 'forces' by which equilibrium is achieved and maintained.

1.3.10. Diffusion processes

An important set of time-development processes in geography can be associated with the concept of diffusion. This usually involves the spatial re-distribution of some components or activities, perhaps emanating from one or more centres. There may also be some population growth in addition. Diffusion of heat into soil (Geiger, 1965), water into soil (Kirkby, 1969) and air pollution (Gustafson and Kortanek, 1972) are examples of diffusion without growth. The diffusion of innovations (Hägerstrand, 1958; Gale, 1972), disease (Bailey, 1957, 1964) and population – as migration (Morrill, 1958) are examples of diffusion with growth. Simulation models can be built of such processes, using the concepts of algebra and elementary probability (Chapters 2, 3, and 7), and also, differential equation models (Chapter 6), or stochastic models (Chapter 8).

1.3.11. Growth and decline: accounting and time development equations

Human geography is almost evenly divided between the study of populations at various scales, and the study of economies and their products (Stone, 1966, Rees and Wilson, 1973, Wilson and Rees, 1974). The time development of such systems can be studied using the concepts of *accounting:* we have to account for system components at the beginning of a period and at the end and to note any change of state. *Rates* of state changes can be calculated, and these used as a basis for projection. For discrete time periods, models can be developed using rates with only the concepts of algebra (Chapters 2 and 3) or matrix algebra (Chapter 4). In some models, time appears explicitly and the models are formulated as sets of either difference equations or differential equations (Chapter 6 though treated below in elementary fashion in Chapter 5) according as to whether time is treated as a discrete variable or a continuous variable. The argument can be extended to embrace the study of animal and plant populations and their spatial distributions (McArthur and Wilson, 1967, Watt, 1968, Metcalf and Eddy, 1972).

Yet another kind of time-development problem which needs somewhat different treatment is the growth of a network. This can be studied in relation to river networks (Haggett and Chorley, 1969) or transport networks (Fine and Cowan, 1971), for example. Stochastic modelling can also be used, as in Chapter 8.

1.3.12. Process studies: flows and mass or energy balance relationships

There are a number of processes with similar characteristics in physical geography which are governed by mass or energy balance relationships – also types of accounting equations. Most of them use a particular differential equation as their basis – the continuity equation. Different systems are represented by different kinds of solutions of this equation (Chapter 6). The range of phenomena which can be modelled in this way include overland flow of water (Kirkby, 1969), hillslope processes (Kirkby, 1969, 1971), and channel flows within a river (Calver, Kirkby and Weyman, 1972).

A similar set of studies of relevance in climatology and ecology can be associated with the concept of energy balance. The most widely used equation is the balance of solar energy at the Earth's surface (cf. Geiger, 1958). In its simplest form, this can be understood using simple algebra, but a fuller analysis involves more complicated functions and the use of differential equations.

1.3.13. Miscellaneous examples

The examples introduced above all relate to major fields of theoretical geography (as outlined in section 1.1) and have applications within several sub-disciplines. It is also worth emphasizing that there is a very wide range of possible applications of mathematics outside the systematic framework – relatively minor, but important, *ad hoc* applications. We will use such examples as appropriate in the rest of the book, and we mention a number of them below.

In Chapter 2, we show how algebraic variables are related in equations. There are many examples of such relationships. Among those we consider at various points are:

(i) trip generation equations (Wilson, 1974, Chapter 9)
(ii) modal split equations (Wilson, 1974, Chapter 9)
(iii) various hydrological relationships – flood frequency analysis (Dalrymple, 1960), ground water (Waltz, 1969).
(iv) scree slopes
(v) linear equations expressing resource capacity constraints.
(vi) sums of geometric progressions as used in d.c.f. calculations for investment appraisal.

Some of these relationships can be exhibited graphically to illustrate the co-ordinate geometry of Chapter 3.

In Chapter 3, we describe the standard functions, and the range of functions which can be constructed out of these. Examples are

(i) power functions in spatial interaction models
(ii) exponential functions in spatial interaction models and elsewhere – exponential growth, decay in time, and so on.
(iii) log function as a slope profile.
(iv) trigonometric functions are used in many contexts: in surveying and photogrammetry, angles of scree slopes, and so on.

These are perhaps the most commonly used functions. Others can be formed as composites of these. Examples are the logistic function (in all kinds of choice models, such as modal split in transport studies), the lognormal functions (income distribution – Mogridge, 1969, distribution of species – MacArthur and Wilson, 1967), exponentially damped power functions (Tanner, 1961) and so on. Two areas which generate a good range of examples of functions are velocity fields (Angel and Hyman, 1970, 1972) and slope profiles (Kirkby, 1969, 1971).

Numerous minor applications illustrate the basic concepts of the calculus (Chapter 5). The differential calculus is used in setting up the underlying assumptions of the intervening opportunities' model, error propagation in photogrammetry (Hallert, 1960). The integral calculus is illustrated by the consumers surplus formula in transport planning (Tressider et al., 1968), the estimation of urban populations for given density curves (Rees, 1973), the integration of simple differential equations such as that which constitutes the underlying assumption of the intervening opportunities model, and the continuous-space spatial interaction model (Angel and Hyman, 1972).

Exercise

1. Write an essay, using examples other than those cited in Chapter 1, on different levels of resolution for the description of geographical systems.

2. Elementary algebra and system description

2.1. Notation and the basic algebraic operations

We assume that the reader is familiar with the basic operations of ordinary arithmetic. Numbers (which may be integers, or expressed using fractions or decimals) may be added, subtracted, multiplied and divided. The most elementary algebra consists of forming expressions using the same four basic operations on letters which stand for numbers. Any letters, usually from the English or Greek alphabets, can be used in this way, and these 'variables' can be combined using the basic operations. If x and y are two variables, we can form the expressions $x + y$, $x - y$, $x \times y$ (sometimes written $x \cdot y$ or, simply, xy) and x/y $\left(\text{or } \frac{x}{y} \text{ or } x \div y\right)$. If x is given the numerical value 4, and y the value 2, then the four expressions have values 6, 2, 8 and 2 respectively. Letters used in this algebra are sometimes called 'constants' or 'parameters'. This means that, although a letter is used, the value is supposed to remain constant for the particular situation being analyzed – though it could be given another value when the same piece of algebra is used for another situation. Alternatively, an actual number may be used in an algebraic expression – as $x + 2$ for example.

It is important to recognize at the outset that the notation which is used for a particular problem is determined merely by convention. *The reader should accept that he or she should never be the prisoner of a particular notation:* it is always possible to change one; it is always possible to read about a mathematical technique or a geographical model in one book, and the same thing in another book but described in a different notation, and to understand that they represent the same thing.

The basis of elementary algebra is the equation. We simply note at this point the existence of the sign '=' and that

$$x = y + z \tag{2.1}$$

for example means that 'the number x is equal to the expression $y + z$'. We shall discuss many examples of the manipulation of equations in section 2.4 below.

The applied mathematician often finds that he runs out of letters for the set of variables he needs for a piece of analysis quite soon. For this reason, he uses subscripts or superscripts to define additional variables. Thus, x_1, x_2 and x_3 may be three such variables (using subscripts) or y^1, y^2, y^3 three more (using superscripts). Subscripts are somewhat more common than superscripts as the latter could just be confused with powers applied to single letter variables (as will become clear later). At this stage the subscripts and superscripts used in this way are simply *labels* used to define variables though we will see in Chapter 4 that they may have greater significance. Note, incidentally, that such labels are nearly always integers.

It is possible and useful to move to a higher level of abstraction and to use letters as subscript or superscript labels. Thus, the sets of variables defined above could equally well be defined as x_i, $i = 1, 2$ or 3; y^i, $i = 1$, 2, or 3. Other letters could be used for the subscripts, and if the numerical range of these subscripts is the same, then this does not change the variable definition in any way. For example, we could define x_j, $j = 1$, 2, or 3 or x_k, $k = 1, 2$, or 3. The *reality* is the set of variables x_1, x_2, x_3 obtained when the subscript letter is given its numerical values.

Any letters can be used as subscript or superscript labels, though it is quite a common convention in many fields of applied mathematics to use *i, j, k, l, m* or *n* where possible. However, these letters are also commonly used as variable names, and so it is important to make definitions very clear in one's own work, and to understand the definitions used by an author.

It is possible to use algebraic expressions as subscripts. Thus, the three variables x_1, x_2, x_3 could be written as x_{i+1}, $i = 0, 1, 2$ or x_{i-1}, $i = 2, 3, 4$.

In other circumstances, a single label does not suffice, and two or more are used. Thus, we might define the nine variables T_{11}, T_{12}, T_{13}, T_{21}, T_{22}, T_{23}, T_{31}, T_{32}, and T_{33} – or as T_{ij}, $i = 1, 2$, or 3, $j = 1, 2$, or 3. (If there is any confusion about two adjacent numbers, e.g. if 12 in T_{12} is read as 'twelve' rather than 'one, two', then a distinguishing comma could be used: $T_{1,2}$ or $T_{i,j}$, though this is only rarely done in practice.) Variables with two subscripts can often be treated as the elements of *matrices*, which will be discussed extensively in Chapter 4. We should also add that it is possible to have more than two subscripts, and two or more superscripts. Further, a set of variables may be defined using both subscripts and superscripts. We shall introduce examples of such variables later.

The use of subscripts (or superscripts, which can be treated the same way in this respect) enables us to develop a convenient notation for sums or products of variables. We use the summation sign, Σ, an uppercase

Greek 's' (sigma), for summation, and Π, an uppercase Greek 'p' (pi) for product. For example, we can write

$$x_1 + x_2 + x_3 = \sum_{i=1}^{3} x_i \tag{2.2}$$

$i = 1$ to 3 is the *range* of the summation (which is assumed to proceed in integer steps), and $\sum_{i=1}^{3} x_i$ means: take the expression following the summation sign, replace i by 1, 2 and 3 in succession and add. Hence, we have the relation (2.2). Similarly:

$$x_1 + x_2 + \ldots + x_{10} = \sum_{i=1}^{10} x_i \tag{2.3}$$

$$x_1 y_1 + x_2 y_2 + x_3 y_3 = \sum_{i=1}^{3} x_i y_i \tag{2.4}$$

$$T_{1j} + T_{2j} + T_{3j} = \sum_{i=1}^{3} T_{ij} \tag{2.5}$$

The product sign is used in exactly the same way, except that the successive operation is multiplication. Thus

$$x_1 x_2 x_3 = \prod_{i=1}^{3} x_i \tag{2.6}$$

Note that when expressions are formed using a summation or product sign, the value of the expression is independent of the letter used for the subscript which appears in the summation (or product). Thus,

$$\sum_{i=1}^{3} x_i = x_1 + x_2 + x_3 \tag{2.7}$$

$$\sum_{j=1}^{3} x_j = x_1 + x_2 + x_3 \tag{2.8}$$

and so on. Such a subscript (or *index* as it is sometimes called)* is thus known as a *dummy* subscript, or dummy index, as opposed to any other kind which is a *free* subscript or free index.

We have so far introduced only the most basic concepts of elementary algebra. However, these prove sufficient for us to show how to describe some of the geographical systems introduced in the previous chapter in algebraic terms. Then we can take the discussion of algebraic methods (and indeed all the other methods) further using these examples.

* 'index' is a generic term which describes either a subscript or a superscript.

2.2. Principles of system description

2.2.1. The algebra of space

Since we are so often concerned with spatial systems, a preliminary discussion of the treatment of space in the discussion of algebraic variables is useful. Space is handled in algebraic terms by using the concepts of co-ordinate geometry. We postpone most of the discussion of this to Chapter 3, but here we do introduce the notion of a co-ordinate system. A Cartesian co-ordinate system is shown in Figure 2.1. There is a vertical axis (the y-axis) and a horizontal axis (the x-axis) and the

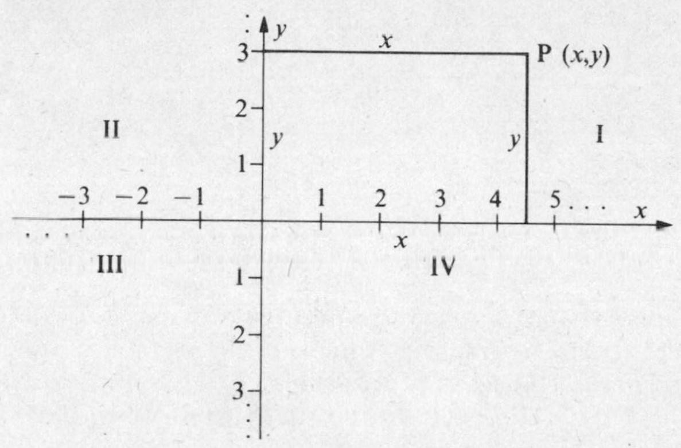

FIG. 2.1. A Cartesian co-ordinate system

possible values of y and x are shown as scales on each. Any point P is then uniquely characterized in space by the values of x and y at the point as shown on the figure. These values are called the *co-ordinates* of the point and are written (x, y). The x-co-ordinate is sometimes called the ***abscissa*** and the y-co-ordinate the *ordinate*. The axes are shown as passing through the point (0, 0), known as the *origin*, and they divide space into four quadrants (I, II, III and IV on the figure). In quadrant I, x and y are both positive, in II, x is negative and y is positive; in III, x and y are both negative; and in IV, x is positive and y is negative.

In a similar way to using integer subscripts as labels, we can use other variables themselves as variable labels. Thus, if H is some variable which varies spatially say with height above the ground, we can define $H(x, y)$ to be 'height at the point (x, y)'.* This is obviously going to be a very

* In Chapter 3, we will see that we will be able to describe H as a *function* of x and y.

useful way of constructing algebraic variables to represent geographical quantities, particularly at a micro-spatial level of resolution, since (x, y) is an exact 'address'.

An alternative way of describing the position of a point is obtained using *polar co-ordinates*, as shown in Figure 2.2. r is the distance from the origin to the point P, and θ is the angle between the horizontal axis OX and OP. It can easily be seen that the pair of polar co-ordinates

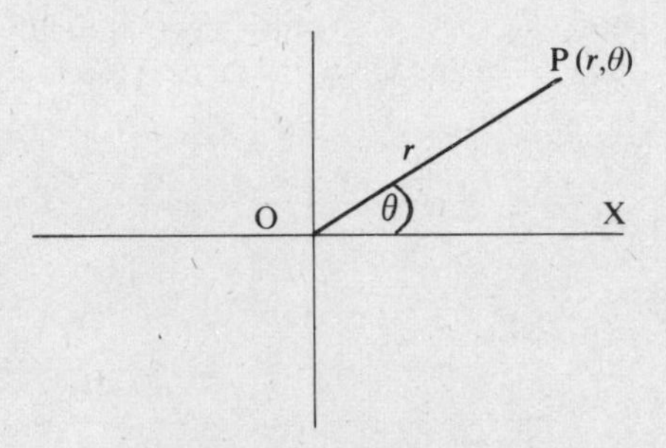

FIG. 2.2. Polar co-ordinates

(r, θ) also uniquely characterizes the position of P in space. We could then use $H(r, \theta)$ as a geographical variable if this was convenient. The quadrants can now be related to the value of θ: I has $0° \leqslant \theta \leqslant 90°$, II has $90° \leqslant \theta \leqslant 180°$, III has $180° \leqslant \theta \leqslant 270°$, and IV has $270° \leqslant \theta \leqslant 360°$. This representation is particularly useful when variables have values displaying angular symmetry – that is, which are independent of θ. They can then be shown with one label only $H(r)$, say, while if a Cartesian system was used, the two labels (x, y) would still be needed. This is also suitable for representation of geographical variables at a micro scale.

We should also note that micro-scale representations can sometimes be devised which do not use a co-ordinate system at all. Consider

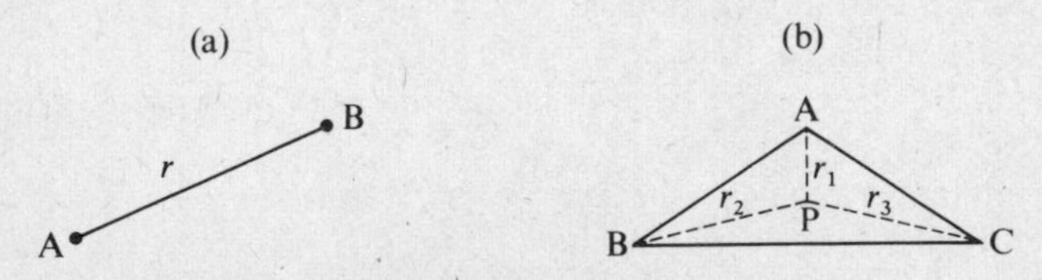

FIG. 2.3. Alternative co-ordinate systems

Figure 2.3. In Figure 2.3 (a) A may be a fixed point in space and B a variable point, and the only variable needed to describe some spatial relationship may be the distance between them, say r. In Figure 2.3 (b), the three points A, B and C may be fixed in space, and P determined in relation to its distance from each of them, shown here as r_1, r_2 and r_3.

Next, we consider the algebraic treatment of space at a meso level of spatial resolution. We will wish to relate quantities to spatial locations – but to locations which represent an area of space rather than a precisely specified address. In other words, we now treat space in a *discrete* way, rather than in the *continuous* representation of Cartesian co-ordinates. This implies that we should divide our study area into *zones*, and label these zones. There is an infinite variety of ways of doing this for any one study area. Some ways are indicated in Figure 2.4.

Figures 2.4 (a)–2.4 (c) show zone boundaries which are uneven. The zone system of Figure 2.4 (b) is 'finer' than that of Figure 2.4 (a) and

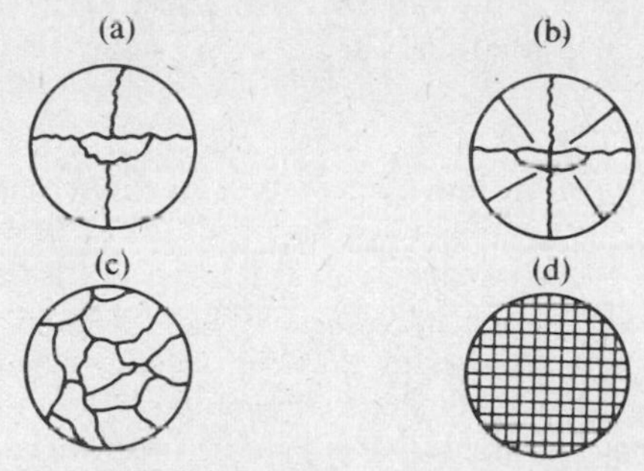

FIG. 2.4. Possible zoning systems

indeed the zones in the former case are subdivisions of the latter. The boundaries in Figure 2.4 (c) bear no relation to those in Figures 2.4 (a) and 2.4 (b). The zone system of Figure 2.4 (d) is different again, and is a uniform square grid system. All such systems can occur in practice in geography and planning. The zones in Figure 2.4 (a) may be wards within a local authority; those of Figure 2.4 (b) subdivisions of these, such as Census Enumeration Districts; those of Figure 2.4 (c) may relate to a different administrative authority, and be, say, postal districts; those of Figure 2.4 (d) may be proposed for research purposes.

Suppose one zone system is chosen. Then, zones must be labelled, so that the labels can be associated with variables which measure geographical quantities. Suppose we wish to construct a variable to represent 'zone population', for example, and to do this by labelling a variable, P. Consider the labelling systems shown in Figure 2.5. Figure 2.5 (a) is a familiar map in which geographical names are used. We could then use P (Headingley) as one population variable. This is clearly inconvenient, and so we number the zones from 1 to 5 (in this case) as shown in Figure 2.5 (b). We can use these zone numbers as subscripts and take the population variables as P_1, P_2, P_3, P_4 and P_5. Finally, we can move to a higher level of abstraction and use i, as a subscript, as a zone label and

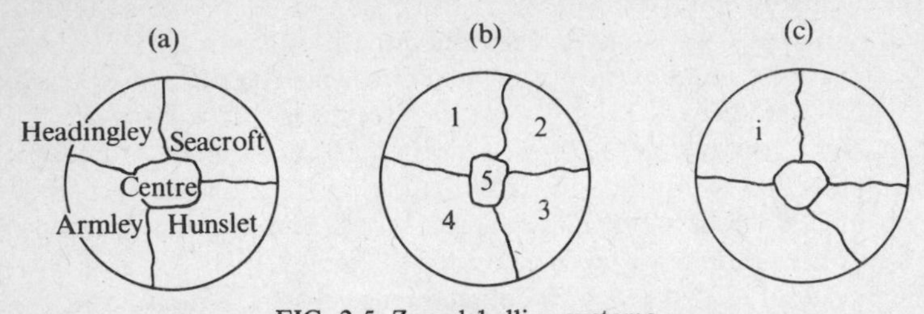

FIG. 2.5. Zone labelling systems

take P_i, $i = 1, 2, 3, 4, 5$ as the variables and refer to P_i as the 'population of any zone, i'. Such a situation is depicted in Figure 2.5 (c).

2.2.2. *Further preliminaries*

Before proceeding to show how to construct algebraic variables to describe geographical systems by working through a range of examples, two final preliminary comments are in order. First, it is not always easy to find a good notation, especially for complicated situations, but it is worthwhile to put in a lot of effort to achieve good results in this respect nonetheless. Any mathematician would endorse the view that 'discovering' a good notation is half the battle in solving a mathematical problem. Second, the analysis from now on may appear to become much more abstract as algebraic variables are defined for geographical systems and further mathematical techniques introduced. *The reader should always be prepared to connect back, to trace back, from algebraic variables to 'concrete' features of the geographical system of interest.* It is nearly always possible to do this, and it is vital if geographical and mathematical understanding are to advance in parallel.

2.2.3. *Von Thünen's and Alonso's problems*

The simplest von Thünen problem is depicted in Figure 2.6. 0 is the centre of a market for agricultural products. Circular symmetry is assumed, so land use is uniform in any circular annulus with 0 as centre. Consider a point P at distance r from 0 as shown. Let i be used as a label for different possible crops; that is, we may take $i = 1$ to represent wheat, $i = 2$ barley, and so on. Let p_i be the market price per unit of crop i, a_i be its production cost, and c_i be the unit transport cost. We need to know the yield of a unit of land for each crop: let y_i be the amount of crop i produced by such a unit of land. Then, the net profit from a unit of land under crop i, discounting transport, would be $(p_i - a_i)y_i$. Transport costs would be ry_ic_i for a plot of land at distance

r from the market 0. We can use these variables to determine the land use pattern, and we do this using elementary algebra and co-ordinate geometry in Chapter 3.

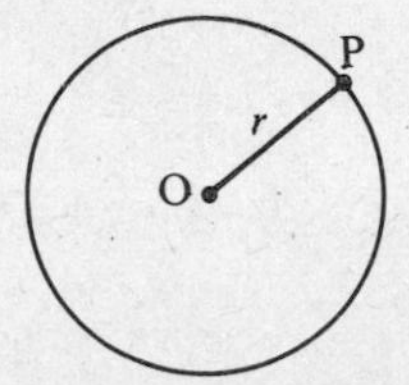

FIG. 2.6. von Thünen's problem

Alonso, in relation to a similar situation, would postulate the amount of utility, $U^k(r)$ say, which an individual k would obtain from a plot land at distance r from 0. This utility is achieved by paying rent for the land, and is also related to other goods purchased. Total purchases must sum to the available income of the individual. Thus we have a constrained maximization problem which leads us to a more general theory of location, which we shall pursue in Chapter 9.

2.2.4. The Weberian problem

A simple Weberian problem has already been hinted at in relation to Figure 2.3 (b) which is repeated here for convenience as Figure 2.7. The position of the point P is determined by the three distances, r_1, r_2 and

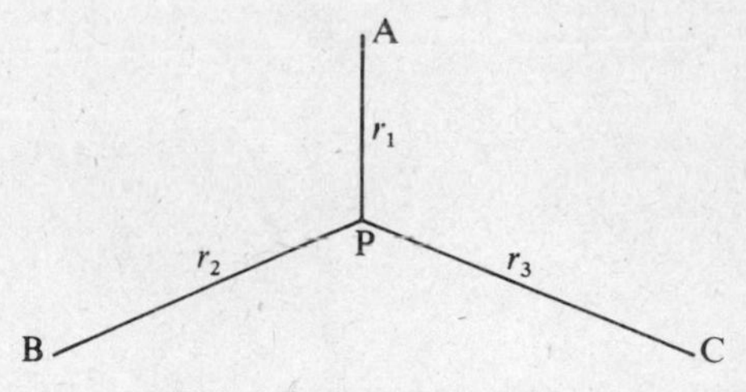

FIG. 2.7. Weber's problem

r_3 measured in relation to fixed points A, B and C as shown. Suppose a firm located at P can make a net profit p, less transport and resource costs, from a unit of some product. As inputs it requires, for each unit of its product quantities of different resources, which we label A, B and C, uniquely available at the points A, B and C respectively. Let these quantities be q_A, q_B and q_C for which it pays p_A, p_B and p_C per unit together with unit transport costs $c_A r_1$, $c_B r_2$, and $c_C r_3$. Then, fairly obviously, the revised net profit per unit of product is:

$$p - (q_A p_A + q_B p_B + q_C p_C) - (c_A q_A r_1 + c_A q_B r_2 + c_A q_C r_3).$$

The problem is to find r_1, r_2 and r_3 so that this is maximized. We shall present the solution to this problem in Chapter 5.

2.2.5. *Quadrat analysis*

Quadrat analysis is concerned with a spatial system such as that depicted in Figure 2.4 (d). However, the grids are not explicitly labelled: rather, we are concerned with the probability of occurrence of the phenomena being located. If there are possibly x such locations within a quadrat, we work with the probability $P(x)$ that this actually occurs using methods outlined in Chapters 7 and 8.

2.2.6. *Meso-spatial interaction and location theory*

At the meso scale, we usually work with an exhaustive system of mutually exclusive zones for the study area which are numbered (i.e. labelled) consecutively from 1 up to N, where N is, say, the total number of zones. All the examples in Figure 2.4 are examples of such zoning systems. We may then need a variable to represent the number of trips from any one zone to any other. We could take this as T, and add two zone labels – one for origin and one for destination as subscripts. Thus T_{12} would then be the number of trips from zone 1 to zone 2, T_{25} the number of trips from zone 2 to zone 5, and so on. Clearly, if there are N zones in all, there are $N \times N = N^2$ such trip

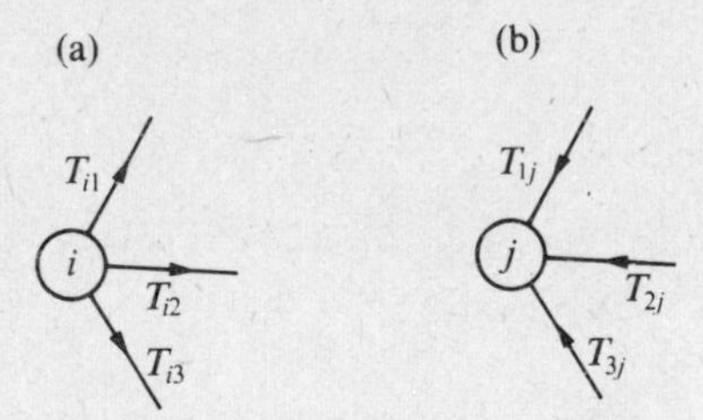

FIG. 2.8. Zonal inflows and outflows

variables. It is convenient, as indicated earlier, to use letters as labels: let i be any zone and j be any zone, and then T_{ij} is the number of trips from i to j, where i can be any number between 1 and N, as can j.

Suppose now we focus on one zone i. It is often convenient to have the sum of all trips leaving i, and to use a separate variable name for this, say O_i (standing for total trip *o*rigins). This sum is shown diagrammatically in Figure 2.8 (a) and algebraically in equation 2.9.

$$O_i = T_{i1} + T_{i2} + T_{i3} + \ldots + T_{iN} \tag{2.9}$$

Similarly we can define D_j to be the sum of all trips entering any zone j. These flows are shown in Figure 2.8 (b) and equation 2.10:

$$D_j = T_{1j} + T_{2j} + T_{3j} + \ldots + T_{Nj} \tag{2.10}$$

We can represent these sums more conveniently using the summation sign introduced earlier. Equations 2.9 and 2.10 can be written

$$O_i = \sum_{j=1}^{N} T_{ij} \tag{2.11}$$

and

$$D_j = \sum_{i=1}^{N} T_{ij} \tag{2.12}$$

respectively. We might also be interested in the total number of trip origins in all zones, and the total number of trip destinations; let these quantities be T and T' respectively.* Then

$$T = O_1 + O_2 + \ldots + O_N = \sum_{i=1}^{N} O_i \tag{2.13}$$

$$T' = D_1 + D_2 + \ldots + D_N = \sum_{j=1}^{N} D_j \tag{2.14}$$

Note that we can substitute for O_i from equation 2.11 into 2.13 and for D_j from equation 2.12 into 2.14 to show that

$$T = \sum_{j=1}^{N} \left(\sum_{i=1}^{N} T_{ij} \right) \tag{2.15}$$

$$T' = \sum_{i=1}^{N} \left(\sum_{j=1}^{N} T_{ij} \right) \tag{2.16}$$

from which it follows (not surprisingly!) that $T = T'$.

In spatial interaction modelling, it will also be necessary to define another doubly subscripted variable as d_{ij}, the *distance* from i to j, t_{ij} the *travel time* from i to j, or as c_{ij} the *cost* of travel from i to j.

Variables like T_{ij} will also be used as the basis of the nodal structure analysis mentioned in section 1.3.6 of Chapter 1.

2.2.7. *Network theory*

Examples of networks are shown in Figure 2.9. A network consists of a

* The prime $'$ is sometimes used to distinguish a variable, as here, with T'.

set of links and nodes. A link of the network connects two nodes. Links are sometimes shown as being *directed,* as in Figure 2.9 (c), implying that travel, or flow, is only permitted in that direction. Unless specifically indicated, links can be assumed to be two way. Figures 2.9 (c) and 2.9 (d) are the sorts of networks which might be traced from maps – a river network and a road network respectively – and the links are not straight lines. Networks are sometimes shown with all links as idealized

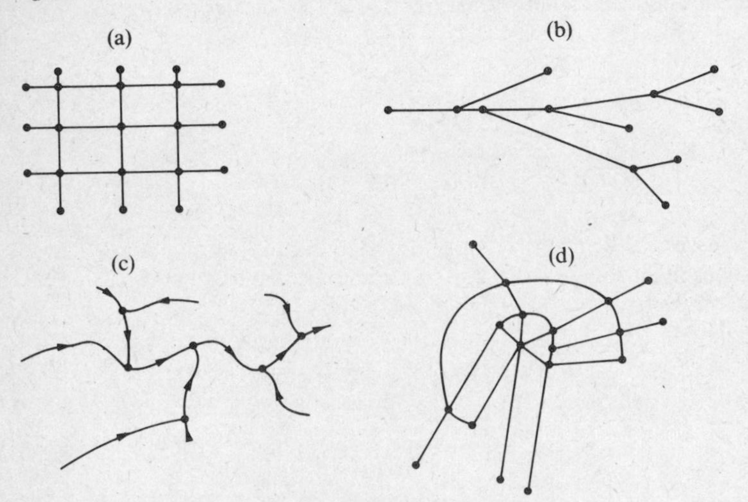

FIG. 2.9. Examples of networks

straight lines, with the emphasis on the topological (connective) structure only rather than the actual spatial arrangement. Figures 2.9 (a) and 2.9 (b) could be taken as examples of such networks.

In order to describe a network, the most important steps are to label the nodes, and to find some way of expressing which nodes are connected – that is, which links exist out of the set of all possible links. There are many ways of carrying out this task. It would be possible for some microanalyses to superimpose a Cartesian co-ordinate system and to label each node with its Cartesian co-ordinates. Whether this is done or not, it is almost always convenient to number the nodes consecutively, from 1 to N, where N is now taken as the total number of nodes. $x(i)$ and $y(i)$, or x_i and y_i, may then be taken as the Cartesian co-ordinates of the ith node. Connectivity can also be specified in a number of ways. We might define a doubly subscripted variable L_{ij}, and set L_{ij} to 1 if node i is directly connected to link j, and to 0 otherwise (which is the basis of the work of Kansky, 1963). Alternatively, we might define two singly subscripted variables (which might be more economical as many L_{ij} values would be zero): n_i may be the number of links connected to zone i and m_j, $j = 1, 2, \ldots n_i$ may be the link numbers of the other end of the link. (Note that m_j is a variable which itself only takes integral values.) We have assumed implicitly in these definitions that we are dealing with one-way links. If a link between i and j is two way, then this can be recorded in one of the above schemes as two links – from i to j and from j to i.

This is a convenient point at which to introduce, informally, the concept of a set. We use letters (possibly subscripted) to label sets of any kind of objects. In this context, we are interested in sets of nodes. Thus, R may be taken as the set of all nodes in some network. If there are N such nodes, we might list them as

$$R = 1, 2, 3, \ldots N \tag{2.17}$$

though '=' is used here to define something rather than as an equation in the usual sense. In particular, we will be interested in the shortest path from any zone i to any zone j. Suppose this is made up of the set of nodes R_{ij}. For example, in Figure 2.9 (d), the shortest path from node 1 to node 4 may be 1–2, 2–3, 3–4. So the set R_{12} is 1, 2, 3, 4.

In order to find shortest paths, we need variables for link characteristics. If d_{ij} is the length of a link from i to j, (and this only has meaning if i and j are directly connected), then, the shortest path from any zone i to any zone j will be that path for which $\Sigma d_{i'j'}$ is a minimum, where the summation is over all links in the path. We can develop a more precise notation for this as follows: let S_{ij}^k be the set of nodes forming the kth path (k simply being a label) from node i to node j. Then we can use the set inclusion sign, $\in$, to develop our notation further. $i' \in S_{ij}^k$ means that the node i' is in the set of nodes which forms the kth path from i to j. Thus $\sum_{i'j' \in S_{ij}^k} d_{i'j'}$ is the total length of the path from i to j. Recall that R_{ij} is the shortest path. So if c_{ij} is the length of the shortest path from i to j

$$c_{ij} = \sum_{i'j' \in R_{ij}} d_{i'j'} \tag{2.18}$$

Thus, in relation to Figure 2.9 (d), if $d_{12} = 1$, $d_{23} = 3$, $d_{34} = 1$, then

$$c_{14} = \sum_{i'j' \in R_{14}} d_{i'j'} \tag{2.19}$$

$$= d_{12} + d_{23} + d_{34} \tag{2.20}$$

$$= 1 + 3 + 1 \tag{2.21}$$

$$= 4 \tag{2.22}$$

A task of Chapter 9 will be to show how the best path can be identified from among all possible paths.

This is an informal sketch only of the descriptive basis of network theory (sometimes also known as graph theory), and there will be further developments in relation to specific examples later.

2.2.8. Diffusion processes

The simplest diffusion processes to describe are those which take place 'vertically' at a point and so do not demand variables which are spatially labelled. Examples are the diffusion of heat in soil and the infiltration of water into soil. Figure 2.10 (a) shows the heat diffusion situation. Here

FIG. 2.10. Diffusion of heat and water into soil

the vertical dimension, expressed by x, the depth from the surface, is important. The main variable is the flow of heat B, and this will be related to x, time t, temperature T (which will itself depend on x and t). The differential equation for this process will be given in Chapter 6.

The water infiltration situation is depicted in Figure 2.10 (b). In this mechanism, the infiltration rate is related to infiltration at the surface, and so we do not use a depth variable. The overall infiltration rate f has two main components – the transmission rate A and the diffusion constant B, which produces a contribution to f which is time dependent. The relationship is expressed in a single equation (which is discussed in Chapter 3) to illustrate the power function.

A third physical example of diffusion concerns the spread of air pollution. An extra dimension (literally) is added to the problem in this case, since the variation of concentration of pollutants with height is important. A three-dimensional version of the Cartesian co-ordinate system first presented in Figure 2.1 is shown in Figure 2.11. The point P is now labelled by three co-ordinates (x, y, z), the first two of which can be considered to give the ground position, and z the height above the ground. We can then assume that the polluting source is located at the origin, (0, 0, 0), that there is a wind blowing in the direction of the x-axis of mean speed u, and we can model the pollutant concentration R, at (x, y, z) at time t, writing it as $R(x, y, z, t)$, using x, y, z and t as labels. We shall give an equation for this in Chapter 3 to illustrate the normal distribution function.

We also considered various examples of spatial diffusion in human geography. Most models of such a process use a quadrat zone system as

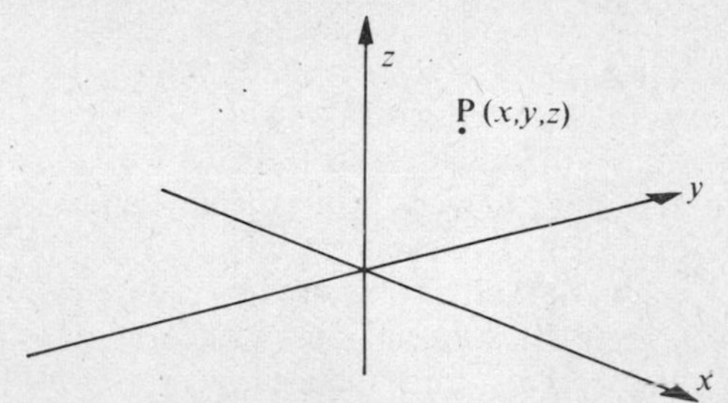

FIG. 2.11. A three-dimensional Cartesian co-ordinate system

in Figure 2.4 (d). In this case, it is useful to label a quadrat by, say, the Cartesian co-ordinates of its centroid. Most models then predict something like $P(x, y)$ which is the probability that the innovation or disease or whatever, has reached the quadrat labelled (x, y), perhaps modified by a quantity variable also. Haggett (1969) uses this sort of probability mechanism, but in a network system rather than a zone system: diffusion takes place along the links of a network. This representation seems appropriate in a regional study (as was the case in his study) where the 'nodes' are towns and villages within the region.

2.2.9. *Accounting and time development*

We introduced the notion of accounting in Chapter 1 in relation to changing states of system components during some time period. Consider a spatially distributed population, using a zone system as in Figure 2.4. Ignore, for the time being, births and deaths. At time t, let $P_i(t)$ be the population of zone i, and at time $t + T$, let $P_i(t + T)$ be the population of zone i. If $P_i(t + T)$ differs from $P_i(t)$, this will have resulted from interzonal migration. An accounting variable can be defined as follows: let $K_{ij}(t, t + T)$ be the number of people who were 'in state i' – that is 'resident in zone i' at time t – but who were in state j at time $t + T$. If there are N zones, then the set of variables $K_{i1}(t, t + T)$, $K_{i2}(t, t + T)$, . . . $K_{iN}(t, t + T)$ 'accounts for' the population $P_i(t)$. It enumerates what could have happened to the population during the time period. [Note that the non-movers are recorded as $K_{ii}(t, t + T)$.] We have

$$P_i(t) = \sum_{j=1}^{N} K_{ij}(t, t+T) \tag{2.23}$$

and also

$$P_i(t+T) = \sum_{j=1}^{N} K_{ji}(t, t+T) \qquad (2.24)$$

We will explore the properties of these kinds of accounting variables in Chapter 4, below, to illustrate matrix concepts. These notions are also relevant to the study of Markov processes (Chapter 8).

Similar two subscript, time-labelled, variables can be defined to describe the transactions of an economy: $Z_{mn}(t, t+T)$ may be taken as the amount of product of industry m used by industry n during period t to $t+T$, if $Y_m(t, t+T)$ is the amount of m consumed directly, and $X_m(t, t+T)$ is the total amount produced, then we have the accounting relationship.

$$\sum_{n=1}^{N} Z_{mm}(t, t+T) + Y_m(t, t+T) = X_m(t, t+T) \qquad (2.25)$$

Models based on these variables will also be explored in Chapter 4. With the above definition, no spatial labels were used. It is also possible to add such labels, the usual zone labels such as the is and js of Figure 2.4, and to record the spatial structure of the economy.

These arguments can be extended to cover animal and plant populations. A model of animal populations will have some of the features of the human population model (spatial redistribution, birth and death), and also of the economic model (prey – predator relationships). Such models use time as a discrete variable or, perhaps more commonly, as a continuous variable, and a simple example is presented in Chapter 5.

2.2.10. Process studies based on balance relationships.

We illustrate system description in another kind of process by reference to mass balance relationships. We explore the method of system descrip-

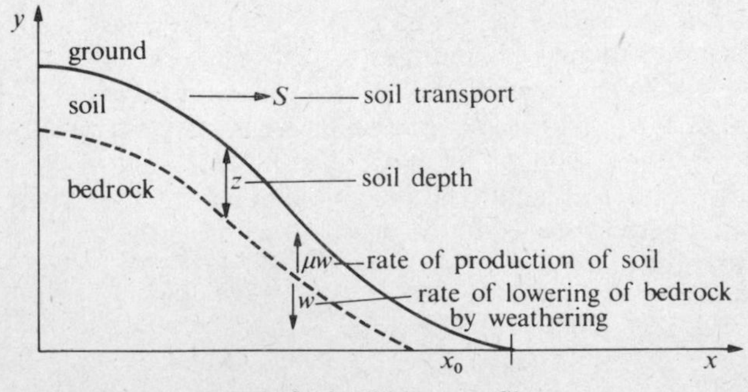

FIG. 2.12. Hillslope profile

tion for such a problem by considering a slope profile and soil transport down the slope. To simplify matters, we consider cross-section through the line of steepest descent as shown in Figure 2.12. The cross-section is shown embedded in a Cartesian co-ordinate system, and the algebraic variables which represent the main physical variables are shown on the figure. These are S, z, μ, W, y and x. Two rates are defined: M and D, the rate of mechanical lowering of the bedrock and the rate of chemical lowering respectively. These rates are related to the main physical variables, and to time t, and differential equations are developed which describe the slope profile. The detailed discussion is in Chapter 6.

2.2.11. Concluding comments

The above examples should illustrate a wide variety of ways in which algebraic variables can be defined to describe a system of interest. The ability to identify and to define such variables in a convenient form is in part a skill which can be acquired by practice, and for this reason it is essential for the reader to tackle the exercises at the end of this chapter, and indeed, in the rest of the book.

2.3. Relating data to algebraic variables

In the previous section, we have shown how to use algebraic variables to describe a system of interest. A *particular* system will be described by assigning appropriate numerical values to these variables. Usually, this numerical data, for geographical systems, will be presented in either map form, or as a table. Suppose, for example, we are given

(a)

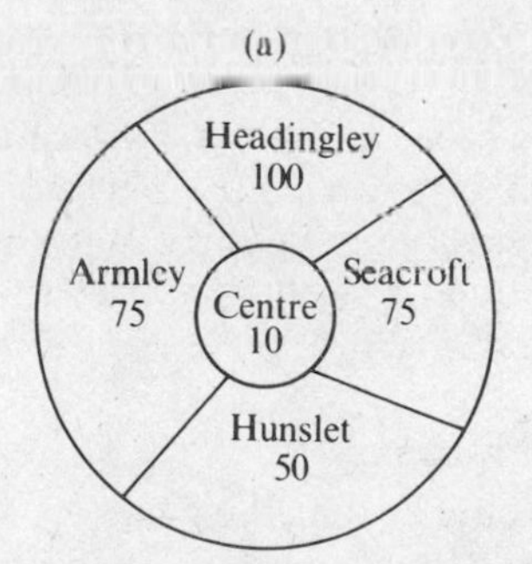

(b)

Zone	Pop (000's)
Centre	10
Headingley	100
Seacroft	75
Hunslet	50
Armley	75

FIG. 2.13. Zone labelling and data presentation

population data for the zones of a city as in Figure 2.5 (a). These are shown presented in map form in Figure 2.13 (a) and in table form in Figure 2.13 (b). The corresponding algebraic variable could be defined as P_i the population of zone i, $i = 1, 2, 3, 4$ and 5, say. Thus, the first task is to assign zone numbers to the zones. Suppose we do this as in

Figure 2.5 (b). These decisions can be recorded in a table as in Figure 2.14 (a). Then, Figures 2.14 (b)–2.14 (d) show different ways of relating the data to the variables, each being equivalent to any other. The tables shown as 2.14 (b) and 2.14 (c) show that any letter can be used to represent the subscript on the population variable at this stage.

(a)		(b)		(c)		(d)	
Zone	Zone No.	i	P_i	j	P_j	Variable	Value
Headingley	1	1	100	1	100	P_1	100
Seacroft	2	2	75	2	75	P_2	75
Hunslet	3	3	50	3	50	P_3	50
Armley	4	4	75	4	75	P_4	75
Centre	5	5	10	5	10	P_5	10

FIG. 2.14. Alternative forms of zonal data presentation

Most other data assignment tasks involving algebraic variables with spatial labels can be handled in a similar way. Other examples were 2 or 3 variable problems without spatial labels, heat conduction in soil, for example, these are straightforward in this respect. Yet others, such as Alonso's location theory, involved space being labelled by a continuous variable, such as distance from the city centre. In such cases, algebraic variables have to be related to data using such devices as contours, or the space must be divided into zones and labelled as discrete units. It is rarely possible to match such 'continuous' variables against data directly.

This short section is intended to give a broad indication of how to relate data to variables. Further discussion is more effective in terms of numerical examples as the rest of our argument proceeds. Exercises are also set which involve numerical substitution for algebraic variables.

2.4. Simple equations

2.4.1. Linear equations in one dependent variable

The simplest possible kind of equation would be something like

$$y = 2 \tag{2.26}$$

which simply assigns the value 2 to the variable y. The variable which appears on the left hand side of such an equation is the *dependent variable.* More interestingly, we may consider an equation such as

$$y = x + \mathrm{a} \tag{2.27}$$

where x is another variable and a is a constant. y is still known as the dependent variable, it 'depends' on x and a, while x is known as an independent variable. There may be more than one independent variable, as in

$$y = \mathrm{a}(x + z) \tag{2.28}$$

where a is again a constant and x and z are both independent variables.

Even simple equations are not always arranged as tidily as these examples, where the dependent variable appears alone on the left hand side. When this is not the case, we say that we can 'solve' the equation for the dependent variable. Thus, if

$$x = \mathrm{a} - y \tag{2.29}$$

we can solve for y as

$$y = \mathrm{a} - x \tag{2.30}$$

or, if

$$\mathrm{a}y = x \tag{2.31}$$

we can solve for y as

$$y = x/\mathrm{a} \tag{2.32}$$

We can describe the process of solving an equation more formally as follows. Consider an equation to be of the form

$$X = Y \tag{2.33}$$

where X and Y are each 'expressions'. Then we can add, multiply, subtract or divide, indeed carry out any algebraic operation on one side of the equation provided we also do the same thing on the other. Thus:

$$X + \mathrm{a} = Y + \mathrm{a} \tag{2.34}$$

$$X - \mathrm{a} = Y - \mathrm{a} \tag{2.35}$$

$$\mathrm{a}X = \mathrm{a}Y \tag{2.36}$$

or

$$X/\mathrm{a} = Y/\mathrm{a} \tag{2.37}$$

Equation 2.30 was obtained from equation 2.29 by adding y to each side and subtacting x from each side, a process sometimes also known as 'taking a variable to the other side of the equation (and changing the sign)'. Equation 2.32 was obtained from 2.31 by dividing each side by a. As a further example, suppose

$$16y + 2 = 6y + 5x \tag{2.38}$$

and we solve for y as follows:
(i) substract $6y$ from each side to give

$$10y + 2 = 5x \tag{2.39}$$

(ii) subtract 2 from each side, to give

$$10y = 5x - 2 \tag{2.40}$$

(iii) divide each side by 10:

$$y = \frac{5x - 2}{10} = 0{\cdot}5x - 0{\cdot}2. \tag{2.41}$$

Usually, of course, these operations are carried out simultaneously and without conscious effort.

Quite simple equations do turn up in the literature. For example, if O_i is the number of non-home-based trip origins generated in the ith district used for the London Transportation Study (Greater London Council, 1968), then

$$O_i = -72{\cdot}10 + 0{\cdot}393U_i + 0{\cdot}935V_i + 0{\cdot}159W_i \tag{2.42}$$

where

U_i = number of cars owned in the district,
V_i = amount of employment in general retail services in the district,
and W_i = amount of employment in services, construction and extractive industries.

If this same relationship was to be used in another city, we might write it in the form

$$O_i = \mathrm{a} + \mathrm{b}U_i + \mathrm{c}V_i + \mathrm{d}W_i \tag{2.43}$$

where a, b, c and d are constants to be determined for the particular city. For London, a = − 72·10, b = 0·393, c = 0·935 and d = 0·159.

Another simple example occurs in flood frequency analysis, If over n years, floods are recorded and ranked in order of magnitude, then the mean recurrence interval, in years of an mth magnitude flood, can be taken as

$$T = \frac{n + 1}{m} \tag{2.44}$$

A further physical example is provided by the relationship of variables of a ground water system (Waltz, 1972). If P is ground water pressure, W is unit weight, then P/W has the dimensions of length and can be taken as 'pressure head'. If the elevation head is Z, then the flow potential, or total head H is given by

$$H = \frac{P}{W} + \mathrm{Z}. \tag{2.45}$$

If we 'measured' H, and wished to solve for P, say, this could be rearranged to give

$$P = (H - \mathrm{Z})W \tag{2.46}$$

Most examples of equations involve functions which will be introduced in Chapter 3, and so we will postpone further discussion until then.

2.4.2 Quadratic equations

y^2 (y 'squared') is y × y, and is a special case of a power function which we will define in Chapter 3. It is also sometimes referred to as y raised to the power 2, instead of 1 as in the linear case. Equations which contain terms in y^2, as well as y, when y is the dependent variable are known as quadratic equations. Consider

$$y^2 = 4 \tag{2.47}$$

Thus, clearly, $y = 2$ is a solution, since $2^2 = 2 \times 2 = 4$. But also -2 is a solution, since $(-2)^2 = (-2) \times (-2) = 4$. Thus

$$y = 2 \text{ or } -2. \tag{2.48}$$

We can also use the square root sign, and write

$$y = \pm\sqrt{4} \tag{2.49}$$

bearing in mind that the root can have either sign.

Suppose now,

$$(y + 1)^2 = 4 \tag{2.50}$$

Then,

$$y + 1 = 2 \text{ or } -2$$

$$\text{so } y = 1 \text{ or } -3. \tag{2.51}$$

Equation 2.50 is

$$y^2 + 2y + 1 = 4 \tag{2.52}$$

$$y^2 + 2y - 3 = 0. \tag{2.53}$$

This suggests that, in general, we can solve quadratic equations of the form 2.53 by re-arranging them in the form of equation 2.50. The trick is to form a perfect square from the y^2 and y terms, and to group all the constants together on the right hand side. Thus, starting with

$$y^2 + 6y - 7 = 0 \tag{2.54}$$

we get

$$(y+3)^2 - 9 - 7 = 0 \tag{2.55}$$

so

$$(y+3)^2 = 16 \tag{2.56}$$

$$y + 3 = 4 \text{ or } -4 \tag{2.57}$$

$$y = 1 \text{ or } -7 \tag{2.58}$$

In general, consider the equation

$$ay^2 + by + c = 0 \tag{2.59}$$

where a, b and c are constants. Write it in the form

$$y^2 + \frac{b}{a}y + \frac{c}{a} = 0 \tag{2.60}$$

by dividing through by a. Then we can proceed as before:

$$\left(y + \frac{b}{2a}\right)^2 - \frac{b^2}{4a^2} + \frac{c}{a} = 0 \tag{2.61}$$

$$\left(y + \frac{b}{2a}\right)^2 = \frac{b^2}{4a^2} - \frac{c}{a} = \frac{b^2 - 4ac}{4a^2} \tag{2.62}$$

$$y + \frac{b}{2a} = \pm \frac{\sqrt{(b^2 - 4ac)}}{4a^2} \tag{2.63}$$

$$= \pm \frac{\sqrt{(b^2 - 4ac)}}{2a} \tag{2.64}$$

so

$$y = -\frac{b \pm \sqrt{(b^2 - 4ac)}}{2a} \tag{2.65}$$

which is the standard formula for finding the two solutions, or roots, of a quadratic equation.

We can construct a simple example as follows. Suppose a quantity y of some resource can be produced at price p per unit, that the price paid per unit is a function of the total produced, in fact as

$$p = 10 - y \tag{2.66}$$

in some suitable units, and that 24 units of cash are available. How much of the resource is produced? If y is produced, the total cost is

$$py = (10 - y)y \tag{2.67}$$

So, in this case, y is the solution of

$$(10 - y)y = 24 \tag{2.68}$$

or, rearranging

$$y^2 - 10y + 24 = 0. \tag{2.69}$$

We can solve by completing the square or using the formula 2.65. Let us use the formula with a = 1, b = −10, c = 24:

$$y = \frac{10 \pm \sqrt{(100 - 96)}}{2} \tag{2.70}$$

So,

$$y = \frac{10 \pm 2}{2} = 6 \text{ or } 4. \tag{2.71}$$

Thus there are, as usual, two solutions which would cost the resource user 24 units. Clearly, in this case, he would take the largest solution, 6. Sometimes both solutions have an obvious relevance; in others, one only, as here.

2.4.3. Cubic and higher order equations

We can write $x \times x \times x = x^3$, and raise x to the third power. Equations containing such cubic terms (and, possibly, quadratic, linear and constant terms) are called, unsurprisingly, cubic equations. Equations containing x^4 and lower order terms are called quartic equations, and so on. Such equations do not have the importance of linear or quadratic equations in geography and planning at the present time, and this is fortunate, because they are less easy to solve. There are algebraic methods for the solution of cubic equations (though we shall not present them here) but no general methods of solution for higher order equations, though in particular cases it may be possible to 'spot' solutions by inspection. In Chapter 3, we shall show how to solve cubic equations graphically and this will suffice for our purposes.

2.4.4. Checking that equations make sense

It is important for the reader to develop a variety of ways of checking that any equation he uses is sensible. There are two main aspects of this task: does the equation make sense in *mathematical* terms, and does the equation make sense in system (that is, *geographical*) terms. The first of these is probably the most important at this stage. For example, the subscript structure should be correct. In equation 2.43 above, each variable contains the subscript i and this is a *free* subscript (as distinct from *dummy* subscripts defined in relation to the summation sign earlier). If one of the is on the right hand side was a j, then this should be checked as being unlikely. Broadly speaking, free subscripts attached to the dependent variable on the left hand side should appear somewhere on the right hand side, and *vice versa.*

In geographical terms, the most obvious check relates to the *dimensions* of the variable: if the right hand side of the equation is 'oranges', or more sensibly, 'trips', then the left hand side should be oranges or trips. This issue has to be approached with caution, however. Equation 2.43 again illustrates this: O_i is trips, while U_i is cars owned. However, the coefficient b, of U_i could be considered to have the dimension of trips per car owned, which sets matters right again.

2.5. Simultaneous equations – more than one dependent variable

So far, we have restricted ourselves to one equation with a single dependent variable. We can also consider the situation where there are many equations (simultaneous equations) with many dependent variables. It is in general true (though there are many important exceptions) that when there are exactly as many equations as there are 'unknowns' (i.e. dependent variables), then the equations can be solved for the unknowns. If there are more equations than unknowns, then the system is usually overdetermined; if there are fewer equations, the system is underdetermined. For our purposes, the second of these 'uneven' cases is the most important, and such equations sets will turn up as 'constraints' in Chapter 9. For the present, however, we will assume that we have the same number of equations as unknowns, and that a unique solution exists.

Consider a two equation – two variable system, where x_1 and x_2 are the dependent variables and y_1 and y_2 are independent variables (or may even be taken as constants). We are now using subscripts to label variables (to facilitate the extension to greater numbers of variables later) and also, x-variables as dependent variables and y-variables as independent, reverse of what we had before. We make no apology for this: there is no agreed convention, and so the reader is invited to consider this as an example of a notation switch!

Consider the following pair of equations:

$$0{\cdot}8x_1 - 0{\cdot}1x_2 = y_1 \tag{2.72}$$

$$-0{\cdot}3x_1 + 0{\cdot}7x_2 = y_2 \tag{2.73}$$

We have to solve these for x_1 and x_2. There are various ways of proceeding.

Method 1. Solve equation 2.72 for x_1 in terms of x_2 and y_2:

$$0{\cdot}8x_1 = 0{\cdot}1x_2 + y_1 \tag{2.74}$$

multiplying by 10/8,

$$x_1 = \frac{1}{8}x_2 + \frac{10}{8}y_1. \tag{2.75}$$

Then substitute this x_1 into equation 2.73 and solve for x_2:

$$-0{\cdot}3\left(\frac{1}{8}x_2 + \frac{10}{8}y_1\right) + 0{\cdot}7x_2 = y_2 \tag{2.76}$$

Multiply by 80 to simplify the coefficients:

$$-3(x_2 + 10y_1) + 56x_2 = 80y_2 \tag{2.77}$$

Thus

$$53x_2 = 80y_2 + 30y_1 \tag{2.78}$$

so

$$x_2 = \frac{80y_2 + 30y_1}{53}. \tag{2.79}$$

Then, we can substitute for x_2 into equation 2.75:

$$x_1 = \frac{1}{8}\left(\frac{80y_2 + 30y_1}{53}\right) + \frac{10}{8}y_1 \tag{2.80}$$

which, after some re-arrangement, gives

$$x_1 = \frac{70y_1 + 10y_2}{53}. \tag{2.81}$$

Method 2: We can operate on each side of either equation at will, in the same way and this, in effect, includes the possibility of adding or subtracting equations. Multiply equation 2.72 by 7 to give

$$5{\cdot}6x_1 - 0{\cdot}7x_2 = 7y_1. \tag{2.82}$$

Then add to equation 2.73

$$5{\cdot}3x_1 = 7y_1 + y_2 \tag{2.83}$$

so

$$x_1 = \frac{7y_1 + y_2}{5{\cdot}3} \tag{2.84}$$

agreeing with equation 2.81. Similarly, multiply equation 2.72 by 3 and equation 2.73 by 8 and add, eliminating x_1:

$$5{\cdot}3x_2 = 3y_1 + 8y_2 \tag{2.85}$$

so

$$x_2 = \frac{3y_1 + 8y_2}{5{\cdot}3} \tag{2.86}$$

agreeing with equation 2.79. In this case, method 2 is the quicker and more elegant, though less direct. This situation is common in mathematics: there is a 'sledgehammer' method of finding a solution (method 1 in this case), the most obvious method which can be guaranteed to work, or a more elegant and quicker method if it can be discovered. In Chapter 4, we will also see how to write this equation system in matrix terms. This leads to another method of solving simultaneous equations.

The equations 2.72 and 2.73, represent a model of a two sector economy. The variables x_1 and y_1 represent total product and final demand respectively for the first sector, and x_2 and y_2 equivalent variables for the second. The model generates the equations in the form 2.72 and 2.73, but we are usually given the final demand variables y_1 and y_2, and so we need to solve, as we have done, for x_1 and x_2.

This two variable–two equation system is a special case of a more general equation system which we will study more deeply in Chapter 4. Such a system, using doubly subscripted letters for the constants, can be written:

$$
\begin{aligned}
(1-a_{11})x_1 - a_{12}x_2 \ldots - a_{1N}x_N &= y_1 \\
-a_{21}x_1 + (1-a_{22})x_2 \ldots - a_{2N}x_N &= y_2 \\
&\vdots \\
-a_{N1}x_1 - a_{N2}x_2 \ldots + (1-a_{NN})x_N &= y_N
\end{aligned}
\qquad (2.87)
$$

The reader can easily check that if we take $N = 2$, $a_{11} = 0{\cdot}2$, $a_{12} = 0{\cdot}1$, $a_{21} = 0{\cdot}3$, and $a_{22} = 0{\cdot}3$, then this system reduces to that given by equations 2.72 and 2.73.

This is a convenient point to show how we can use a slightly more abstract mathematical notation to write such an equation system in a much more compact form. First, we write down the ith equation:

$$-a_{i1}x_1 - a_{i2}x_2 \ldots + (1-a_{ii})x_i \ldots - a_{iN}x_N = y_i \quad i = 1, 2, \ldots N \qquad (2.88)$$

where $i = 1, 2, \ldots$ N indicates that such an equation holds for each i value. Thus the whole system (2.87) can be described as in 2.88. We can represent it in an even more compact form if we define a special two subscript set of coefficients known as *Kronecker deltas:*

$$\delta_{ij} = \begin{cases} 1 \text{ if } i = j \\ 0 \text{ if } i \neq j \end{cases} \qquad (2.89)$$

Thus, $\delta_{11} = 1$, $\delta_{12} = \delta_{23} = \delta_{31} = 0$, etc. Then equation 2.88 can be written

$$\sum_{j=1}^{N} (\delta_{ij} - a_{ij})x_j = y_i, \quad j = 1, 2, \ldots N \qquad (2.90)$$

The reader can check this by writing the sum on the left hand side explicitly and using the definition 2.89.

So far, we have only considered linear simultaneous equations – the dependent variables only appear linearly, to the power 1. We could also consider non-linear, say quadratic or cubic simultaneous equations. Solution procedures are then much more difficult.

2.6. Inequalities

Four kinds of inequality relationships are commonly used:

$<$ means 'less than'

$\leqslant$ means 'less than or equal to'

$>$ means 'greater than'

$\geqslant$ means 'greater than or equal to'

These are then used to constrain variables. Thus

$$x < 2 \tag{2.91}$$

means 'x is less than 2'. Unlike an equation, we cannot solve for x, but we can note that the variable x takes a restricted range of values. It is possible to manipulate inequalities in much the same way as equations. We can add and subtract the same quantity to each side. For example, if

$$x + y \leqslant 4 \tag{2.92}$$

we can also write this as

$$x \leqslant 4 - y \tag{2.93}$$

Or, if

$$x^2 + 2x - y^2 - 2y \geqslant 0 \tag{2.94}$$

we can write this as

$$x^2 + 2x \geqslant y^2 + 2y \tag{2.95}$$

if this is more convenient.

We can also multiply or divide each side by the same quantity, *with one important proviso:* if the number which is being multiplied or divided is negative, then the inequality is reversed. Thus, if

$$x \geqslant 2 \tag{2.96}$$

then clearly

$$2x \geqslant 4 \tag{2.97}$$

but if we multiply by -2, we do *not* have

$$-2x \geqslant -4$$

(which implies $2x \leqslant 4$, or $x \leqslant 2$, the opposite to our starting result), but

$$-2x \leqslant -4 \tag{2.98}$$

which is equivalent to the original assumption.

At this stage, we content ourselves with the introduction of the concepts only. Their meaning will be illustrated in graphical terms in the discussion of co-ordinate geometry in Chapter 3, and they will play a substantial role in the discussion of mathematical programming in Chapter 9. In each of these cases, geographical examples will be given.

2.7. Some simple series

2.7.1. The arithmetic progression

We conclude this chapter on elementary algebra by introducing the concept of series, and giving the sums of some standard finite series. We can define a sequence, $u_n, n = 1, 2, \ldots N$, say, and then

$$S_N = u_1 + u_2 + \ldots + u_N = \sum_{n=1}^{N} u_n \qquad (2.99)$$

as the sum of N terms of the series.

Consider an *arithmetic progression:* each successive term is formed from the previous one by adding a constant difference, or:

$$u_n = u_{n-1} + \mathrm{d} \qquad (2.100)$$

and S_N is

$$S_N = a + (a + \mathrm{d}) + (a + 2\mathrm{d}) + \ldots + [a + (N - 1)\mathrm{d}] \qquad (2.101)$$

where a is the first term. Note that a occurs N times, and S_N can be written

$$S_N = Na + \mathrm{d}(1 + 2 + \ldots + N - 1) \qquad (2.102)$$

Put

$$s = 1 + 2 + \ldots + N - 1 \qquad (2.103)$$

We now use a trick; write the terms in reverse order:

$$s = N - 1 + 2 + \ldots + 1 \qquad (2.104)$$

Add equations 2.103 and 2.104:

$$2s = N + N + \ldots + N = N(N - 1) \qquad (2.105)$$

(since there are $N - 1$ terms) and so

$$s = N(N - 1)/2 \qquad (2.106)$$

Thus, substituting for $1 + 2 + \ldots + N - 1$ in equation 2.102:

$$S_N = Na + \frac{\mathrm{d}N(N-1)}{2} \qquad (2.107)$$

2.7.2. The geometric progression

The other common standard series is the *geometric progression* formed by multiplying successive terms by a constant factor r:

$$u_n = \mathrm{r}u_{n-1} \tag{2.108}$$

$$S_N = a + a\mathrm{r} + a\mathrm{r}^2 + \ldots + a\mathrm{r}^{N-1} \tag{2.109}$$

if a is the first term. We use another trick to find this sum. Multiply by r:

$$\mathrm{r}S_N = a\mathrm{r} + a\mathrm{r}^2 + \ldots + a\mathrm{r}^N \tag{2.110}$$

Subtract equation 2.110 from 2.109:

$$(1-\mathrm{r})S_N = a - a\mathrm{r}^N \tag{2.111}$$

and so

$$S_N = \frac{a - a\mathrm{r}^N}{1-\mathrm{r}} = \frac{a(1-\mathrm{r}^N)}{(1-\mathrm{r})} \tag{2.112}$$

This formula is useful in economic evaluation. If a project is implemented which, over 10 years say, yields net benefits B in each successive year, and R is the rate of interest, then the present value is

$$\mathrm{P} = \mathrm{B} + \frac{\mathrm{B}}{1+\mathrm{R}} + \frac{\mathrm{B}}{(1+\mathrm{R})^2} + \ldots + \frac{\mathrm{B}}{(1+\mathrm{R})^9} \tag{2.113}$$

This is a geometric progression with first term $a = \mathrm{B}$, and multiplier $\mathrm{r} = \frac{1}{1+\mathrm{R}}$. There are 10 terms, so $N = 10$. We can substitute into equation 2.112 to give

$$\mathrm{P} = S_{10} = \frac{\mathrm{B}\left[1 - \frac{1}{(1+\mathrm{R})^N}\right]}{1 - \frac{1}{1+\mathrm{R}}} \tag{2.114}$$

The geometric progression also allows us to introduce the concept of an infinite series. Let $N \to \infty$ (which means, let N become infinitely large) in equation 2.109, so

$$S = S_\infty = a + a\mathrm{r} + \ldots \tag{2.115}$$

$$= a(1 + r + r^2 + \ldots) \tag{2.116}$$

If * $|r| < 1$, this infinite series converges to a finite answer. In the formula 2.112, $r^N \rightarrow 0$ as $N \rightarrow \infty$ if $|r| < 1$, so

$$S = \frac{a}{1-r} \tag{2.117}$$

is the sum to infinity. Conversely, we note that this implies that

$$\frac{1}{1-r} = 1 + r + r^2 + r^3 \ldots \tag{2.118}$$

In other words, the right-hand side can be considered to be a series expansion of the left hand side. We shall find this particular result useful later in offering an interpretation of the economic base model.

Exercises

Section 2.1

1. Write expressions for the following, using a summation sign:

(a) $t_1 + t_2 + t_3 + t_4$, (b) $t_3 + t_4 + t_5$,

(c) $T_{i1} + T_{i2} + T_{i3} + T_{i4}$, (d) $T_{1j} + T_{2j} + T_{3j} + T_{4j}$

2. Write expressions for the following *without* summation signs:

(a) $\sum_{i=1}^{3} y_i$, (b) $\sum_{k=1}^{3} y_k$, (c) $\sum_{i=1}^{5} T_{ij}$, (d) $\sum_{j=1}^{3} S_{ijk}$,

(e) $\sum_{i=1}^{3} \sum_{j=1}^{4} U_{ij}$

Section 2.2

3. Draw a Cartesian co-ordinate system with scales on the x- and y-axes running from -10 to 10 in each case and plot the following points:

(0, 0), (0, 2), (3, 0), (1, 3), (1, 10), (10, 3), (−2, 6), (2, −6), (−2, −6).

4. Draw a polar co-ordinate system, allowing r to have a maximum of 10 units, and plot the following points (r, θ): (2,30°), (1,90°), (0, 0), (0,10°), (10,120°), (8,150°), (6,180°), (5,210°), (7,270°), (3,330°), (4,360°).

5. Sketch two zone systems for a hypothetical (or real!) study area to illustrate different levels of resolution. Does the finer system aggregate to the less fine? If not, design a third system which is an aggregate of the finer system.

* $|r|$ means 'the absolute value of r' – i.e. ignoring any negative sign.

6. Give some hypothetical values to the variables defined for von Thünen's problem in section 2.2.3. and calculate net profit for your example.

7. For some hypothetical network, label zones and assign distances to individual links in the manner of section 2.2.7. Find, by inspection, the sets R_{ij} for each (i, j) pair, and hence find the c_{ij} s using equation 2.18.

8. At time t, region 1 contains 100 people and region 2,200. In the period t to $t + T$, there are no births or deaths, but 20 people migrate from region 1 to region 2, and 80 people from region 2 to region 1. What are the terms $K_{11}(t, t + T)$, $K_{12}(t, t + T)$, $K_{21}(t, t + T)$ and $K_{22}(t, t + T)$ relevant to equation 2.23 for this problem? Hence find $P_i(t + T), i = 1, 2$, using equation 2.24

Section 2.3

9. Assume the following data:

i	x_i	y_i
1	5	2
2	4	6
3	6	3
4	10	8
5	2	7

and $W_1 = 0{\cdot}5$, $W_2 = 0{\cdot}5$, $W_3 = 1{\cdot}0$, $W_4 = 2{\cdot}0$, $W_5 = 2{\cdot}0$ and that y_i $T_{ij} = x_i + y_i$ and $S_{ijk} = T_{ij}W_k$. Find S_{122} (i.e. $S_{ijk}, i = 1, j = 2, k = 2$), S_{245}, S_{333}, and S_{451}.

Section 2.4

10. Solve the following equations for x:

(a) $2x + 3 = 5$, (b) $2x + a = b$, (c) $ax + b = c$, (d) $a/x + y = b$, (e) $P = (H - z)x$, (f) $x^2 - 7x + 12 = 0$, (g) $x^2 - x - 12 = 0$, (h) $2x^2 + x - 1 = 0$, (i) $4x^2 - 1 = 0$, (j) $ax^2 + 2bx + c = 0$, (k) $a_1x^2 + a_2x + a_3 = 0$, (l) $x^3 = 8$.

Section 2.5

11. Solve the following equations for x and y:
    (a) $0{\cdot}9x - 0{\cdot}2y = 0{\cdot}5,\ -0{\cdot}4x + 0{\cdot}6y = 0{\cdot}8$;
    (b) $3x + 2y = 12,\ 5x + y = 13$; (c) $5x - y = 2,\ x + y = 4$;
    (d) $10x + 3y = 6,\ x + 5y = 2$.

12. If δ_{ij} is a Kronecker delta, write the equation system

$$\sum_{j=1}^{3} (\delta_{ij} - a_{ij})x_j = y_i, \quad j = 1, 2, 3$$

in full, using neither the summation sign, Kronecker deltas or i or j subscripts.

Section 2.7

13. If $S = 2 + 5 + 8 + 11 + \ldots + 32$

find the sum of the terms on the right-hand side using the formula for an arithmetic progression.

14. If

$$S = 2 + 6 + 18 + 54 + 162$$

find the sum of the terms on the right-hand side using the formula for a geometric progression. Check by direct addition that the result is true.

3. Functions, graphs, co-ordinate geometry, and more equations

3.1. Introduction

IN this chapter, we complete our presentation of elementary mathematics and introduce concepts which are essential for all further development. The reader who reaches the end of this chapter, and completes the exercises, will have the tools to read much of the existing literature on the basis of the first three chapters alone, and so what follows is particularly important.

The argument is organized as follows. In Chapter 2, we introduced the notion of a relationship, expressed by an equation, between two variables x and y one of which was the dependent variable the other the independent. We begin by generalizing this concept of a *functional* relationship by introducing a formal general definition of a function in section 3.2. Next, we introduce the concepts of graphs and co-ordinate geometry, as this provides a useful tool for the visual presentation of particular functions. In section 3.4, we can then introduce a range of standard functions, and we show how to extend this range by combining functions in section 3.5. Some results on series expansion of functions are presented in section 3.6, followed by an account of functions of more than one independent variable. Up to this point, relatively few examples are introduced. Since the reader should by this time be equipped to handle much more algebraic manipulation, a range of examples are presented in section 3.8, which illustrate a wider variety of equations than was possible in Chapter 2, together with a number of the functions introduced earlier.

3.2. Some formal definitions and general notation

3.2.1. Definition of a function

Formally, a function can be defined as follows. Let A be a set of objects and B a set of objects, and then a function f, defined in the set A, is a rule (or relationship) associating with any element of set A, say a, an element of set B, say b. It can be written

$$b = \mathrm{f}(a) \tag{3.1}$$

This rule is sometimes called a *mapping* of the elements of the set A onto the set B. This is shown diagrammatically in Figure 3.1. This kind of diagram is one of the simplest ways of expressing the rule. If one of the sets is large, it is a particularly inconvenient way.

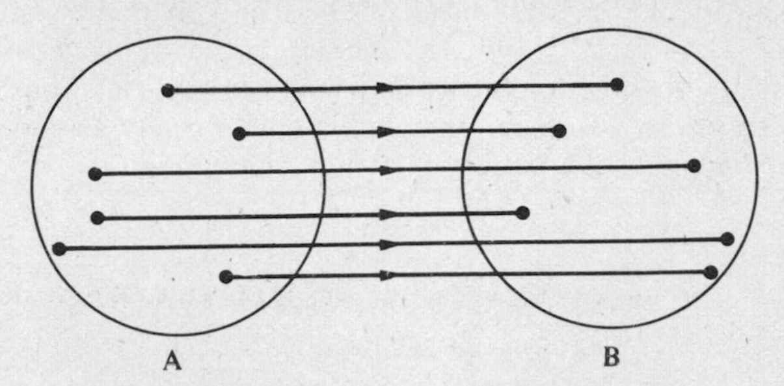

FIG. 3.1. A function relating the elements of two sets

This is an appropriate place to recall (cf. section 2.2.7) the sign for inclusion within a set: ∈. Thus,

$$a \in \mathrm{A} \tag{3.2}$$

is read: 'a is a member of the set A', and then equation 3.1 could be written more fully as

$$b = \mathrm{f}(a), \ a \in \mathrm{A}, \ b \in \mathrm{B} \tag{3.3}$$

meaning that the functional relationship holds for a in set A and b in set B.

Notice that here we are using the letters A and B to denote *sets* and not algebraic variables. We have not so far specified the nature of the objects in each set. Usually, of course, for our purposes, they will be numbers. For example, if A and B are each the set of all real numbers, a simple rule would be to say that 'b is the number which is twice a', and so in algebraic terms

$$b - 2a \tag{3.4}$$

In this case, f in equation 3.1 is the function 'multiplication by 2'. A similar functional relationship between y and x would be $y = 2x$, or in general

$$y = \mathrm{f}(x) \tag{3.5}$$

Following on from the Chapter 2 definitions, we would clearly define y to be the *dependent* variable and x to be the *independent* variable.

3.2.2. Notation

Relations such as 3.1 and 3.5 are functional relationships in a general notation. We use such a notation when we wish to say that y and x are functionally (say, 'causally', within a model) related, but we do not wish to specify the function (or we may not even know it). Thus, equation 3.5 reads: 'y is some function f of x'. We may use another letter to represent an alternative functional relationship:

$$y = \mathrm{g}(x) \tag{3.6}$$

for example. Or sometimes we may use the dependent variable itself:

$$y = y(x) \tag{3.7}$$

Note also that we may have a function of a function. For example, if

$$y = \mathrm{f}(x) \tag{3.8}$$

and

$$x = \mathrm{g}(z) \tag{3.9}$$

then

$$y = \mathrm{f}[\mathrm{g}(z)] \tag{3.10}$$

so that we can now exhibit y as a function of z directly. In such a case, x is an intermediate variable which can be eliminated. For example, if f is the function 'multiplication by 2' and y is the function 'squared', then

$$y = 2x \tag{3.11}$$

$$x = z^2 \tag{3.12}$$

and so

$$y = 2z^2 \tag{3.13}$$

3.2.3. Implicit and inverse functions

We have assumed so far that the dependent variable, say y, can always be shown explicitly in terms of the independent variable, say x, as in equations 3.8 or 3.11. Consider a relationship such as

$$x^2 + y^2 = a^2 \tag{3.14}$$

This gives the relationship *implicitly*. A general relationship of this form may be written

$$\mathrm{f}(x, y) = 0 \tag{3.15}$$

or, for the particular example,

$$x^2 + y^2 - a^2 = 0 \tag{3.16}$$

In this case, we can obtain an explicit form:

$$y = \pm\sqrt{(a^2 - x^2)} \tag{3.17}$$

though this is not always possible. Consider

$$y^5 + xy^2 + x^2y - 6 = 0 \tag{3.18}$$

for example. This is a quintic equation in y which cannot be solved explicitly for y, though it does express a relationship between y and x of the form 3.15.

Another useful concept is that of an *inverse* function. Suppose

$$\mathrm{f}(y) = x \tag{3.19}$$

Then, if such an equation is 'solved' for y, the solution is written

$$y = \mathrm{f}^{-1}(x) \tag{3.20}$$

where f^{-1} is the function which is the inverse of f. Note that f^{-1} is a whole symbol. It is *not* $1/\mathrm{f}$! Such a reciprocal would be written $[\mathrm{f}(x)]^{-1}$. For example, if f is the function 'squared', then

$$y^2 = x \tag{3.21}$$

so that

$$y = \sqrt{x} \tag{3.22}$$

and f^{-1} is the function 'square root'. We shall find in section 3.4 below that the concept of an inverse function will help to extend our range of functions since for any function, we can also investigate its inverse.

3.3. Graphs and co-ordinate geometry

3.3.1. Graphs

We introduced the basic concepts of Cartesian co-ordinate systems in section 2.2.1 of the previous chapter, and, in particular, in Figure 2.1. It should be immediately clear that this gives us the basis for 'plotting graphs' of functions: we simply plot each point (x, y) for which y and x are functionally related. Usually, in the cases in which we are interested, this process will generate a continuous curve, in some cases with occasional discontinuities. The simplest such relationship is

$$y = x \tag{3.23}$$

and a little thought or experimentation shows this to be a straight line through the origin at 45° to either axis, as in Figure 3.2. Any point on the line satisfies the relation 3.23. Equation 3.11 leads to a steeper

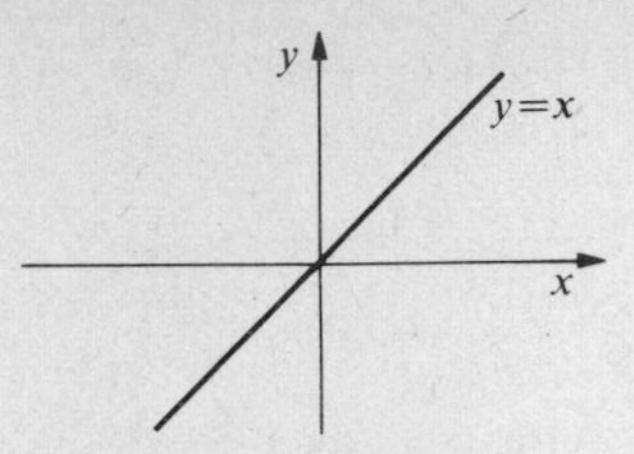

FIG. 3.2. Graph of $y = x$

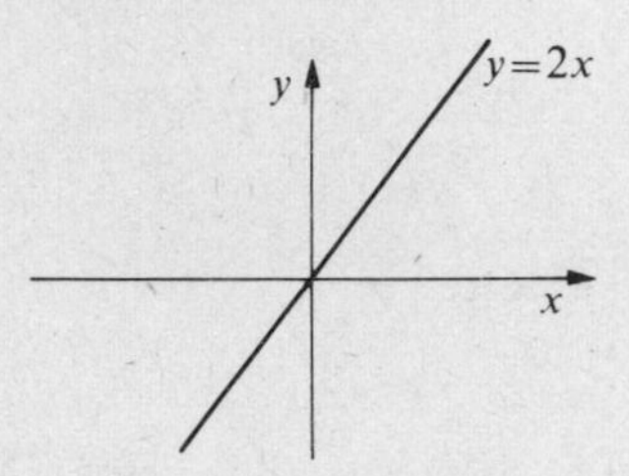

FIG. 3.3. Graph of $x = z^2$

straight line, as shown in Figure 3.3. The relationship between x and z in equation 3.12 is shown in Figure 3.4.

Implicitly defined functions, such as that in equation 3.14 can also be represented graphically. This particular case can be seen to represent

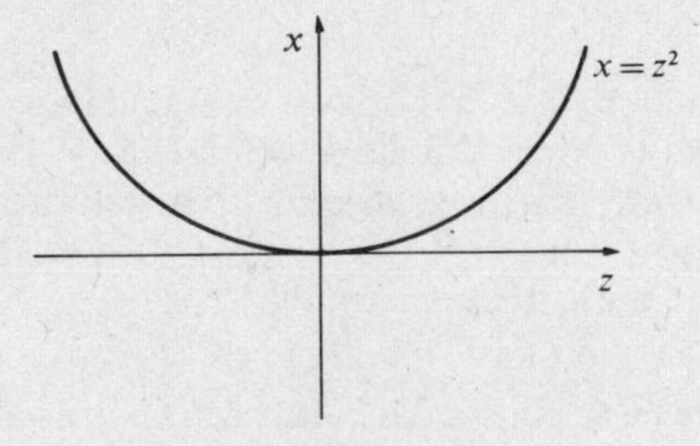

FIG. 3.4. Graph of $x = z^2$

the set of points on a circle about the origin of radius a, as in Figure 3.5.

So far, we have only considered well-defined algebraic relationships between two variables, and this produces *smooth* lines and curves. We can also plot graphs of empirical relationships between pairs of variables, and indeed this is one of the most common ways of representing such data. From an experiment, or survey, only a finite number of observations will be available which could be plotted as in Figure 3.6.

Such relationships are rarely smooth, if only because of errors in measurement as well as the complicated nature of natural or social

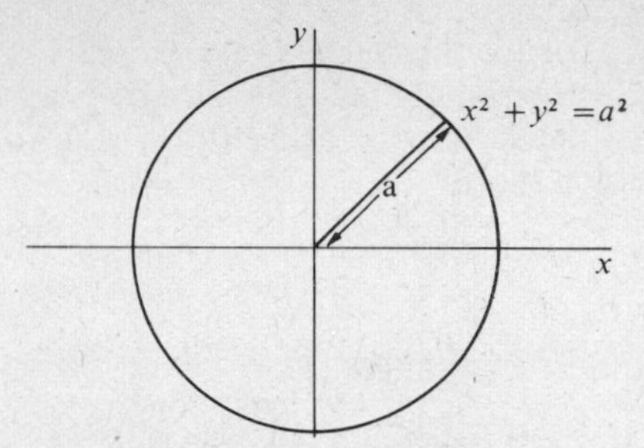

FIG. 3.5. Graph of $x^2 + y^2 = a^2$

systems. However, in Figure 3.6 (a), we might think that the data exhibit a more or less linear relationship, while in Figure 3.6 (b) a quadratic one. Many other examples are suggested for the reader's investigation in the

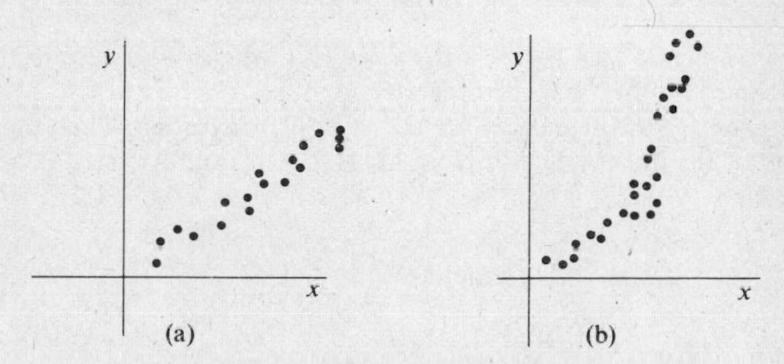

FIG. 3.6. Graphs of empirical data

exercises at the end of the chapter. The test of fitting smooth curves to empirical data is one of the main tasks of statistics, and we shall return to this, for the straight line case in Chapter 5. These informal concepts will suffice for the time being.

3.3.2. The geometry of the straight line

A relationship of the form

$$ay + bx + c = 0 \tag{3.24}$$

where a, b and c are constants is known as a linear relationship, since only linear (first-order) terms in y and x are included. Without loss of generality (by dividing through by a, renaming the constants, and rearranging), it can be written in the form

$$y = mx + c \tag{3.25}$$

We begin with the assertion that such linear relationships always represent straight lines when plotted, and conversely, any straight line has a corresponding linear equation of the form 3.25. We shall be able to justify the assertion shortly.

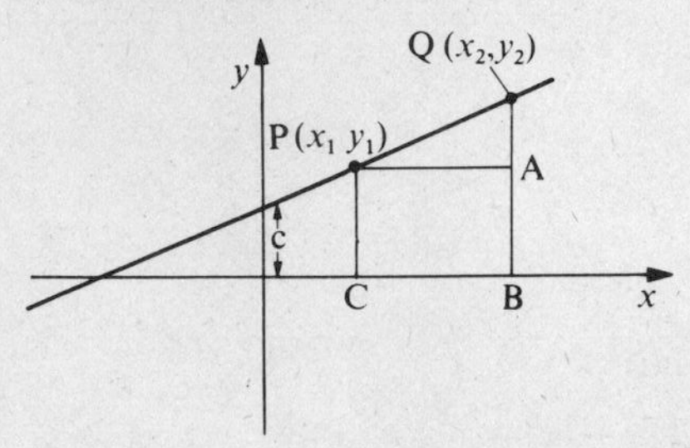

FIG. 3.7. The straight line $y = mx + c$

It can easily be seen that c is the value of y when $x = 0$, and so it is the *intercept* on the y axis. To investigate m, consider Figure 3.7. Let $P(x_1, y_1)$ be any point on the line and $Q(x_2,y_2)$ any other. Then the *gradient* of the line is AQ/AP. Now AQ is $y_2 - y_1$, and AP is $x_2 - x_1$. So, we have

$$\text{gradient} = \frac{y_2 - y_1}{x_2 - x_1} \tag{3.26}$$

and we must have

$$y_2 = mx_2 + c \tag{3.27}$$

$$y_1 = mx_1 + c \tag{3.28}$$

since both points are on the line, so, using these equations

$$y_2 - y_1 = mx_2 - mx_1 = m(x_2 - x_1) \tag{3.29}$$

We can now substitute from equation 3.29 into the numerator of the right hand side of 3.26:

$$\text{gradient} = \frac{m(x_2 - x_1)}{x_2 - x_1} = m \tag{3.30}$$

Thus, we see that m is the gradient of the line.

A straight line, by definition, has a constant gradient, and the above algebra thus bears out our assertion of the one-to-one connection between straight lines and linear equations.

This is a convenient point to introduce δ as the notation for an increment in a variable.* For points P and Q in Figure 3.7, we might have

$$\delta y = y_2 - y_1 \tag{3.31}$$

$$\delta x = x_2 - x_1 \tag{3.32}$$

These quantities are shown in Figure 3.8: The gradient m can now be seen to be

$$\text{m} = \frac{\delta y}{\delta x} \tag{3.33}$$

We now see, from another viewpoint, why the line $y = 2x$ in Figure 3.3 is 'steeper' than $y = x$ in Figure 3.2. Also, the definition 3.26 shows that if m is positive, the line is forward sloping, as in Figures 3.2 and 3.3, while if it is negative, the line is backward sloping. We can construct an example of a backward sloping line as follows. Suppose a total

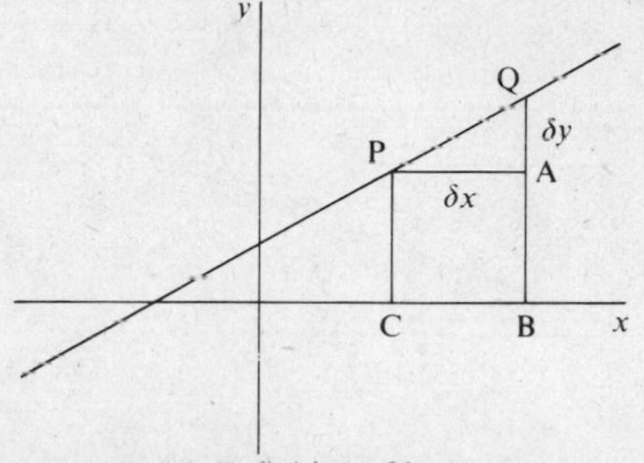

FIG. 3.8. Definition of increments

amount B, is the available budget for transport investment in a given year, and that y can be spent on road building and x on public transport facilities. Then,

$$y + x = \text{B} \tag{3.34}$$

or, if we use the standard form 3.25

$$y = -x + \text{B} \tag{3.35}$$

we see that this is a straight line with gradient -1 and intercept B as shown in Figure 3.9.

3.3.3. The intersection of two straight lines

Unless two lines are parallel, they intersect at a point. We can begin to

* Occasionally a capital greek delta, Δ, is used for this purpose.

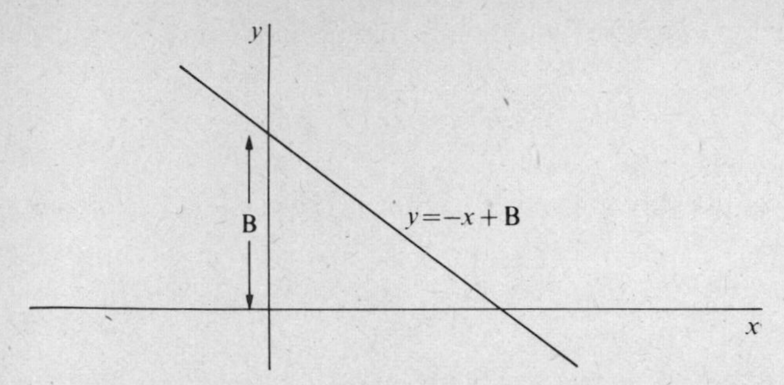

FIG. 3.9. A backward sloping straight line

show how algebra can be represented geometrically, or *vice versa*, by investigating the intersection of two straight lines from both viewpoints.

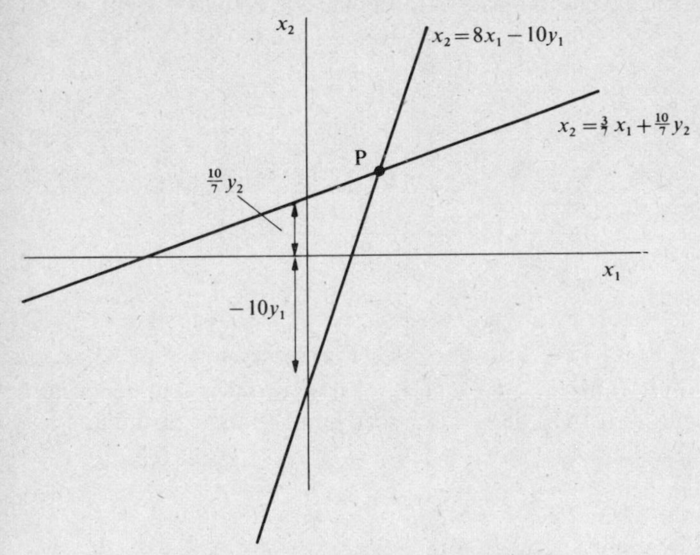

FIG. 3.10. Intersection of two straight lines

Consider the two equations 2.72 and 2.73 in Chapter 2, repeated here for convenience:

$$0{\cdot}8x_1 - 0{\cdot}1x_2 = y_1 \tag{3.36}$$

$$-0{\cdot}3x_1 + 0{\cdot}7x_2 = y_2 \tag{3.37}$$

These can be written

$$x_2 = 8x_1 - 10y_1 \tag{3.38}$$

$$x_2 = \tfrac{3}{7}x_1 + \tfrac{10}{7}y_2 \tag{3.39}$$

and treated as straight lines in the (x_1, x_2) plane, the first having gradient 8 and intercept $-10y_1$, the second, gradient $\frac{3}{7}$ and intercept $\frac{10}{7}y_2$. They are 'plotted' (notionally – because we do not have numerical values for the constants y_1 and y_2) in Figure 3.10. The point P which is the intersection of the two lines is, of course, on *each* line, and so its co-ordinates are simply the solution of the simultaneous equations already known from 2.79 and 2.81 to be $\dfrac{80y_2 + 30y_1}{53}, \dfrac{70y_1 + 10y_2}{53}$. If y_1 and y_2 had numerical values, we now see that the equations could have been solved graphically: the lines could have been plotted, and the solution 'read-off' as the co-ordinates of the point of intersection. Conversely, given any two straight lines, say

$$y = m_1x + c_1 \tag{3.40}$$

and

$$y = m_2x + c_2 \tag{3.41}$$

we can find the co-ordinates of the point of intersection by solving them as simultaneous equations. Substitute for y from 3.41 into 3.40:

$$m_2x + c_2 = m_1x + c_1 \tag{3.42}$$

so

$$x = -\frac{c_2 - c_1}{m_2 - m_1} \tag{3.43}$$

Then, using equation 3.40

$$y = -m_1\frac{(c_2 - c_1)}{m_2 - m_1} + c_1 \tag{3.44}$$

$$= -m_1\frac{(c_2 - c_1)}{m_2 - m_1} + \frac{c_1(m_2 - m_1)}{m_2 - m_1} \tag{3.45}$$

so

$$y = \frac{c_1m_2 - m_1c_2}{m_2 - m_1} \tag{3.46}$$

3.3.4. The distance between two points

It is convenient to have a formula for the distance between two points. Consider the points $P(x_1, y_1)$ and $Q(x_2, y_2)$. Pythagoras' theorem tells us that, in Figure 3.7

$$PQ^2 = AP^2 + AQ^2 \tag{3.47}$$

so

$$PQ = \sqrt{(AP^2 + AQ^2)} \tag{3.48}$$

Thus, our formula is

$$PQ = \sqrt{((x_2 - x_1)^2 + (y_2 - y_1))} \tag{3.49}$$

3.3.5. Equations of particular straight lines

The standard form of equation of a straight line is 3.25, and since this contains two constants m and c, we can find these, and hence the equation, given any two pieces of information about the line. We consider some of the most important cases in turn.

(a) *Given gradient and intercept:* in this case (which is not the most common in practice) m and c are given directly.

(b) *Given two points on the line,* say (x_1, y_1) and (x_2, y_2). The equation can be obtained in this case by a trick. Consider Figure 3.11.

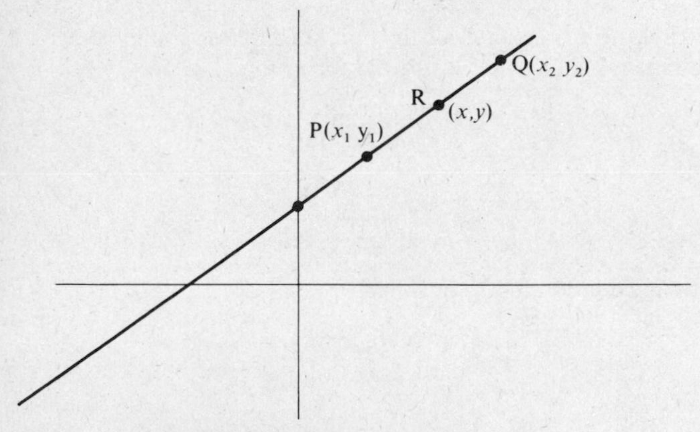

FIG. 3.11. A line through two given points

If $R(x, y)$ is any point on the line, then the gradient of PR must equal the gradient PQ. Thus

$$\frac{y - y_1}{x - x_1} = \frac{y_2 - y_1}{x_2 - x_1} \tag{3.50}$$

This can be rearranged as

$$y - y_1 = (x - x_1)\frac{(y_2 - y_1)}{x_2 - x_1} \tag{3.51}$$

and so

$$y = x\frac{(y_2 - y_1)}{(x_2 - x_2)} + y_1 - x_1\frac{(y_2 - y_1)}{x_2 - x_1} \tag{3.52}$$

which can be written

$$y = x\frac{(y_2 - y_1)}{(x_2 - x_1)} + \frac{x_2 y_1 - x_1 y_2}{x_2 - x_1} \tag{3.53}$$

and so

$$\mathrm{m} = \frac{y_2 - y_1}{x_2 - x_1} \tag{3.54}$$

(which is not surprising), and

$$\mathrm{c} = \frac{x_2 y_1 - x_1 y_2}{x_2 - x_1} \tag{3.55}$$

(c) *Given the gradient, m, and one point* (x_1, y_1) on the line. We must have

$$y = \mathrm{m}x + \mathrm{c} \tag{3.56}$$

and we can find c since

$$y_1 = \mathrm{m}x_1 + \mathrm{c}, \tag{3.57}$$

giving

$$\mathrm{c} = y_1 - \mathrm{m}x_1 \tag{3.58}$$

(d) Given the gradient m of a *perpendicular* line and a point. That is, the task is to find the equation of the *normal* to a line at a point (x_1, y_1) say. If the given gradient is m, the gradient of a perpendicular line is $-\frac{1}{\mathrm{m}}$, so the line is

$$y = -\frac{x}{\mathrm{m}} + \mathrm{c} \tag{3.59}$$

and we can find c as before since we must have

$$y_1 = -\frac{x_1}{m} + c \qquad (3.60)$$

giving

$$c = y_1 + \frac{x_1}{m} \qquad (3.61)$$

3.3.6. The interpretation of linear inequalities

Linear inequalities take the same form as linear equations, but with inequality signs replacing equality signs. The main concepts were introduced in section 2.6 of the previous chapter. We can now use our knowledge of the geometry of the straight line to give a graphical interpretation to inequalities. In equation 3.34, which we used to illustrate a backward sloping straight line, amounts of expenditure on highways x and public transport y were related to the total budget, B. It would be more usual if this relationship was represented by an inequality, with B as an upper limit:

$$x + y \leqslant B \qquad (3.62)$$

Any such inequality can be given a graphical interpretation with the help of the corresponding straight line, in this case

$$x + y = B \qquad (3.63)$$

which was plotted in Fig. 3.9. Any point satisfying the inequality lies 'below' the line and (since also $x \geqslant 0$ and $y \geqslant 0$ because negative expenditures are impossible) this is the shaded area in Figure 3.12. Such an area is called a feasible region – it consists of 'possible' (x, y) points.

Suppose we now modify the example slightly. Let x be the amount of investment in factory X in a town, and y that in factory Y. Suppose 100 jobs per unit of investment in X are created, and 50 in Y, so that $100x + 50y$ jobs are created, and suppose 20 units of expenditure for local shops are created per unit for X and 40 units per unit of Y, giving $20x + 40y$ in all. Then if B is a budget constraint, we may require

$$x + y \leqslant B \qquad (3.64)$$

but also

$$100x + 50y \geqslant E \qquad (3.65)$$

and

$$20x + 40y \geqslant S \qquad (3.66)$$

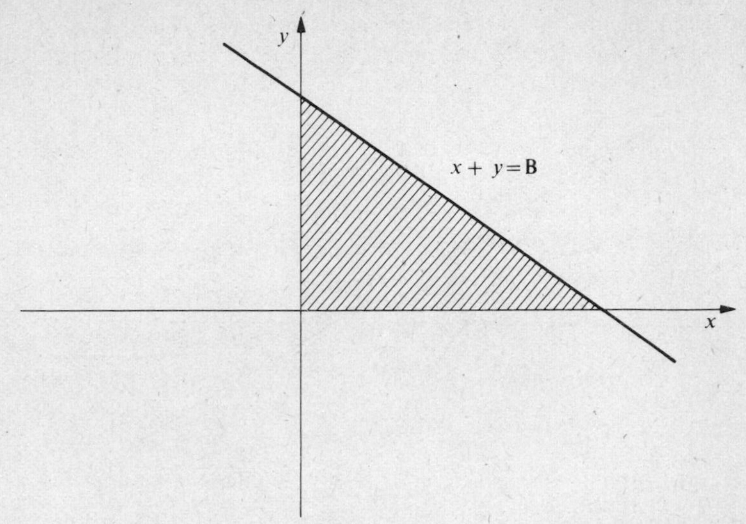

FIG. 3.12. Interpretation of a linear inequality – 1

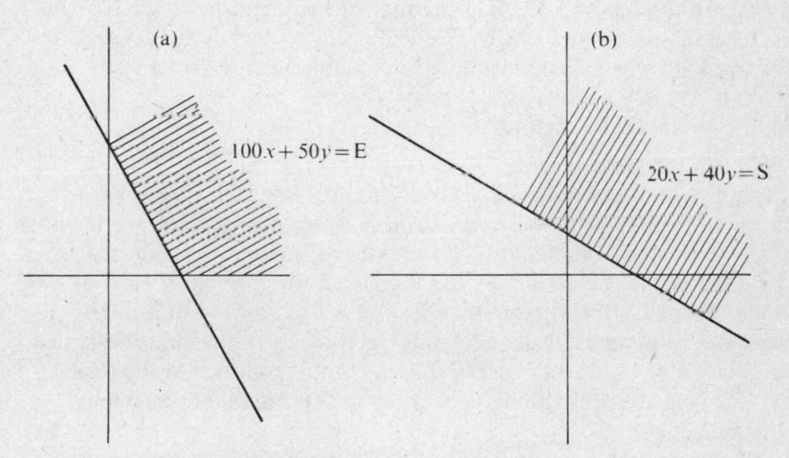

FIG. 3.13. Interpretation of a linear inequality – 2

to generate some minimum number of jobs E and some minimum purchasing power S. The straight lines associated with 3.65 and 3.66 are

$$100x + 50y = \mathrm{E} \tag{3.67}$$

and

$$20x + 40y = \mathrm{S} \tag{3.68}$$

and the inequalities are satisfied for points *above* these lines, as in Figure 3.13. Since for this problem all three inequalities must hold

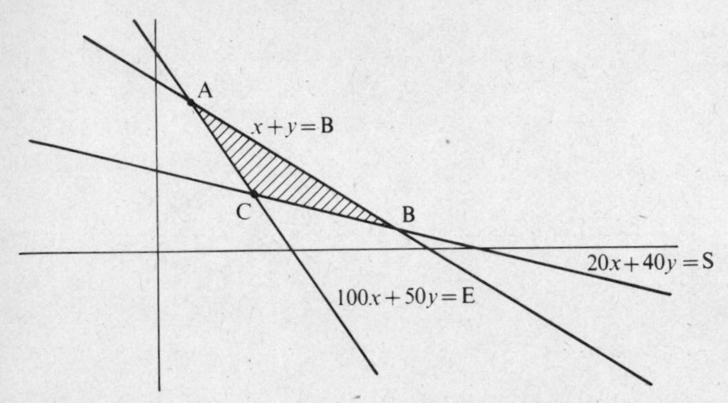

FIG. 3.14. Interpretation of three simultaneous linear inequalities

simultaneously, we need to identify the feasible region which is common to Figures 3.12, 3.13 (a) and 3.13 (b) and this is the triangle ABC exhibited in Figure 3.14. We will need to identify feasible regions of this kind when we discuss linear programming in Chapter 9.

3.4. The standard functions

3.4.1. Introduction

Although it is in some sense arbitrary to decide which functions are 'standard' and which not, in this section we give an account of the most commonly occurring functions, and refer to these as the standard set. Where possible, we define the functions in a first-principles kind of way, and we indicate their nature graphically. A full account of how to produce such graphs really demands the tools of the calculus described in Chapter 5, to identify such things as turning points, but the reader will be able to check the sense of the graphs presented by plotting a few points himself.

We begin with a detailed discussion of the *power* function. This is important in its own right, but also as a basis for the construction of logarithmic and exponential functions as implicit or inverse power functions. We then discuss the main trigonometric functions and, in passing, present the bases of trigonometry, and we note the existence of inverse trigonometric functions. We will then show how to construct a wide range of further functions from this standard set in section 3.5.

3.4.2. The power function

We have already introduced the concepts of x^2 as $x \times x$, x^3 as $x \times x \times x$, and so on. x^2 is read as 'x squared', or 'x raised to the power 2 (or index 2)'. Thus, we have defined the function

$$y = x^n \tag{3.69}$$

for positive integer values of n, and such a function is called a power function. We now show how to extend the definition so that x^n has a meaning for other values of n. First, note that

$$y = x \tag{3.70}$$

could be written

$$y = x^1 \tag{3.71}$$

In other words, x is the power function for $n = 1$, though the 1 is rarely written explicitly as in equation 3.71. Further, by convention, we can define

$$y = x^0 = 1 \tag{3.72}$$

giving x^n a meaning for $n = 0$. That is, anything raised to the power zero is 1. x^n for n as a negative integer can be dealt with if we note that

$$1/x = x^{-1} \tag{3.73}$$

$$1/x^2 = x^{-2} \tag{3.74}$$

and, generally,

$$1/x^n = x^{-n} \tag{3.75}$$

The reader can easily check that, with these definitions, the following rules for combining power functions hold:

$$x^m \times x^n = x^{m+n} \tag{3.76}$$

$$x^m/x^n = x^{m-n} \tag{3.77}$$

$$(x^m)^n = x^{mn} \tag{3.78}$$

These relationships hold for all integer values of m and n.

We now proceed to deal with non-integer values of n. Suppose we define the function y by

$$y^n = x \tag{3,79}$$

where n is an integer. That is, we use an implicit definition. y is that number which, when raised to the nth power, is equal to x. y is known as the nth root of x, and is sometimes written explicitly as

$$y = \sqrt[n]{x} \tag{3.80}$$

The most familiar example of such a relationship is the square root. If

$$y^2 = x,$$

then

$$y = \sqrt[2]{x} = \sqrt{x} \tag{3.81}$$

(where in this case the 2 in the root sign is usually omitted). We can also write

$$y = x^{1/2} \tag{3.82}$$

or more generally, in relation to the definition 3.79

$$y = x^{1/n} \tag{3.83}$$

Thus, this value of y is the value given to x^m, where $m = 1/n$, and n is an integer. Thus, we can now have fractional powers. Since

$$1/x^{1/n} = x^{-1/n} \tag{3.84}$$

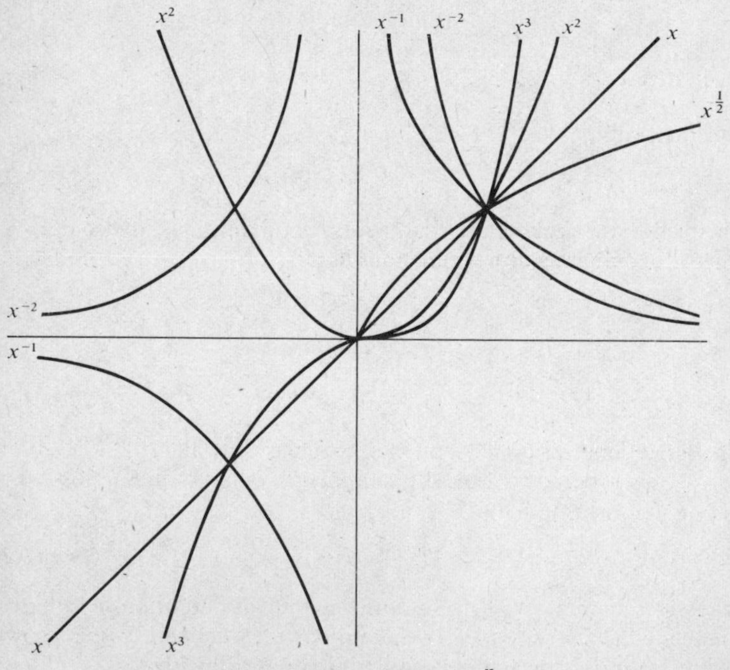

FIG. 3.15. Graphs of $y = x^n$

we can have negative fractions, and since

$$(x^{1/n})^m = x^{m/n} \tag{3.85}$$

where m and n are both integers, the definition can easily be extended to non-integers greater than 1. Figure 3.15 shows $y = x^n$ plotted for various values of n.*

3.4.3. The logarithmic function

We can define the logarithmic function with another implicit definition. For some constant a, let

$$a^y = x \tag{3.86}$$

That is, when a is raised to the power y, the result is x. Then y is known as the logarithm of x to the base a and is written explicitly as

$$y = \log_a x \tag{3.87}$$

The most common values of a are 10, e (the base of 'natural' logarithms, which we discuss in section 3.4.4 below) or 2. Values of $\log_{10}x$ are given in standard tables. Some examples, from tables, are given in Table 3.1 below.

TABLE 3.1

Some logarithms

x	$y = \log x$	because
1	0	$10^0 = 1$
2	0·3010	$10^{0\cdot3010} = 2$
3	0·4771	$10^{0\cdot4771} = 3$
4	0·6021	$10^{0\cdot6021} = 4$
5	0·6990	$10^{0\cdot6990} = 5$
.	.	.
.	.	.
10	1	$10^1 = 10$

Most of the interesting and distinctive properties of logarithms are independent of the base, and so they will be stated here for base a

* In this, and subsequent, figures, the 'plots' are indicative only and should not be taken as exact.

Since

$$a^0 = 1 \tag{3.88}$$

$$\log_a 1 = 0 \tag{3.89}$$

and

$$a^1 = a \tag{3.90}$$

so

$$\log_a a = 1 \tag{3.91}$$

Perhaps the most important property is

$$\log_a x_1 + \log_a x_2 = \log_a(x_1 x_2) \tag{3.92}$$

That is: the sum of the logarithms of two numbers is equal to the logarithm of the product. It is this property which facilitates multiplication using log tables, or the operation of slide rules. This result can be proved as follows: put

$$\log_a x_1 = y_1 \tag{3.93}$$

so

$$a^{y_1} = x_1 \tag{3.94}$$

and similarly define y_2 such that

$$\log_a x_2 = y_2 \tag{3.95}$$

so

$$a^{y_2} = x_2 \tag{3.96}$$

Thus,

$$x_1 x_2 = a^{y_1} a^{y_2} = a^{y_1 + y_2} \tag{3.97}$$

using 3.76. By definition

$$\log(x_1 x_2) = y_1 + y_2 \tag{3.98}$$

and so equations 3.93, 3.95 and 3.98 prove the result. Similarly, it can be shown that

$$\log_a x_1 - \log_a x_2 = \log_a(x_1/x_2) \tag{3.99}$$

and

$$\log_a x^m = m \log_a x \tag{3.100}$$

Finally, we show the relationship between logarithms to different bases. If

$$a^{y_1} = x \tag{3.101}$$

$$y_1 = \log_a x \tag{3.102}$$

and

$$b^{y_2} = x \tag{3.103}$$

$$y_2 = \log_b x \tag{3.104}$$

We must have

$$a^{y_1} = b^{y_2} \tag{3.105}$$

Take logs to base a:

$$\log_a a^{y_1} = \log_a b^{y_2} \tag{3.106}$$

and so, using equation 3.100 on each side, and noting that $\log_a a = 1$,

$$y_1 = y_2 \log_a b \tag{3.107}$$

so

$$\log_a x = \log_a b \times \log_b x \tag{3.108}$$

$\log_a x$ is only defined for $x > 0$ and is plotted in Fig. 3.16. As expected, it increases much more slowly than $y = x$.

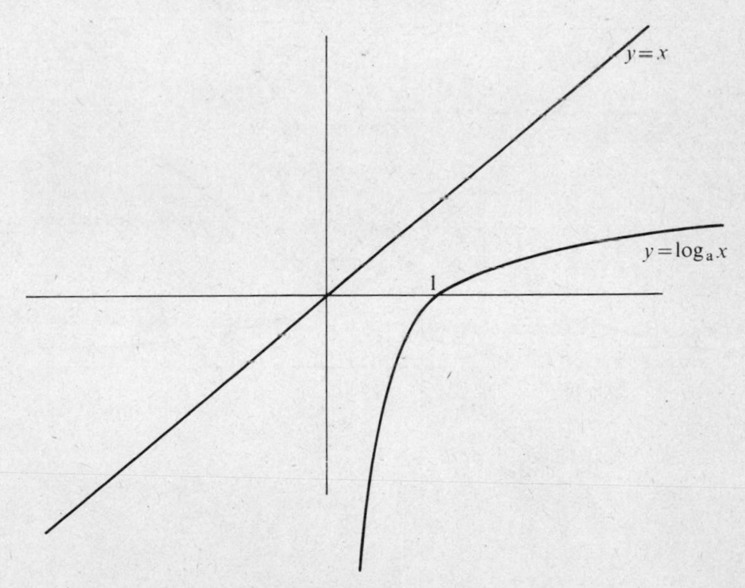

FIG. 3.16. Graphs of $y = \log_a x$

3.4.4. The exponential function

A particularly important natural number (of the same nature as π) is e. It has the following value to 8 decimal places:

$$e = 2{\cdot}71828182 \tag{3.109}$$

We shall learn more about its significance in Chapter 5 when the tools of the calculus are available, and we will also have a method of calculating it. Meanwhile, we can define

$$y = e^x \tag{3.110}$$

as the exponential function. It is sometimes written

$$y = \exp(x) \tag{3.111}$$

Note that, using the earlier definition of a logarithm, this implies that

$$x = \log_e y \tag{3.112}$$

so the exponential function is the inverse of one of the logarithmic functions.

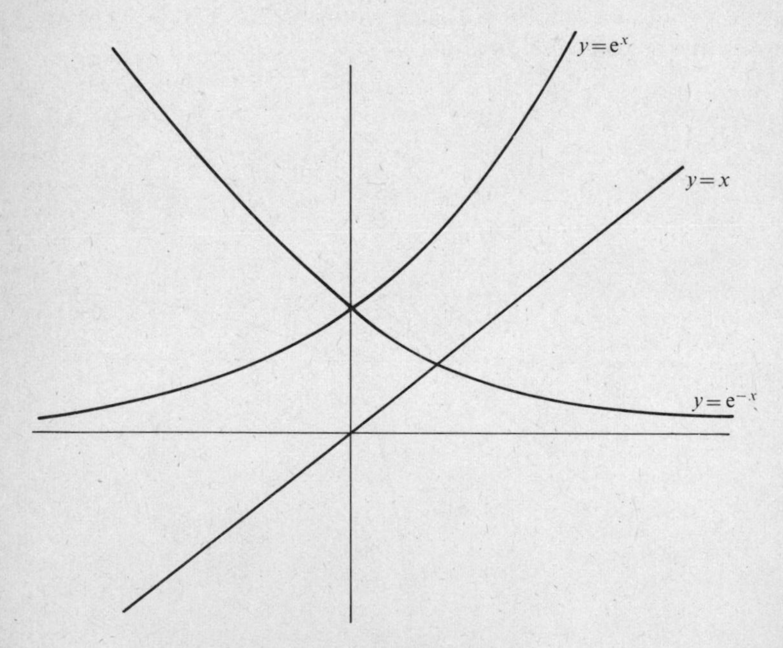

FIG. 3.17. Exponential functions

We can also define

$$y = e^{-x} \tag{3.113}$$

as the negative exponential function. Both of these exponential functions are plotted, relative to $y = x$, in Figure 3.17. More generally, we can introduce a constant a (which can take either sign) and define

$$y = e^{ax} \tag{3.114}$$

and varying a varies the rate of growth. Some exponential functions for

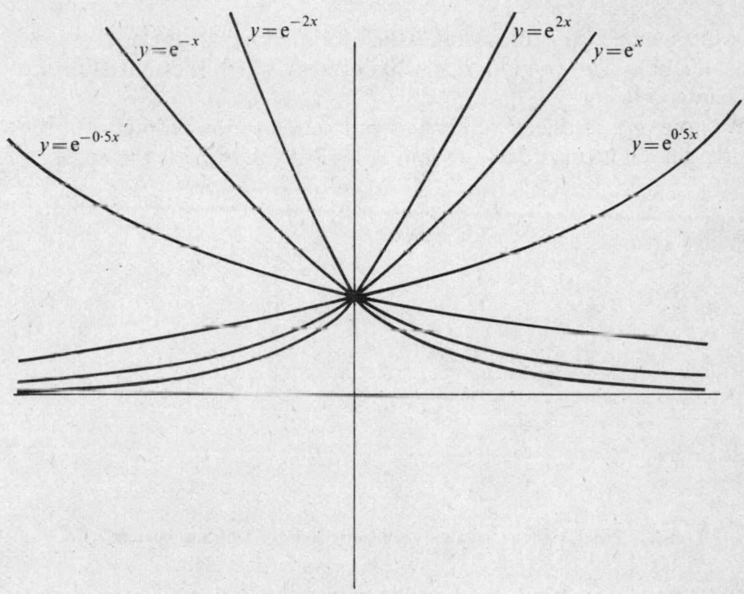

FIG. 3.18. Graphs of $y = e^{ax}$

a range of values of a are plotted in Figure 3.18.

3.4.5. Trigonometric functions

Consider the angle θ as defined in relation to a Cartesian co-ordinate system in Figure 3.19. The angle between OP and OA is θ. Let r be the length of OP, so using Pythagoras' theorem,

$$r^2 = x^2 + y^2 \tag{3.115}$$

or

$$r = \sqrt{(x^2 + y^2)} \tag{3.116}$$

The principal trigonometric functions are then defined by the following ratios:

$$\sin\theta = \frac{y}{r} \tag{3.117}$$

$$\cos\theta = \frac{x}{r} \tag{3.118}$$

$$\tan\theta = \frac{y}{x} \tag{3.119}$$

We will assume, in the discussion which follows that all angles are measured in *radians* (π radians = 180 degrees), which is essential for use in Chapter 5 below.

We note one particular result here in relation to the geometry of the straight line. It is easy to see, from Figure 3.19, that if θ is the angle

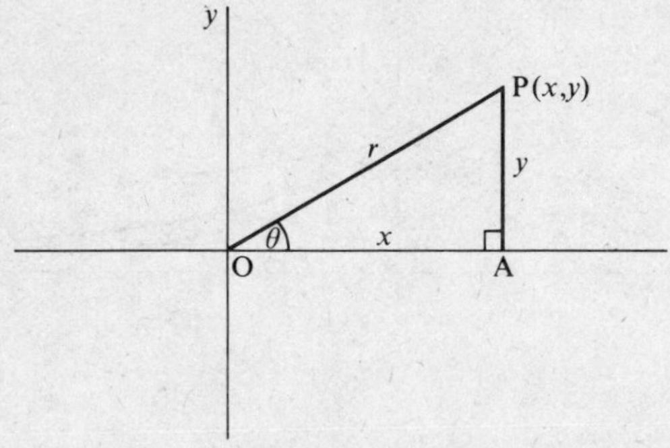

FIG. 3.19. Angle θ, related to a Cartesian co-ordinate system

between any line $y = mx + c$ and the x-axis, then since m is the gradient of the line, it can be identified with $\tan\theta$.

The sin, cos, and tan functions take a range of values as θ moves from 0 to 2π, and then they repeat. In other words, they are *cyclic* functions. Plots are shown in Figure 3.20, and the reader can check these from the above definitions.

Although these functions have been defined in terms of an angle, they have more general uses. In terms of a general relation $y = f(x)$, between y and x, it will sometimes be appropriate to take f as one of these functions for example

$$y = \sin x \tag{3.120}$$

when an oscillating function is required.

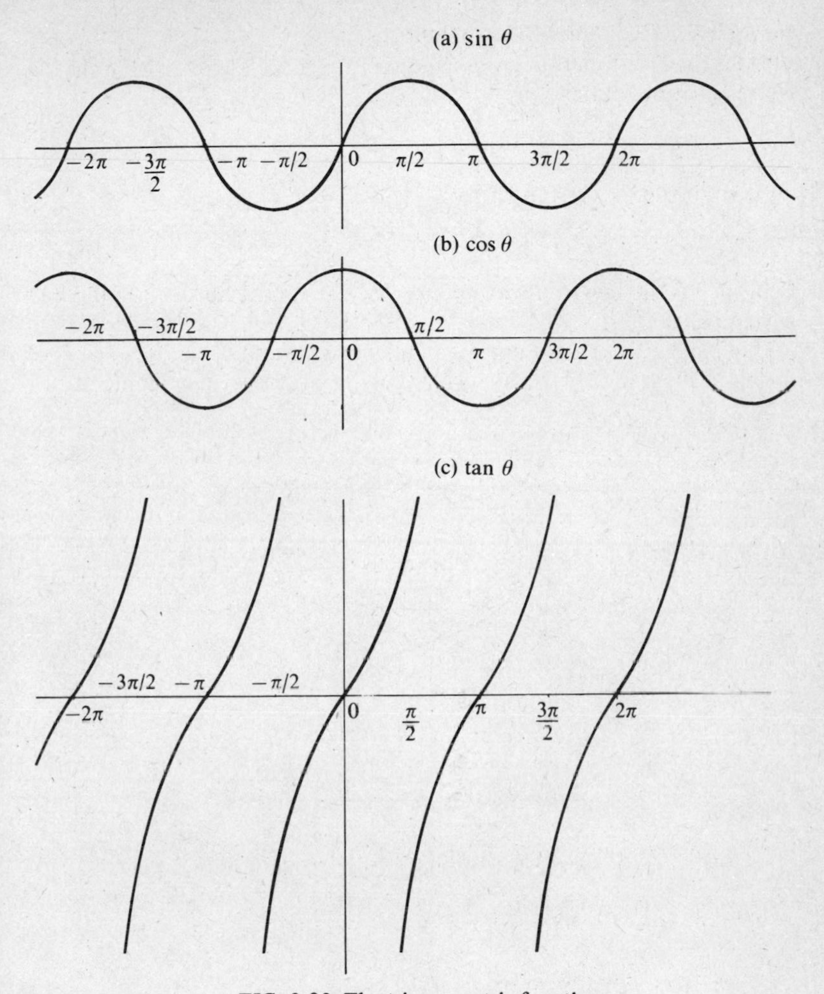

FIG. 3.20. The trigonometric functions

The reciprocals of the trigonometric functions are sometimes given special names, and are defined as follows:

$$\operatorname{cosec}\theta = 1/\sin\theta \qquad (3.121)$$

$$\sec\theta = 1/\cos\theta \qquad (3.122)$$

$$\operatorname{cotan}\theta = 1/\tan\theta \qquad (3.123)$$

3.4.6. Inverse trigonometric functions

Consider the sine function given in equation 3.120. To investigate the inverse function, define y by

$$\sin y = x \qquad (3.124)$$

and then, formally, we can write

$$y = \sin^{-1} x \qquad (3.125)$$

[N.B. $\sin^{-1}$ is the inverse function, *not* $1/\sin x$, that reciprocal should be written $(\sin x)^{-1}$.]

Thus, in 3.125, y is the angle, in radians, whose sine is x. It is plotted in Figure 3.21. (Note, incidentally, that this is an example of

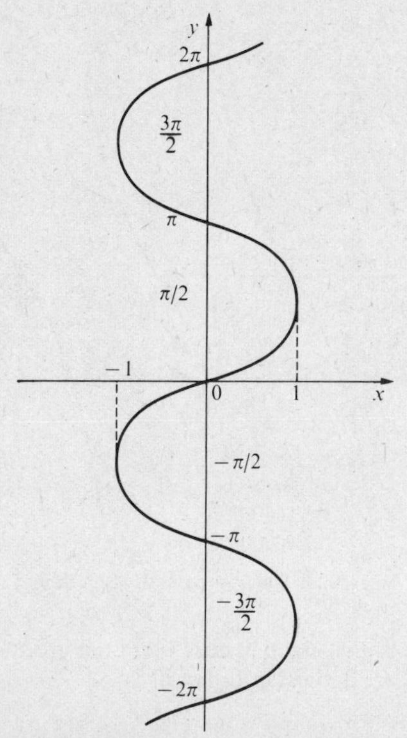

FIG. 3.21. An inverse trigonometric function: $y = \sin^{-1}x$

a many-valued function. Most of our functions give a single y corresponding to any given x, but in this case, for $-1 \leqslant x \leqslant 1$, there are many possible y values.)

In an analogous manner we could define

$$y = \cos^{-1}x \tag{3.126}$$

and

$$y = \tan^{-1}x \tag{3.127}$$

and plot these. These functions do not have direct applications at present, but are important in relation to various standard integrals in Chapter 5.

3.5. A wider range of functions

3.5.1. Introduction

A wide range of functions can be obtained by combining the standard functions in various ways using any of the standard algebraic operations, or the 'function of a function' principle. Thus, we might form

$$y = x^2 + x + 1 \tag{3.128}$$

$$y = x^3 - 3x^2 \tag{3.129}$$

$$y = (\sin x)/x \tag{3.130}$$

$$y = x \log_e x \tag{3.131}$$

using standard operations, or

$$y = \sin(x^2) \tag{3.132}$$

using the 'function of a function' concept. We can go on to form ever more elaborate combinations of standard functions if required. In the rest of this section, we present a number of combinations of standard functions which are almost standard, or which have particular geographical and planning applications.

3.5.2. Polynomial and rational functions; graphical solution of equations

A function which is formed as a sum of power functions, say

$$y = x^2 + 2x + 1 \tag{3.133}$$

is known as a polynomial. It is said to be of the degree of the highest power present, so the above example is quadratic, or second order. In general terms, a polynomial of the nth degree may be written

$$y = a_n x^n + a_{n-1} x^{n-1} + \ldots + a_0 \tag{3.134}$$

$$= \sum_{m=0}^{n} a_m x^m \qquad (3.135)$$

A special case of polynomial is the bimonial distribution which will turn up in Chapter 7.

A *rational* function is the ratio of two polynomials:

$$y = \frac{a_n x^n + a_{n-1} x^{n-1} + \ldots + a_0}{b_m x^m + b_{m-1} x^{m-1} + \ldots + b_0} \qquad (3.136)$$

Polynomial functions are often used to 'fit' empirical data, and rational functions can be used to approximate to other functions (Baker and Gemmel, 1970) as so-called Padé approximants.

The most common polynomial functions are quadratic and cubic. They can always be plotted graphically, and this is a convenient point to indicate the 'shape' of some of them, and we can show in passing how equations can be solved graphically. Consider

$$y = x^2 + 6x - 7 \qquad (3.137)$$

This is plotted in Figure 3.22. Note that the curve crosses the x-axis at

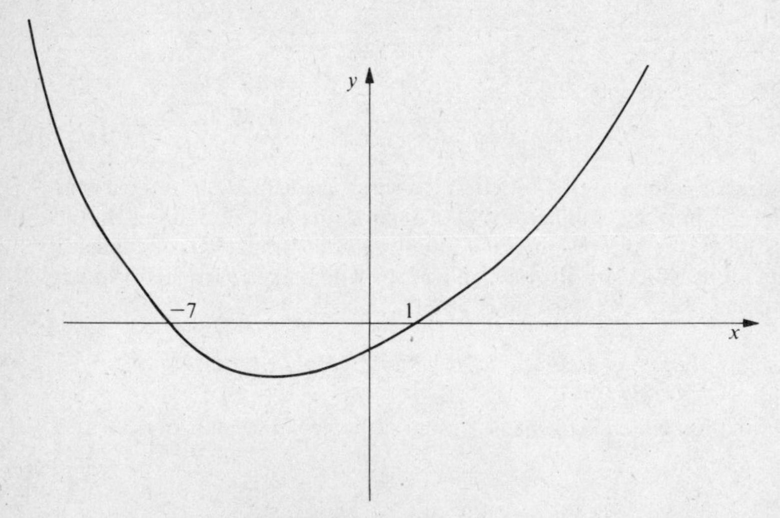

FIG. 3.22. Graph of $y = x^2 + 6x - 7$

$x = 1$ and $x = -7$, since $y = 0$ at the x-axis, this means that the axis crossings are the solutions of

$$x^2 + 6x - 7 = 0 \qquad (3.139)$$

and so we have solved a quadratic equation graphically. For obvious reasons, graphical solutions are less accurate than algebraic ones and so the latter methods are preferable (see the discussion following equation 2.54 in Chapter 2). However, for higher order equations, as we noted earlier, it may be impossible to obtain an exact algebraic solution, and then graphical methods can be helpful. (The other alternative is to use iterative methods on a computer). Consider a cubic polynomial

$$y = 2x^3 + 9x^2 - 32x + 21 \qquad (3.139)$$

and suppose we want the values of x for which $y = 0$. This function is plotted in Figure 3.23. There are now three solutions, $x = -7, 1$, or $1{\cdot}5$

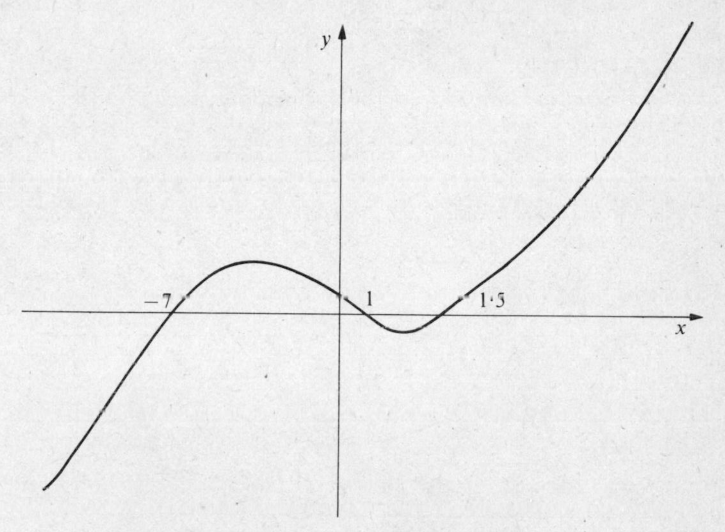

FIG. 3.23. Graph of $y = 2x^3 + 9x^2 - 32x + 21$

for which $y = 0$. The solution can be checked in this case, since the polynomial factors.

$$y = (x^2 + 6x - 7)(2x - 3) \qquad (3.140)$$

but typically such factoring would not be possible.

3.5.3. *The hyperbolic functions*

A well-known group of functions which turn out to have properties in *calculus* not unlike the trignometric functions, are the *hyperbolic* functions which are constructed out of exponential functions as follows:

$$y = \sinh x = (e^x - e^{-x})/2 \tag{3.141}$$

$$y = \cosh x = (e^x + e^{-x})/2 \tag{3.142}$$

$$y = \tanh x = \sinh x/\cosh x \tag{3.143}$$

$$= (e^x - e^{-x})/(e^x + e^{-x}) \tag{3.144}$$

and for completeness we can also define the reciprocal functions

$$\text{sech}\, x = 1/\cosh x \tag{3.145}$$

$$\text{cosech}\, x = 1/\sinh x \tag{3.146}$$

$$\coth x = 1/\tanh x \tag{3.147}$$

We could also define inverse hyperbolic functions, such as

$$y = \sinh^{-1} x \tag{3.148}$$

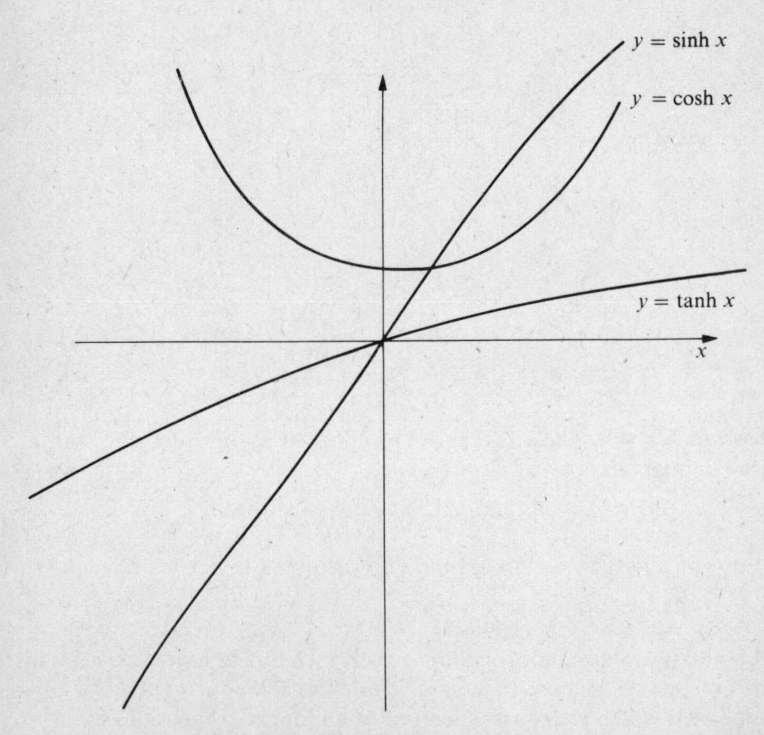

FIG. 3.24. The hyperbolic functions

by starting with

$$\sinh y = x \tag{3.149}$$

as an implicit definition of y.

Plots of the three main functions are given in Figure 3.24. While these functions have few, if any, direct application in geography and planning at present, we present them here as available for future use, and because they have important properties in the calculus. The best known property of any of them is perhaps that $y = \cosh x$ is the equation of a hanging chain!

3.5.4. The logistic function, and a note on asymptotes

As we shall see later, the logistic function occurs frequently in geography and planning. As a preliminary, it is useful to introduce more formally than hitherto the notion of an *asymptote*. Consider the plot of $y = e^{-x}$ in Figure 3.17. e^{-x} is never zero, but as x increases (as $x \to \infty$) it becomes increasingly near to zero. A line such as the x-axis, the line $y = 0$, here is called an asymptote. Now, suppose in some situation we know that for large negative values of some variable, the function is very near some lower bound asymptote, while for large positive values,

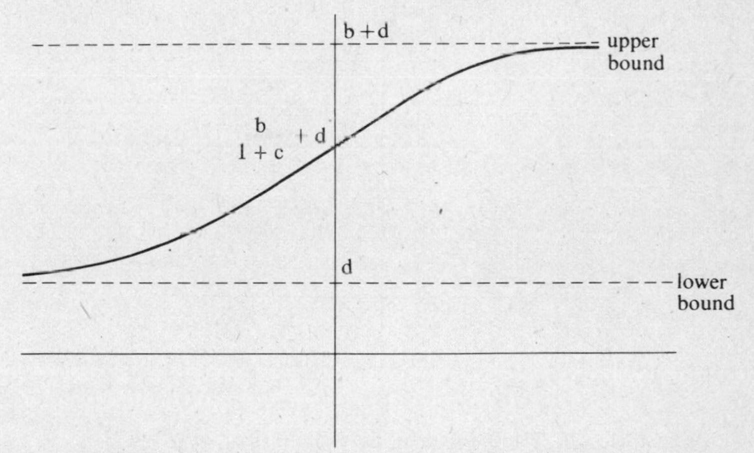

FIG. 3.25. The logistic function

it is very near to some upper bound asymptote, and there is steady growth in between. Such a curve is shown in Figure 3.25, and is the logistic function. Consider

$$y = \frac{b}{1 + ce^{-ax}} + d \tag{3.150}$$

For large negative x, $e^{-ax} \to \infty$, so $b/(1 + ce^{-ax}) \to 0$, and $y \to d$, the lower asymptote. As $x \to \infty$, $e^{-ax} \to \infty$, and so $b/(1 + ce^{-ax}) \to b$, and $y \to b + d$, the upper asymptote. The axis is crossed at $y = b/(1 + c) + d$, as shown. If a is negative, the curve will slope the other way, with the upper asymptote reached at negative x and *vice-versa.* This function often turns up in modal split models (Wilson, 1974, Chapter 9).

3.5.5. Miscellaneous examples

As indicated in the introduction, a wide variety of functions can be generated, and above we have introduced some of the more important ones. Others can be mentioned briefly. If we use the function-of-a-function concept to take the exponential of the quadratic term $-ax^2$, to give

$$y = e^{-ax^2} \tag{3.151}$$

we obtain a function which has the shape of the familiar normal (or Gaussian) distribution. It is plotted in Figure 3.26. This is in very common use, of course.

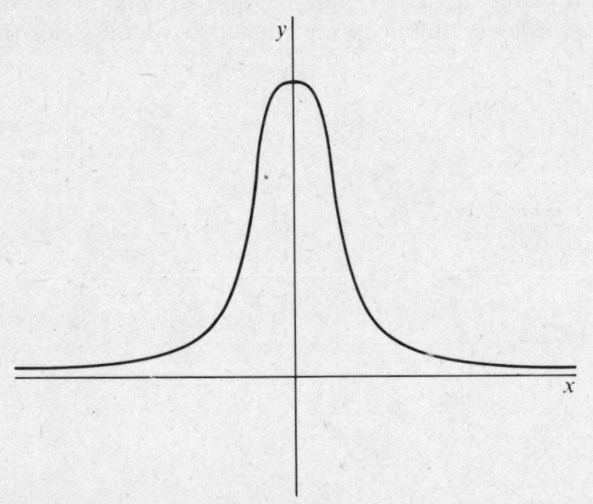

FIG. 3.26. The Gaussian, or normal, distribution: $y = e^{-ax^2}$

Sometimes we may be unsure as to whether to fit a negative exponential function, e^{-ax}, or a power function, x^{-a}, to a particular set of backward-sloping points obtained as empirical data. For roughly equivalent values of a and n, the power function 'dips' more steeply near the origin. For such purposes, a composite function

$$y = x^{-n}e^{-ax} \tag{3.152}$$

may be convenient. An even more general function has been proposed by March (1971) for similar purposes:

$$y = x^{-n} e^{-ax^b} \qquad (3.153)$$

Many further examples of composite functions will arise naturally in the rest of the book, and turn up in geographical and planning literature.

3.6. Series expansion of functions

We discovered by coincidence (in exploring the geometric progressoin) in Chapter 2 that some functions can be represented as infinite series. A reference back to equation 2.118 shows that

$$y = \frac{1}{1-x} = 1 + x + x^2 + \ldots = \sum_{n=0}^{\infty} x^n \qquad (3.154)$$

This series expansion is true for $|x| < 1$, where $|x|$, called the 'modulus of x' or 'mod x', denotes the positive value of x: that is

$$|x| = \begin{cases} x & \text{if } x > 0 \\ -x & \text{if } x < 0 \end{cases} \qquad (3.155)$$

Some of the other standard functions have simple series expansions, and we note their existence here. We shall give a form of proof in Chapter 5 when the tools of calculus are available.

We begin with the power function, giving an expansion, not for x^a, but $(1+x)^a$, which is valid for $|x| < 1$:

$$(1+x)^a = 1 + ax + \frac{a(a-1)}{2!}x^2 + \frac{a(a-1)(a-2)}{3!}x^3 + \ldots \qquad (3.156)$$

which is

$$(1+x)^a = \sum_{n=0}^{\infty} \frac{a(a-1)\ldots(a-n+1)}{n!} x^n \qquad (3.157)$$

in a more concise notation. Note the use of the *factorial*: $n!$ (which reads 'n factorial') is defined by

$$n! = 1.2.3.4\ldots(n-2)(n-1)n \qquad (3.158)$$

We can also give an expansion for a logarithmic function, in this case for $\log_e(1+x)$, valid for $|x| < 1$:

$$\log_e(1+x) = x - \frac{x^2}{2} + \frac{x^3}{3} \ldots \qquad (3.159)$$

$$= \sum_{n=1}^{\infty} (-1)^{n-1} \frac{x^n}{n} \tag{3.160}$$

Note the use of the factor $(-1)^{n-1}$, which is of alternating sign as n increases, to represent this series using a summation sign.

The series expansion of e^x is true for all values of x and is given by

$$e^x = 1 + \frac{x}{1!} + \frac{x^2}{2!} + \frac{x^3}{3!} \dots \tag{3.161}$$

$$= \sum_{n=0}^{\infty} \frac{x^n}{n!} \tag{3.162}$$

Note that this gives a method for calculating e by putting $x = 1$.

Next, we consider the trigonometric functions. There are expansions of $\sin x$ and $\cos x$, in each case for all values of x:

$$\sin x = x - \frac{x^3}{3!} + \frac{x^5}{5!} \dots \tag{3.163}$$

$$= \sum_{n=1}^{\infty} (-1)^{n+1} \frac{x^{2n-1}}{(2n-1)!} \tag{3.164}$$

Note the device which has been used here to represent the series using the summation sign.

$$\cos x = 1 - \frac{x^2}{2!} + \frac{x^4}{4!} \dots \tag{3.165}$$

$$= \sum_{n=0}^{\infty} (-1)^n \frac{x^{2n}}{2n!} \tag{3.166}$$

In relation to the first term, note that, by convention,

$$0! = 1 \tag{3.167}$$

We will not give here the expansion of $\tan x$, which is much more complicated (see, for example, Phillips, 1930, p. 351). The same comment applies to the reciprocal trigonmetric functions.

It is easy to see from their definitions, and from 3.161 that there are simple expansions for $\sinh x$ and $\cosh x$:

$$\sinh x = x + \frac{x^3}{3!} + \frac{x^5}{5!} + \dots \tag{3.168}$$

$$= \sum_{n=1}^{\infty} \frac{x^{2n}}{(2n-1)!} \tag{3.169}$$

$$\cosh x = 1 + \frac{x^2}{2!} + \frac{x^4}{4!} + \ldots \tag{3.170}$$

$$= \sum_{n=0}^{\infty} \frac{x^{2n}}{(2n)!} \tag{3.171}$$

Again, it is much more difficult to give an expansion for $\tanh x$ and the reciprocal functions, and the inverse trigonometric functions.

The series expansions presented above are in terms of power functions. Their main utility for our purposes lies in obtaining approximate forms of functions for small values of x, such as

$$\sin x \simeq x - \frac{x^3}{3!} \tag{3.172}$$

for small x, and where '$\simeq$' denotes 'approximately equal to'.

It is also possible to expand functions in terms of other base functions. An example is the use of the trigonometric functions in Fourier series of the form

$$f(x) = \tfrac{1}{2}a_0 + \sum_{n=1}^{\infty} (a_n \cos nx + b_n \sin nx) \tag{3.173}$$

Functions are, in effect, considered as sums of waves of different lengths. This is of some importance to geographers because of the use of this concept in *trend surface mapping* (Haggett, 1965). The methods for calculating the coefficients a_n and b_n for any function $f(x)$ involve integral calculus, which we do not reach until Chapter 5. But as an illustration, we present a series expansion for $y = e^x$, valid for $0 < x < 2\pi$:

$$e^x = \left[\frac{e^{2\pi}-1}{\pi}\right]\left[\frac{1}{2} + \frac{1}{1^2+1}\cos x + \frac{1}{2^2+1}\cos 2x + \ldots - \frac{1}{1^2+1}\sin x - \frac{1}{2^2+1}\sin 2x \ldots\right] \tag{3.174}$$

3.7. Functions of more than one independent variable

3.7.1. Notation and examples

So far, we have concentrated on functions of one independent variable, However, situations arise where more than one is needed. If z is a dependent variable, and x and y now both independent variables, we may have

$$z = ax + by \tag{3.175}$$

where a and b are constants (and compare equation 2.43 of Chapter 2). We can easily extend our general notation and write, when appropriate,

$$z = f(x, y) \tag{3.176}$$

to indicate that z is some function f of x and y. We can have as many independent variables as required. If y is taken as dependent variable, and $x_1, x_2 \ldots x_N$ as N independent variables, then a functional relationship could be expressed as

$$y = f(x_1, x_2, x_3, \ldots x_N) \tag{3.177}$$

We should perhaps also note that *implicit* definitions are also possible. For example, we might take

$$ye^{-yx_1x_2} = a \tag{3.178}$$

where a is a constant, to define a functional relationship between y, x_1, and x_2. In general, we could write

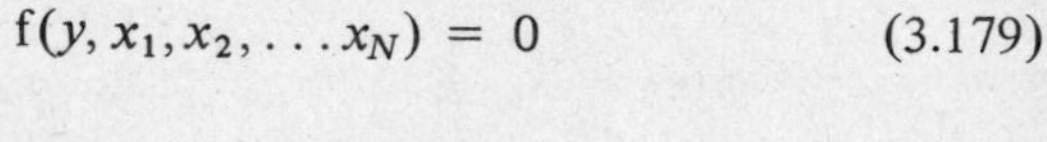

$$f(y, x_1, x_2, \ldots x_N) = 0 \tag{3.179}$$

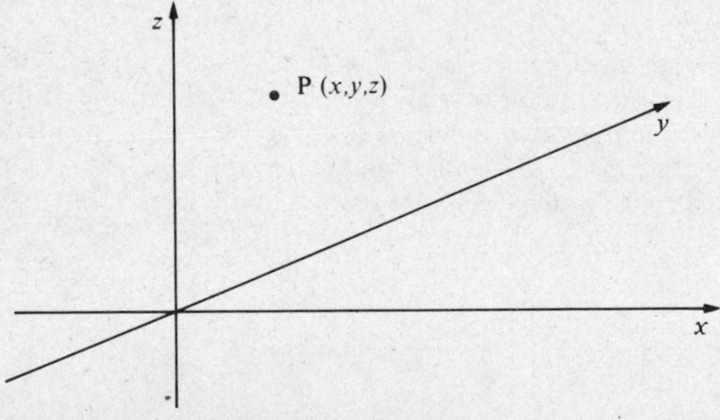

FIG. 3.27. A three-dimensional Cartesian co-ordinate system

3.7.2. Higher dimensional co-ordinate geometry

We showed earlier how to exhibit a relationship between variables y and x using a two-dimensional cartesian co-ordinate system. For three variables, say z, y, and x, we can build a three-dimensional Cartesian co-ordinate system, as we indicated in another context with Figure 2.11 of Chapter 2, which, for convenience, is repeated here as Figure 3.27.

The x and y axes are at right angles and are considered to form a plane perpendicular to this page, while the z axis is vertical, at right angles to that plane, through the origin. Any point P in three-dimensional space is then uniquely characterized by its co-ordinates (x, y, z).

A functional relationship, such as that given in equation 3.176 can then be exhibited. Any pair of values (x, y) locate a point in the (x, y) plane, and $z = \mathrm{f}(x, y)$ then gives the corresponding z value, and is thus characterized as a point a height z 'above' (x, y). It can easily be seen that, as a relation between x and y in two dimensions appears as a *curve*, a relation between x, y, and z in three dimensions appears as a *surface*.

We mention two particular cases which have parallels in two dimensions. First, a linear relationship (which would be a straight line in two dimensions) such as

$$\mathrm{a}x + \mathrm{b}y + \mathrm{c}z + \mathrm{d} = 0 \tag{3.180}$$

is a *plane* in three dimensions. Second, we consider the three-dimensional equivalent of

$$x^2 + y^2 = \mathrm{a}^2 \tag{3.181}$$

which, in two dimensions, is a circle of radius a. The corresponding concept in three dimensions is

$$x^2 + y^2 + z^2 = \mathrm{a}^2 \tag{3.182}$$

and any attempt to plot the points (bearing in mind Pythagoras' theorem) will soon convince the reader that this represents a *sphere* of radius a.

The concepts can be extended to even higher dimensions. (x_1, x_2, x_3, x_4), for example, can be considered to be a 'point' in a four-dimensional Cartesian space, but the three-dimensional nature of the world we live in inhibits the visual presentation of such a space, and so it remains a mathematical abstraction. The field of N-dimensional geometry is a well-developed one, but any further consideration of it would take us beyond the scope of this book.

3.8. More examples of equations, functions and associated algebraic manipulation

3.8.1. Introduction

The principles used for solving equations, and for associated algebraic manipulation, were outlined in section 2.4 of Chapter 2. Equations which relate to real world examples often contain functions of the types which have been defined earlier in this chapter. Thus, it is now possible to examine many more examples and to illustrate the kind of algebra and associated geometrical interpretation that the student of modern geography and planning frequently has to deal with. By the end of the chapter, and on completion of the exercises, the reader should be sufficiently well equipped to tackle much of the literature with a reasonable degree of understanding.

3.8.2. Infiltration rates

In section 2.2.8 of Chapter 2 we introduced examples of diffusion processes, one of which was concerned with the infiltration of water into soil. The main variables were defined in relation to Figure 2.10 (b): f was the instantaneous rate of infiltration, and it was made up of two components, a transmission rate A which produces a contribution to f which is time independent, and a diffusion constant B, which produces a contribution to f related to $t^{-1/2}$. These variables are then related by the following equation (Philip, 1957):

$$f = A + B \cdot t^{-1/2} \tag{3.183}$$

Thus, if A and B are taken as constants, f is a function of t, and can be plotted as in Figure 3.28. For large values of t, $f \simeq A$, and is asymptotic as shown.

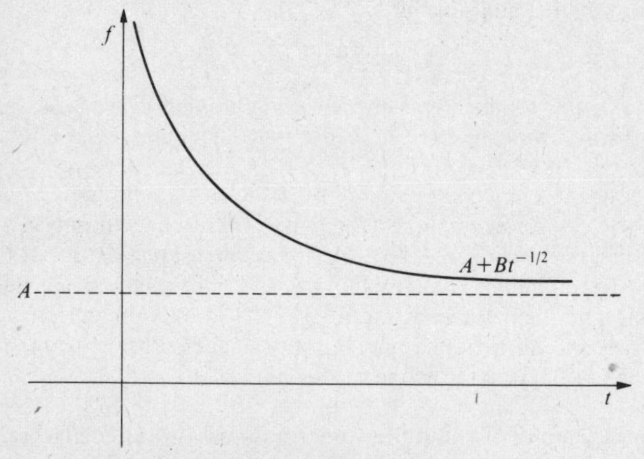

FIG. 3.28. Graphs of infiltration rate function

We give two examples of algebraic manipulation associated with this formula. First, suppose we want to find the time t_0 associated with a particular rate f_0:

$$f_0 = A + Bt_0^{-1/2} \tag{3.184}$$

We can solve for t as follows:

$$t_0^{-1/2} = \frac{f_0 - A}{B} \tag{3.185}$$

$$t_0^{1/2} = \frac{B}{f_0 - A} \tag{3.186}$$

and so

$$t_0 = \left(\frac{B}{f_0 - A}\right)^2 \tag{3.187}$$

Second, suppose we are given two pairs of measurements: $f = f_0$ when $t = t_0$ and $f = f_1$ when $t = t_1$, and we wish to determine A and B from these. Clearly

$$f_0 = A + Bt_0^{-1/2} \tag{3.188}$$

$$f_1 = A + Bt_1^{-1/2} \tag{3.189}$$

We can treat these as linear simultaneous equations in A and B. Subtract equation 3.188 from 3.189:

$$f_1 - f_0 = B(t_1^{-1/2} - t_0^{-1/2}) \tag{3.190}$$

so

$$B = \frac{f_1 - f_0}{t_1^{-1/2} - t_0^{-1/2}} \tag{3.191}$$

Then, substitute for B in equation 3.188:

$$f_0 = A + t_0^{-1/2}\left[\frac{f_1 - f_0}{t_1^{-1/2} - t_0^{-1/2}}\right] \tag{3.192}$$

and we can solve for A as follows:

$$A = f_0 - t_0^{-1/2}\left[\frac{f_1 - f_0}{t_1^{-1/2} - t_0^{-1/2}}\right] \tag{3.193}$$

$$= \frac{f_0(t_1^{-1/2} - t_0^{-1/2}) - t_0^{-1/2}(f_1 - f_0)}{(t_1^{-1/2} - t_0^{-1/2})} \tag{3.194}$$

$$= \frac{f_0 t_1^{-1/2} - f_1 t_0^{-1/2}}{t_1^{-1/2} - t_0^{-1/2}} \tag{3.195}$$

3.8.3. Scree slopes

Figure 3.29 (a) depicts a cliff before erosion begins, and Figure 3.29 (b) the cliff after a period of cliff retreat. If the cliff height is h_1, the height

FIG. 3.29. The geometry of scree slopes

CQ is h_2, and the angle of the scree slope, α, which can all be measured how can we find the width AD = a? The area ABCD must be equal to the area PCQ. Further, since

$$\frac{QC}{PC} = \tan \alpha \tag{3.196}$$

$$PC = QC/\tan \alpha = h_2/\tan \alpha \tag{3.197}$$

Thus,

$$ABCD = PCQ \tag{3.198}$$

implies that

$$h_1 a = \frac{1}{2} h_2 \frac{h_2}{\tan \alpha} \tag{3.199}$$

and we can solve this for a to give

$$a = \frac{h_2^2}{2h_1 \tan \alpha} \tag{3.200}$$

3.8.4. Horton's network laws

Figure 3.30 shows a hypothetical river system with its branches classified into the orders of a hierarchy in the manner of Strahler (1957). This is a fourth-order basin, as there are four levels in the hierarchy. Let N_i be the number of branches at level i. Then Horton's (1945) law is that this sequence forms a geometrical progression so that

$$\frac{N_{i-1}}{N_i} = \mathrm{p} \tag{3.201}$$

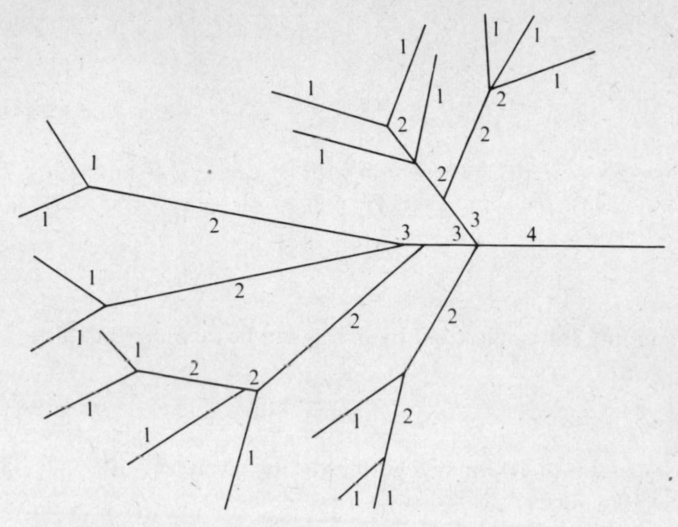

FIG. 3.30. A hypothetical river network with stream orders

say, where p is some constant, in this case obviously greater than one. If we put $i = 4$, equation 3.201 can be arranged to give

$$N_3 = \mathrm{p}N_4 \tag{3.202}$$

and similarly

$$N_2 = \mathrm{p}N_3 \tag{3.203}$$

and

$$N_1 = \mathrm{p}N_2 \tag{3.204}$$

Thus

$$N_1 = \mathrm{p}(\mathrm{p}N_3) = \mathrm{p}^2(\mathrm{p}N_4) = \mathrm{p}^3 N_4 \tag{3.205}$$

since

$$N_4 = 1 \tag{3.206}$$

$$N_1 = \mathrm{p}^3 \tag{3.207}$$

The reader can easily check that this result can be stated in general terms as

$$N_i = \mathrm{p}^{(k-i)} \tag{3.208}$$

where k is the order of the basin (which is 4 in our example).

Notice that the total of branches is

$$S = \sum_{i=1}^{k} N_i \tag{3.209}$$

$$= \sum_{i=1}^{k} p^{(k-i)} \tag{3.210}$$

and this is a geometrical progression with first term p^{k-1} and ratio p^{-1}. Thus, we can use the formula 2.112 to give

$$S = \frac{p^{k-1}(1-p^{-k})}{(1-p^{-1})} \tag{3.211}$$

If we multiply top and bottom by p, this can be rearranged to give

$$S = \frac{p^k - 1}{p - 1} \tag{3.212}$$

How could p be determined geometrically, given measurements of N_i? Take logs in equation 3.208:

$$\log N_i = \log p^{(k-i)} = (k-i)\log p \tag{3.213}$$

putting

$$y_i = \log N_i \tag{3.214}$$

$$y_i = k \log p - i \log p \tag{3.215}$$

Figure 3.31 shows a 'best fit' straight line of $\log N_i$ against i (and we shall show how to find this using regression analysis in Chapter 5 –

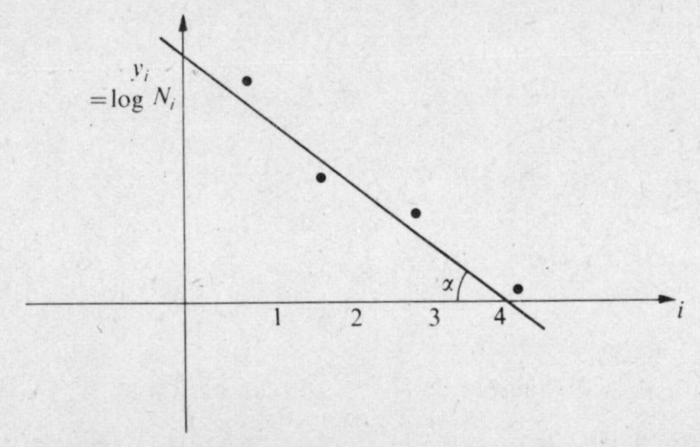

FIG. 3.31. Graphical solution for a stream branching constant

section 5.9.2) through 4 points. As would be anticipated from equation 3.211, this is a backward sloping straight line with gradient $-\log p$. Thus, in relation to α as defined in the figure,

$$\log p = \tan \alpha \tag{3.216}$$

so

$$p = 10^{\tan \alpha} \tag{3.217}$$

if logs had been taken to the base 10.

3.8.5. A simple population model

Consider a spatial system of N regions, $i = 1, 2, \ldots N$, and let $P_i(t)$ be the population of the ith region at time t. Let $b(t, t+T)$ and $d(t, t+T)$ be birth and death rates respectively for the period t to $t+T$, assumed to be uniform over the whole system. Let m_{ij} be fixed and unchanging rates of migration for each period of length T. Then we may suppose that

$$\begin{aligned} P_i(t+T) = & (1 + b(t, t+T) - d(t, t+T))P_i(t) \\ & + \sum_{j \neq 1} (m_{ji} - m_{ij})P_j(t) \end{aligned} \tag{3.218}$$

This is an example of a recursive model: the value of some variables at time $t + T$ are calculated from the values of the variables for an earlier period. Then, for the next period for example, we could continue with

$$\begin{aligned} P_i(t+2T) = & (1 + b(t+T, t+2T) - d(t+T, t+2T))P_i(t+T) \\ & + \sum_{j \neq i} (m_{ji} - m_{ij})P_j(t+T) \end{aligned} \tag{3.219}$$

and so on.

To make things more realistic, we might suppose that birth rates and deaths rates are each declining over time according to a logistic function (see Figure 3.25 earlier and related discussion) as shown in Figure 3.32.

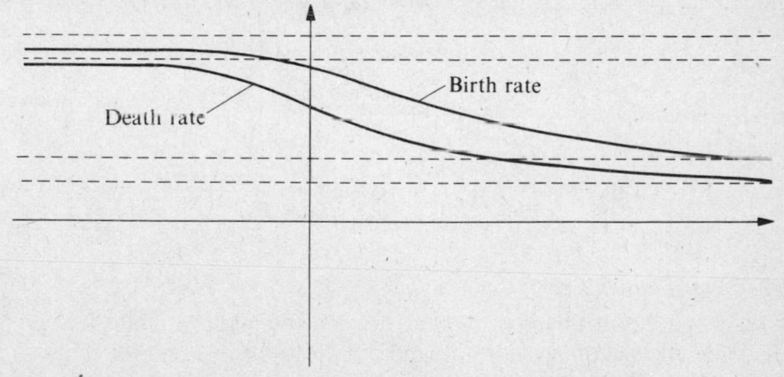

FIG. 3.32. Graphs of hypothetical birth and death rates

Let a point on the curve represent $b(t, t+T)$ for some t, and similarly for $d(t, t+T)$. Then equation 3.150 and related discussion shows that suitable functional forms are:

$$b(t, t+T) = \mathrm{q}_b/(1 + \mathrm{r}_b e^{-\mathrm{p}_b t} + \mathrm{s}_b) \qquad (3.220)$$

$$d(t, t+T) = \mathrm{q}_d/(1 + \mathrm{r}_d e^{-\mathrm{p}_d t} + \mathrm{s}_d) \qquad (3.221)$$

where p_b, q_b, r_b, s_b, p_d, q_d, r_d and s_d are constants to be determined. These values can be substituted into equation 3.218 to give $P_i(t)$ as a function of t explicitly.

How would we set about finding the values of the constants from a set of observations? Consider the birth rate given by equation 3.220. The upper and lower asymptotes would have known values and are, in terms of the constants $\mathrm{q}_b + \mathrm{s}_b$ and s_b respectively, and so this obviously enables q_b and s_b to be determined. Suppose we now 'solve' for $\mathrm{r}_b e^{-\mathrm{p}_b t}$ multiply through by $(1 + \mathrm{r}_b e^{-\mathrm{p}_b t})$ and rearrange.

$$b(t, t+T)(1 + \mathrm{r}_b e^{-\mathrm{p}_b t}) = \mathrm{q}_b + \mathrm{s}_b(1 + \mathrm{r}_b e^{-\mathrm{p}_b t}) \qquad (3.222)$$

so

$$(b(t, t+T) - \mathrm{s}_b)\mathrm{r}_b e^{-\mathrm{p}_b t} = \mathrm{q}_b + \mathrm{s}_b - b(t, t+T) \qquad (3.223)$$

so

$$\mathrm{r}_b e^{-\mathrm{p}_b t} = \frac{\mathrm{q}_b + \mathrm{s}_b - b(t, t+T)}{b(t, t+T) - \mathrm{s}_b} \qquad (3.224)$$

We can now obtain r_b by putting $t = 0$, giving, since $e^0 = 1$,

$$\mathrm{r}_b = \frac{\mathrm{q}_b + \mathrm{s}_b - b(0, T)}{b(0, T) - \mathrm{s}_b} \qquad (3.225)$$

and then p_b from any other known value of $b(t, t+T)$, say $b(t_0, t_0 + T)$:

$$e^{-\mathrm{p}_b t_0} = \frac{\mathrm{q}_b + \mathrm{s}_b - b(t_0, t_0 + T)}{\mathrm{r}_b\,[b(t_0, t_0 + T) - \mathrm{s}_b]} \qquad (3.226)$$

so

$$\mathrm{p}_b = -\frac{1}{t_0}\log_e\left\{\frac{\mathrm{q}_b + \mathrm{s}_b - b(t_0, t_0 + T)}{\mathrm{r}_b\,[b(t_0, t_0 + T) - \mathrm{s}_b]}\right\} \qquad (3.227)$$

3.8.6. von Thünen's rings

In section 2.2.3 of Chapter 2 we introduced the main variables for von Thünen's analysis of agricultural land use around some market centre (cf. Figure 2.6). At distance r from the centre O transport costs would be ry_ic_i, where y_i is the volume of the ith crop produced in a unit area of land and c_i the transport cost per mile. If such a crop sells at price p_i and the fixed costs of production are a_i, the net profit is

$$R_i = (p_i - a_i)y_i - ry_ic_i \tag{3.228}$$

For each crop i, this gives a backward sloping straight line relationship as shown in Figure 3.33 for three such crops. In this case, crop 1 has the highest intercept value $(p_1 - a_1)y_1$ but the steepest gradient $-y_1c_1$.

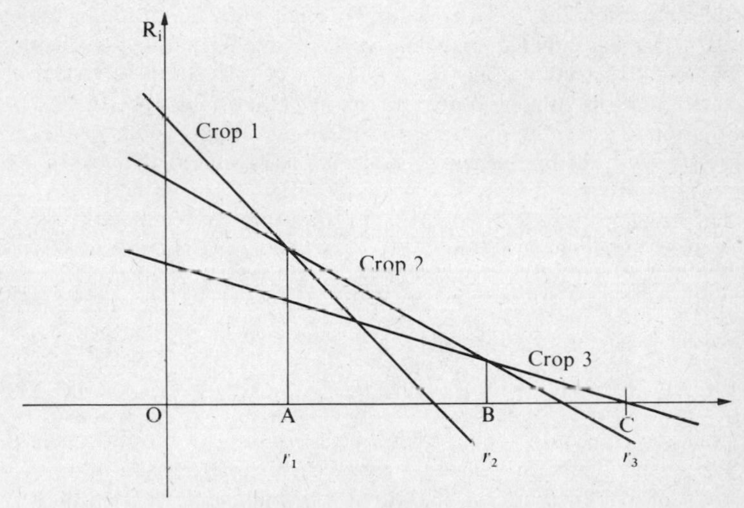

FIG. 3.33. Land rents according to von Thünen

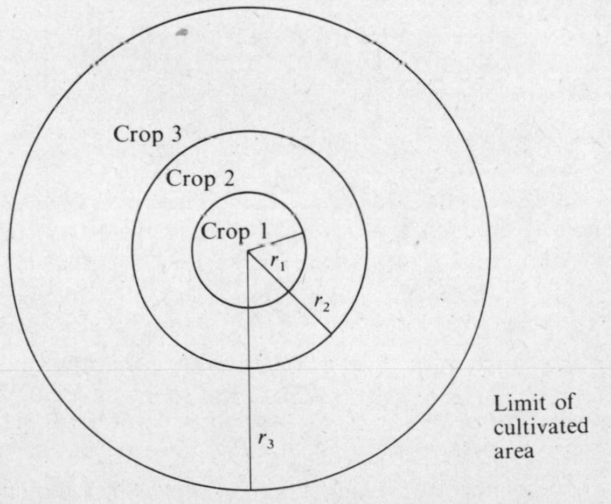

FIG. 3.34. von Thünen's rings

R_1 exceeds R_2 and R_3 up to the point A, $r = r_1$; for $r_1 < r \leqslant r_2$, R_2 is highest, and for $r_2 < r \leqslant r_3$, R_3 is highest. There would be no cultivation for $r > r_3$ because all the Rs are negative. Thus, this set of constants, generating these straight lines, would produce the land use allocation shown in Figure 3.34.

3.8.7. Meso-spatial interaction and location models

We introduced the main variables for spatial interaction and location models in section 2.2.6 of Chapter 2. We begin with the shopping model (Huff, 1964, Lakshmanan and Hansen, 1965, see Wilson, 1974, Chapter 4 for a detailed account). Let S_{ij} be the flow of cash from the residences of zone i to shops in zone j for some spatial system. Let P_i be the population of zone i, e_i mean expenditure on shopping goods per head in zone i, W_j the attractiveness of shops in zone j, and c_{ij} the cost of travel from i to j.

The usual model can be constructed from three proportionality hypotheses. In general, if

$$y \propto x \tag{3.229}$$

where '$\propto$' means 'is proportional to', then

$$y = \mathrm{k}x \tag{3.230}$$

where k is a constant which is said to be a 'constant of proportionality', or a 'normalizing factor' (since it may be calculated so that a set of y values sum to a known total). For the shopping model, we hypothesize that

$$S_{ij} \propto e_i \mathrm{P}_i \tag{3.231}$$

$$S_{ij} \propto W_j^{\alpha} \tag{3.232}$$

$$S_{ij} \propto c_{ij}^{-\beta} \tag{3.233}$$

That is S_{ij} is proportional to the total spending power of zone i, the attractiveness of shops in zone j, raised to some power, and a decreasing function of inter-zonal travel cost, c_{ij} ($\beta > 0$). This implies that

$$S_{ij} = \mathrm{K} e_i P_i W_j^{\alpha} c_{ij}^{-\beta} \tag{3.234}$$

where K is the constant of proportionality. Note that the sum of S_{ij} over j, $\sum_j S_{ij}$, is the total of cash leaving i, and so must be $e_i \mathrm{P}_i$. That is,

$$\sum_j S_{ij} = e_i \mathrm{P}_i \tag{3.235}$$

It turns out to be more convenient then, to replace K in equation 3.234 by K_i, and to calculate it so that 3.235 is satisfied. (We shall concentrate mainly on the algebra of this here – for a detailed discussion of the nature of the model, see the references cited earlier). Thus

$$S_{ij} = K_i e_i P_i W_j^{\alpha} c_{ij}^{-\beta} \qquad (3.236)$$

and we substitute into equation 3.331 and solve for K_i:

$$\sum_j K_i e_i P_i W_j^{\alpha} c_{ij}^{-\beta} = e_i P_i \qquad (3.237)$$

Note that

$$\sum_i ax_i = ax_1 + ax_2 + \ldots \qquad (3.238)$$

$$= a(x_1 + x_2 + \ldots) \qquad (3.239)$$

$$= a\sum_i x_i \qquad (2.240)$$

That is, if a term inside a summation sign is independent of the index being summed, it can be taken outside the summation sign as a common factor. $K_i e_i P_i$ are independent of j and so equation 3.236 can be written

$$K_i e_i P_i \sum_j W_j^{\alpha} c_{ij}^{-\beta} = e_i P_i \qquad (3.241)$$

$e_i P_i$ now cancels from each side:

$$K_i \sum_j W_j^{\alpha} c_{ij}^{-\beta} = 1 \qquad (3.242)$$

and so

$$K_i = 1/\sum_j W_j^{\alpha} c_{ij}^{-\beta} \qquad (3.243)$$

So, the full statement of the model is given by equation 3.236 and 3.243. It is possible to substitute explicitly for K_i from 3.243 into 3.235 to give

$$S_{ij} = e_i P_i \frac{W_j^{\alpha} c_{ij}^{\beta}}{\sum_j W_j^{\alpha} c_{ij}^{\beta}} \qquad (3.244)$$

It may be useful at this stage to exhibit the workings of this kind of model in relation to some hypothetical figures. Let us consider the spatial system shown in Fig. 3.35. The distances are shown on the figure, and so if we take these as a measure of 'cost', $c_{11} = 1{\cdot}0$, $c_{22} = 2{\cdot}585$, $c_{33} = 2{\cdot}0$, and all the other c_{ij}s are $5{\cdot}0$. The other data is given in Table 3.2.

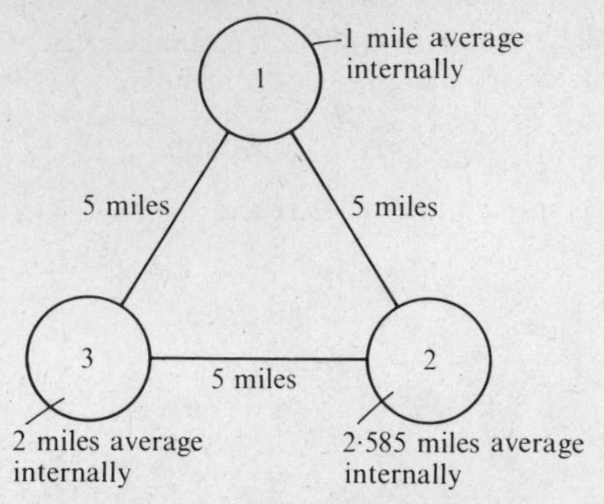

FIG. 3.35. A three zone spatial system

It is computationally convenient to use the two-equation form of the model given by (2.236) and (3.243). The first step is then to calculate the K_is for $i = 1, 2$, and 3. To be completely explicit, we write equation (3.243) for each i value (though this would not normally be necessary), putting $\alpha = 1, \beta = 1$:

$$K_1 = 1/(W_1c_{11}^{-1} + W_2c_{12}^{-1} + W_3c_{13}^{-1}) \tag{3.245}$$

$$K_2 = 1/(W'_1c_{21}^{-1} + W_2c_{22}^{-1} + W_3c_{23}^{-1}) \tag{3.246}$$

TABLE 3.2

Hypothetical data for spatial interaction example

Zone number (i)	e_i	P_i	e_iP_i	W_j
1	2	50	100	10
2	1	1000	1000	100
3	1	500	500	20

We can take $\alpha = 1, \beta = 1$.

$$K_3 = 1/(W_1c_{31}^{-1} + W_2c_{32}^{-1} + W_3c_{33}^{-1}) \tag{3.247}$$

Thus, substituting from the table,

$$K_1 = 1/(10 + 20 + 4) = 1/34 \tag{3.248}$$

$$K_2 = 1/(2 + 38{\cdot}6 + 4) = 1/44{\cdot}6 \tag{3.249}$$

$$K_3 = 1/(2 + 20 + 10) = 1/32 \tag{3.250}$$

We can now substitute in equation 3.236 for each S_{ij}, $i = 1, 2, 3, j = 1, 2, 3$:

$$S_{11} = K_1 e_1 P_1 W_1 c_{11}^{-1} = \frac{1}{34} \cdot 100 \cdot 10 = 29 \cdot 4 \quad (3.251)$$

$$S_{12} = K_1 e_1 P_1 W_2 c_{12}^{-1} = \frac{1}{34} \cdot 100 \cdot 100 \cdot \frac{1}{5} = 58 \cdot 8 \quad (3.252)$$

$$S_{13} = K_1 e_1 P_1 W_3 c_{13}^{-1} = \frac{1}{34} \cdot 100 \cdot 20 \cdot \frac{1}{5} = 11 \cdot 8 \quad (3.253)$$

$$S_{21} = K_2 e_2 P_2 W_1 c_{21}^{-1} = \frac{1}{44 \cdot 6} \cdot 1000 \cdot 10 \cdot \frac{1}{5} = 44 \cdot 8 \quad (3.254)$$

$$S_{22} = K_2 e_2 P_2 W_2 c_{22}^{1} = \frac{1}{44 \cdot 6} \cdot 1000 \cdot 100 \cdot \frac{1}{2 \cdot 585} = 864 \cdot 0 \quad (3.255)$$

$$S_{23} = K_2 e_2 P_2 W_3 c_{23}^{-1} = \frac{1}{44 \cdot 6} \cdot 1000 \cdot 20 \cdot \frac{1}{5} = 89 \cdot 6 \quad (3.256)$$

$$S_{31} = K_3 e_3 P_3 W_1 c_{31}^{-1} = \frac{1}{32} \cdot 500 \cdot 10 \cdot \frac{1}{5} = 31 \cdot 3 \quad (3.257)$$

$$S_{32} = K_3 e_3 P_3 W_2 c_{33}^{1} = \frac{1}{32} \cdot 500 \cdot 100 \cdot \frac{1}{5} = 313 \cdot 0 \quad (3.258)$$

$$S_{33} = K_3 e_3 P_3 W_3 c_{33}^{-1} = \frac{1}{32} \cdot 500 \cdot 20 \cdot \frac{1}{2} = 156 \cdot 5 \quad (3.259)$$

and this set of flows represents the model output for this example. We can calculate total sales in each shopping centre as follows:

$$\sum_i S_{i1} = S_{11} + S_{21} + S_{31} = 29 \cdot 4 + 44 \cdot 8 + 31 \cdot 3 = 105 \cdot 5 \quad (3.260)$$

$$\sum_i S_{i2} = S_{12} + S_{22} + S_{32} = 58 \cdot 8 + 864 \cdot 0 + 313 \cdot 0 = 1235 \cdot 8 \quad (3.261)$$

$$\sum_i S_{i3} = S_{13} + S_{23} + S_{33} = 11 \cdot 8 + 89 \cdot 6 + 156 \cdot 5 = 257 \cdot 9 \quad (3.262)$$

The main characteristic of the spatial interaction model of shopping outlined above is that the flow totals are only constrained – by equation 3.235 – at one end of the interaction. It is thus known as a singly constrained model. It is this feature which allows it to act as a *location* model and predict the spatial distribution of shopping sales, as in equations (3.260)–(3.262). It is also possible to construct a doubly constrained model, which is then solely an interaction model, for such phenomena as the journey to work. We present such a model here to illustrate a new feature of the algebra – the iterative solution of a set of non-linear simultaneous equations.

Let T_{ij} be the flow of work trips from zone i to zone j in some spatial system, and let O_i be the total number of work trip origins in zone i (i.e. the number of resident workers) and D_j the total number of work trip destinations in zone j (i.e. the number of jobs). Thus, in this case, we must have

$$\sum_j T_{ij} = O_i \tag{3.263}$$

$$\sum_i T_{ij} = D_j \tag{3.264}$$

We can construct a model as before by hypothesizing that

$$T_{ij} \propto O_i \tag{3.265}$$

$$T_{ij} \propto D_j \tag{3.266}$$

and

$$T_{ij} \propto c_{ij}^{-\beta} \tag{3.267}$$

In order to ensure that the two sets of constraints (3.263) and (3.264) are satisfied (and the model is thus called *doubly constrained*), instead of the single set of proportionality factors K_i for the shopping model, we now need a double set, which we write $\mathrm{A}_i\mathrm{B}_j$, to give

$$T_{ij} = \mathrm{A}_i\mathrm{B}_j O_i D_j c_{ij}^{-\beta} \tag{3.268}$$

Substitute from (3.268) into (3.263):

$$\sum_j \mathrm{A}_i\mathrm{B}_j O_i D_j c_{ij}^{-\beta} = O_i \tag{3.269}$$

so

$$\mathrm{A}_i = 1/\sum_j \mathrm{B}_j D_j c_{ij}^{-\beta} \tag{3.270}$$

and substituting into (3.360) and solving for B_j,

$$\sum_i \mathrm{A}_i\mathrm{B}_j O_i D_j c_{ij}^{-\beta} = D_j \tag{3.271}$$

giving

$$\mathrm{B}_j = 1/\sum_i \mathrm{A}_i O_i c_{ij}^{-\beta} \tag{3.272}$$

Thus, the model is now given by equations 3.268, 3.270, and 3.272. With the shopping model, we solved for K_i and substituted in the main model equation. Now, we might expect to solve for A_i and B_j and substitute into (3.268). The difficulties which now arise can perhaps be seen more clearly by the reader who is unfamiliar with this sort of algebra if we write out equations 3.270 and 3.272 without the summation signs:

$$A_i = 1/(B_1 D_1 c_{i1}^{-\beta} + B_2 D_2 c_{i2}^{-\beta} + \ldots) \qquad (3.273)$$

$$B_j = 1/(A_1 O_1 c_{ij}^{-\beta} + A_2 O_2 c_{2j}^{-\beta} + \ldots) \qquad (3.274)$$

Thus, each equation for A_i depends on all the B_js, and each equation for B_j on all the A_is. Thus, we have, if there are N zones, a set of $2N$ simultaneous equations. Further, because of the reciprocals on the right hand side, they are non-linear.

These equations are solved by an iterative procedure. Some starting value is assumed for B_j, say

$$B_j = 1 \qquad (3.275)$$

for all j. This is then substituted in equation 3.270 to give us A_is. These A_is are substituted into equation 3.272 and a new set of B_js are calculated. This process is repeated until convergence is achieved – that is, until there is no significant change in the values of the A_is and B_js from one iteration to the next. Such a calculation is usually carried out on a computer, of course. By hand, it would be very time consuming.

In the above examples, we showed the cost function as a power function. It can in practice be any decreasing function of c_{ij}, and another common function is the negative exponential function. We use this to show another kind of algebraic technique used to estimate β in a model. Consider the following slightly simplified (by taking $\alpha = 1$) shopping model:

$$S_{ij} = \frac{e_i P_i W_j e^{-\beta c_{ij}}}{\sum_j W_j e^{-\beta c_{ij}}} \qquad (3.276)$$

We can calculate

$$C(\beta) = \sum_i \sum_j S_{ij} c_{ij} \qquad (3.277)]$$

as the total cost of shopping trips, and we show this as a function of β. Suppose S_{ij}^{obs} is a set of observed values of S_{ij} from some survey. Then we can calculate

$$C^{\text{obs}} = \sum_i \sum_j S_{ij}^{\text{obs}} c_{ij} \qquad (3.278)$$

as an observed total cost. It can be shown that the best value of β is that which satisfies the equation

$$C(\beta) = C^{\text{obs}} \tag{3.279}$$

If we substitute for S_{ij} from 3.276 into 3.277, and then for $C(\beta)$ in 3.279, we get

$$\sum_i \sum_j \left[\frac{e_i P_i W_j e^{-\beta c_{ij}}}{\sum_j W_j e^{-\beta c_{ij}}} c_{ij} \right] = C^{\text{obs}} \tag{3.280}$$

This is obviously a very complicated equation in β, and we cannot hope to solve for β analytically. It is, however, straightforward to obtain a solution graphically. Figure 3.36 shows a plot of $C(\beta)$ against β, and also

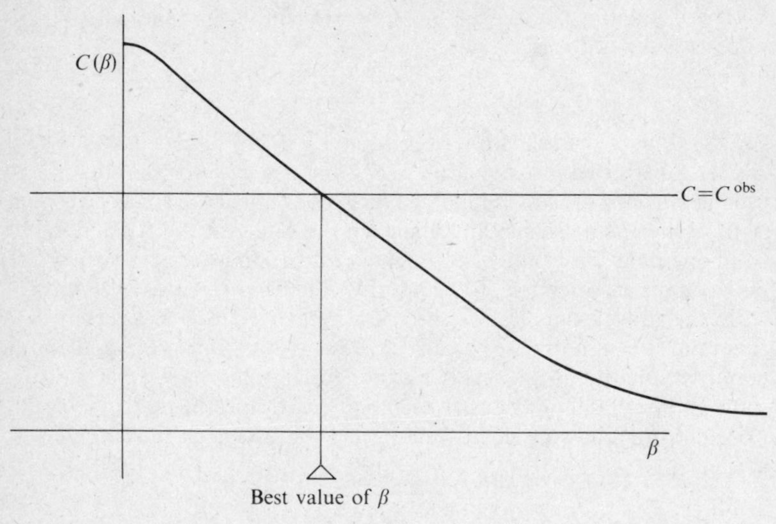

FIG. 3.36. Graphical estimation of spatial interaction model parameter

the straight line $C = C^{\text{obs}}$. $C(\beta)$ turns out to be a monotonically decreasing function of β, and so is equal to C^{obs} at a unique point. In practice, such a procedure can be carried out to any desired degree of accuracy on the computer.

We noted above that $c_{ij}^{-\beta}$ and $e^{-\beta c_{ij}}$ are two most commonly used functions in spatial interaction models. It is interesting to note that there is a simple relationship between them. If c_{ij} in $e^{-\beta c_{ij}}$ is replaced by $\log c_{ij}$, which we might interpret by saying that 'perceived cost' increases less than linearly, it becomes $e^{-\beta \log c_{ij}}$, and then

$$e^{-\beta \log c_{ij}} = e^{\log c_{ij}^{-\beta}} = c_{ij}^{-\beta} \tag{3.281}$$

3.8.8. Diffusion from a polluting source

Finally, we illustrate the concepts introduced earlier of a function of several variables in relation to three-dimensional co-ordinate geometry by studying the diffusion of pollutants from a single source (Gustafson and Kortanek, 1972). Consider the three-dimensional co-ordinate system of Figure 3.37.

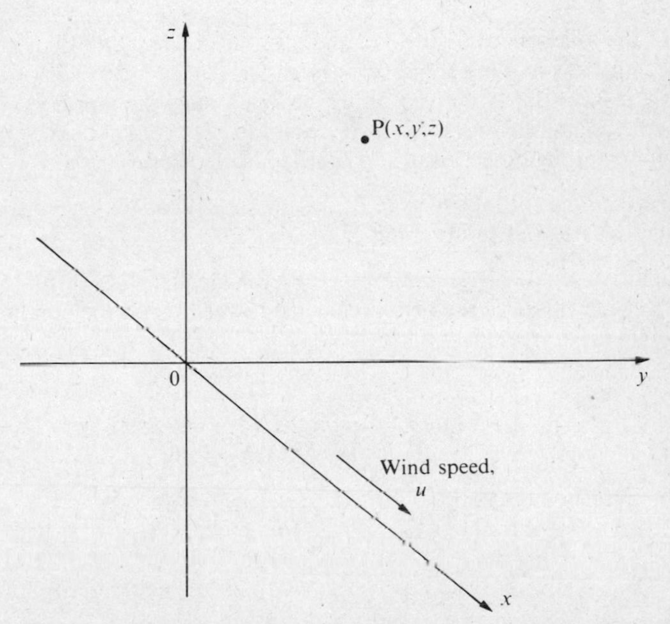

FIG. 3.37. Diffusion of pollutants from a source: frame of reference

Suppose there is a polluting source of strength Q grammes at the origin O, and that there is a mean wind speed u in the direction of the x-axis as shown. Then, it can be shown that, after time t, the amount of pollutant at the point (x, y, z) is $R(x, y, z, t)$ given by

$$R(x, y, z, t) = \frac{Q(2\pi)^{-3/2}}{abc} \exp\left\{\left[\frac{(x-ut)^2}{2a^2} + \frac{y^2}{2b^2} + \frac{z^2}{2c^2}\right]\right\} \qquad (3.282)$$

where a, b, and c are constants. This is, in fact, a three-dimensional normal distribution.

Exercises

Section 3.2

1. Write the equation $y = f(x)$ when f is the following: (a) 'multiplica-

tion by 2', (b) 'squaring', (c) 'add 2, and square', (d) 'square and add 2'.

2. Write the equation $y = f[g(x)]$ when f and g are the following: (a) 'squaring', 'squaring', (b) 'multiplication by 2', 'squaring', (c) 'squaring', 'add 2', respectively.

Section 3.3

3. Plot the following straight lines and give the gradient and the intercept on the y-axis in each case: (a) $y = 3x$, (b) $2x + 3y - 3 = 0$, (c) $y = 4x + 3$, (d) $x + y = 5$, (e) the line joining the points (1, 2) and (2, 5), (f) the line through the point (1, 2) with gradient 3, (g) the line through the point (3, 4) perpendicular to the line $x + y = 5$.

4. Find the point of intersection of the lines $x + y = 5$, $x - y = 3$, and demonstrate your result graphically.

5. Find the feasible region determined by the inequalities $x + y \leqslant 5$, $x - y \leqslant 3$, $x \geqslant 2$. Is the region changed if we also add the condition $y \geqslant 0$?

Section 3.4

6. Plot, roughly, the following functions: (a) $y = 2x^2$, (b) $y = (x + 2)^2$, (c) $y = \log_{10}(2x + 3)$, (d) $y = e^{(x+3)}$, (e) $y = \sin 2x$

7. Obtain in a more concise form expressions for: (a) $x^2x^{2\cdot5}x^3$, (b) $x^3/x^{2\cdot5}$, (c) $\log (x^5)$, (d) $x^{1/3}x^3$, (e) $(x^3)^{1/3}$, (f) $\log x^2 + \log x^2$, (g) $\log x^3 - \log x$

Section 3.5

8. Plot, roughly, the following functions: (a) $y = x^2 + x + 1$, (b) $y = (\sin x)/x$, (c) $y = x \log_e x$, (d) $y = (x + 1)/(x - 1)$, (e) $y = (x^2 + x + 1)/(x + 1)$, (f) $y = x^2 - 7x + 12$, (g) $y = x^2 - x - 12$, (h) $y = 2x^2 + x - 1$, (i) $y = 4x^2 - 1$, (j) $y = x^3 - 8$, (k) $y = 1/(1 + e^{-x}) + 2$, (l) $y = e^{-0\cdot5x^2}$

Section 3.7

10. Explain, using Pythagoras' theorem, why the equation $x^2 + y^2 + z^2 = a^2$ represents a sphere of radius a in three dimensions.

Section 3.8

11. If f and t are related by $f = A + Bt^{-1/2}$, and if $f = 3$ when $t = 0$ and $f = 5$ when $t = 1$, find A and B.

12. If, in relation to Fig. 3.28, $h_1 = 1$, $a = 0{\cdot}5$, $h_2 = 0{\cdot}5$, find α.

13. In a fourth-order river basin, $N_1 = 8$. How many branches are there altogether?

14. Estimate roughly 'sensible' hypothetical values for the coefficients in equations 3.220 and 3.221, and the terms m_{ij} and m_{ji} in equation 3.218, for some two-region system for a five year period. Estimate values for $P_1(0)$ and $P_2(0)$ and hence use your estimated values in equation 3.218 to find $P_1(5)$ and $P_2(5)$ (where time is measured in years).

15. Guess some numerical values for the coefficients in equation 3.228 and then produce the equivalent of Fig. 3.33 and 3.34 for the example you have thus created.

16. If either (i) $T_{ij} = K\mathrm{O}_i\mathrm{D}_j e^{-\beta c_{ij}}$, *or* (ii) $T_{ij} = K\mathrm{O}_i\mathrm{D}_j c_{ij}^{-\beta}$, show, by taking logs of each side, you could, given all the other terms, solve for β. Indicate how β could be obtained graphically if a set of measurements were available for T_{ij}, O_i, D_j and c_{ij} for varying i and j.

17. Invent some alternative data for Table 3.2 and Fig. 3.35, and calculate S_{ij}, where $S_{ij} = \mathrm{K}e_i P_i W_j^{\alpha} c_{ij}^{-\beta}$. Retain the assumption that $\alpha = \beta = 1$. Further, calculate the S_{ij}s given by $S_{ij} = \mathrm{A}_i e_i P_i W_j^{\alpha} c_{ij}^{-\beta}$, where $\mathrm{A}_i = 1/\sum_j W_j^{\alpha} c_{ij}^{-\beta}$ and compare the results with the previous calculation. What are the advantages of the second model?

18. Invent data (O_i, D_j and c_{ij}) for a *two*-region system and the model given by equations 3.268, 3.270 and 3.272. Assume $\beta = 1$. Carry out three iterations of the A_i, B_j calculation, and then find the T_{ij}s. Do you think your iterative process had nearly converged?

4. Matrix algebra

4.1. Definitions

WE have already had several examples of variables which have two subscripts. In section 2.2.6 of Chapter 2, T_{ij} was taken to be the number of trips from the ith zone to the jth zone in some spatial system. In section 2.2.9, we had examples of both population and economic accounts: K_{ij} was the number of people resident in zone i at the beginning of a time period and resident in zone j at the end; Z_{mn} was the amount of the product of industry m used by industry n during some time period. In section 2.2.7, a rather different kind of two subscript variable was defined to describe a network structure. If i and j are any two nodes of some network, then L_{ij} is 1 if there is a direct connection, and zero otherwise. Variations of these examples, and some new ones, will turn up at various stages in the rest of the book. For example, in Chapter 8, p_{ij} will be the probability that a river network has i branches at the jth level of a hierarchy.

Any variable with two subscripts is known as a *matrix*. Let A_{ij}, $i = 1, 2, \ldots m$, $j = 1, 2, \ldots n$ be such a matrix. It is given the name **A**, but is represented in *bold* type, as here. The whole set of A_{ij}s are known as the elements of the matrix, and can be written out as an array as in equation 4.1 below:

$$\mathbf{A} = \begin{bmatrix} A_{11} & A_{12} & A_{13} & \cdots\cdots & A_{1n} \\ A_{21} & A_{22} & A_{23} & \cdots\cdots & A_{2n} \\ \cdots & \cdots & \cdots & \cdots\cdots & \cdots \\ A_{m1} & A_{m2} & A_{m3} & \cdots\cdots & A_{mn} \end{bmatrix} \tag{4.1}$$

That is, the elements are arranged in an array of rows and columns, such that the first index gives the row number, and the second the column number. Thus, the element* A_{23} is located in the 2nd row and 3rd

* The reader should recall that A_{23} reads A two–three, and not A twenty three. If there is any doubt, a comma or a clear space should be inserted between the subscripts.

column; A_{ij} is the element in the ith row and jth column.

The full matrix is sometimes written $\{A_{ij}\}$, which stands for 'the matrix whose (i, j)th element is A_{ij}'. Such a matrix, with m rows and n columns, is known as an $m \times n$ matrix. An important special case is the $m \times 1$ matrix, which is known as a *vector.* With two subscripts, such a matrix would be

$$\mathbf{A} = \begin{bmatrix} A_{11} \\ A_{21} \\ A_{31} \\ \vdots \\ A_{m1} \end{bmatrix} \tag{4.2}$$

In this case, of course, the second index is redundant, and the vector could be written, as is more customary,

$$\mathbf{A} = \begin{bmatrix} A_1 \\ A_2 \\ A_3 \\ \vdots \\ A_m \end{bmatrix} \tag{4.3}$$

Thus, this is the vector whose elements are A_i. That is, a *singly* subscripted variable, such as A_i, is a vector, and is a special case of a matrix.

We might also note, formally, that a non-subscripted variable, or even a single number, can be thought of as 1×1 matrix. It is known as a *scalar.*

It is convenient at the outset to define the transpose of a matrix. Given the $m \times n$ matrix $\mathbf{A}$ defined above, its transpose is denoted by $\mathbf{A}'$, or $\mathbf{A}^T$, and is the $n \times m$ matrix whose (i, j)th element is A_{ji}. Thus

$$\mathbf{A}' = \begin{bmatrix} A_{11} & A_{21} & A_{31} & \cdots & A_{m1} \\ A_{12} & A_{22} & A_{32} & \cdots & A_{m2} \\ A_{13} & A_{23} & A_{33} & \cdots & A_{m3} \\ \cdots & \cdots & \cdots & \cdots & \cdots \\ A_{1n} & A_{2n} & A_{3n} & \cdots & A_{mn} \end{bmatrix} \tag{4.4}$$

Note that the transpose of an $m \times 1$ vector A_i (known as a *column* vector) is a row vector:

$$\mathbf{A}' = [A_1, A_2, \ldots A_m] \tag{4.5}$$

Finally, we recall equations 2.11 and 2.12 of Chapter 2, and we can now see these as the row and column sums respectively of the matrix T_{ij}. If we let X_i be the sum of the ith row of $\{A_{ij}\}$ and Y_j be the sum of the jth column, then

$$X_i = \sum_{j=1}^{n} A_{ij} \tag{4.6}$$

$$Y_j = \sum_{i=1}^{m} A_{ij} \tag{4.7}$$

Note that these row and column sums can themselves be considered as vectors $\{X_i\}$ and $\{Y_j\}$.

4.2. The basic operations of matrix algebra

In this section, we present the main concepts of the algebra of matrices. In section 2.1 of Chapter 2 we did the same thing for ordinary algebraic variables. Now, we consider algebraic operations applied to the matrices themselves. We have to give a meaning to such concepts as **A** + **B**, matrix addition, and **AB**, matrix multiplication. We shall find some surprises – for example, it is usually the case that

$$\mathbf{AB} \neq \mathbf{BA} \tag{4.8}$$

that is, matrix multiplication is non-commutative.

We define the basic operations by setting up a new matrix **C** which is the combination required. Thus, matrix *addition*

$$\mathbf{C} = \mathbf{A} + \mathbf{B} \tag{4.9}$$

is defined by

$$C_{ij} = A_{ij} + B_{ij} \tag{4.10}$$

That is, **C** is the matrix whose elements are the sum of the corresponding elements of **A** and **B**. For example, if

$$\mathbf{A} = \begin{bmatrix} a_{11} & a_{12} \\ a_{21} & a_{22} \end{bmatrix} \tag{4.11}$$

$$\mathbf{B} = \begin{bmatrix} b_{11} & b_{12} \\ b_{21} & b_{22} \end{bmatrix} \tag{4.12}$$

$$\mathbf{C} = \begin{bmatrix} a_{11}+b_{11} & a_{12}+b_{12} \\ a_{21}+b_{21} & a_{22}+b_{22} \end{bmatrix} \tag{4.13}$$

Or, if

$$\mathbf{A} = \begin{bmatrix} a & b \\ c & d \end{bmatrix} \tag{4.14}$$

$$\mathbf{B} = \begin{bmatrix} x & y \\ z & n \end{bmatrix} \tag{4.15}$$

$$\mathbf{C} = \begin{bmatrix} a+x & b+y \\ c+z & d+n \end{bmatrix} \tag{4.16}$$

Or, if

$$\mathbf{A} = \begin{bmatrix} 1 & 2 \\ 1 & 0 \end{bmatrix} \tag{4.17}$$

$$\mathbf{B} = \begin{bmatrix} 3 & 4 \\ 1 & 2 \end{bmatrix} \tag{4.18}$$

$$\mathbf{C} = \begin{bmatrix} 4 & 6 \\ 2 & 2 \end{bmatrix} \tag{4.19}$$

Similarly, we can define matrix subtraction by saying that if

$$\mathbf{C} = \mathbf{A} - \mathbf{B} \tag{4.20}$$

$$C_{ij} = A_{ij} - B_{ij} \tag{4.21}$$

We should perhaps emphasize for both addition and subtraction that the matrices being combined must be of the same size – say both $m \times n$ matrices, and then **C** is an $m \times n$ matrix also.

The next operation has no counterpart in ordinary algebra – *multiplication by a scalar.* If **A** is any matrix, and k a scalar, then

$$\mathbf{C} = k\mathbf{A} \tag{4.22}$$

is the matrix whose (i, j)th element is

$$C_{ij} = kA_{ij} \tag{4.23}$$

That is, each element is multiplied by the scalar in turn.

Next, we turn to *matrix multiplication.* The definition will at first sight seem a little odd: if

$$\mathbf{C} = \mathbf{AB} \tag{4.24}$$

then

$$C_{ij} = \sum_{k=1}^{n} A_{ik} B_{kj} \tag{4.25}$$

where **A** is any $m \times n$ matrix and **B** is any $n \times p$ matrix. Thus, the number of columns in **A** must be equal to the number of rows in **B** but there is no other restriction. **C** is then an $m \times p$ matrix. Equation 4.25, in words, says that the (i, j)th element of the product is formed by taking the kth element of the ith row of **A**, multiplying it by the kth element of the jth column of **B**, and adding all such elements.

For example, if **A** is the 2×2 matrix

$$\mathbf{A} = \begin{bmatrix} a_{11} & a_{12} \\ a_{21} & a_{22} \end{bmatrix} \tag{4.26}$$

and **B** is the 2×3 matrix

$$\mathbf{B} = \begin{bmatrix} b_{11} & b_{12} & b_{13} \\ b_{21} & b_{22} & b_{23} \end{bmatrix} \tag{4.27}$$

then we can form each element of the product matrix **C** using equation 4.25 as follows:

$$C_{11} = \sum_{k=1}^{2} a_{1k} b_{k1} = a_{11}b_{11} + a_{12}b_{21} \tag{4.28}$$

$$C_{12} = \sum_{k=1}^{2} a_{1k} b_{k2} = a_{11}b_{12} + a_{12}b_{21} \tag{4.29}$$

and so on, to give, in full,

$$\mathbf{C} = \mathbf{AB} = \begin{bmatrix} a_{11}b_{11} + a_{12}b_{21} & a_{11}b_{12} + a_{12}b_{21} & a_{11}b_{13} + a_{12}b_{23} \\ a_{21}b_{11} + a_{22}b_{21} & a_{21}b_{12} + a_{22}b_{22} & a_{21}b_{13} + a_{22}b_{23} \end{bmatrix} \tag{4.30}$$

Note that, since **A** and **B** are 2 × 2 and 2 × 3 respectively, **C** is a 2 × 3 matrix. A numerical example is as follows:

$$\mathbf{A} = \begin{bmatrix} 2 & 3 \\ 1 & 1 \\ 1 & 0 \end{bmatrix} \tag{4.31}$$

$$\mathbf{B} = \begin{bmatrix} 1 & 1 & 1 \\ 1 & 0 & 2 \end{bmatrix} \tag{4.32}$$

The, using equation 4.25,

$$C_{11} = 2 \times 1 + 3 \times 1 = 5 \tag{4.33}$$

$$C_{12} = 2 \times 1 + 3 \times 0 = 2 \tag{4.34}$$

and, proceeding similarly,

$$\mathbf{C} = \mathbf{AB} = \begin{bmatrix} 5 & 2 & 8 \\ 2 & 1 & 3 \\ 1 & 1 & 1 \end{bmatrix} \tag{4.35}$$

Note that, since **A** is a 3 × 2 matrix, and **B** a 2 × 3 matrix, the product matrix is a 3 × 3 one. The reader should practice matrix multiplication using the exercises until he or she is completely familiar with it.

Note that having defined the product of two matrices, we can define the product of any number of matrices **A**, **B**, **C**, . . . but, because in general $\mathbf{AB} \neq \mathbf{BA}$, order must always be preserved. We speak of *pre-multiplication* of **A** by **B** to form **BA**, and *post-multiplication* to form **AB**.

It is convenient to define division using the concept of an inverse. If A and B are numbers, then A/B could, of course, be represented as AB^{-1}. B^{-1} is the inverse of B. Thus, $5^{-1} = 1/5$ is the inverse of 5. So 'division' is 'multiplication by the inverse'. The inverse of a matrix can only exist for square, $n \times n$, matrices, and then not always. First, we define the unit $n \times n$ matrix. It is

$$\mathbf{I} = \begin{bmatrix} 1 & 0 & 0 \ldots. & 0 \\ 0 & 1 & 0 \ldots. & 0 \\ 0 & 0 & 1 \ldots. & 0 \\ \ldots & \ldots & \ldots & \ldots \\ 0 & 0 & 0 \ldots. & 1 \end{bmatrix} \tag{4.36}$$

That is, it is the $n \times n$ matrix with ones in the diagonal and zeros elsewhere. Then, the inverse, if it exists, of a square matrix $\mathbf{B}$ can be designated $\mathbf{B}^{-1}$, and is the matrix such that

$$\mathbf{B}^{-1}\mathbf{B} = \mathbf{B}\mathbf{B}^{-1} = \mathbf{I} \tag{4.37}$$

(Note that, in this case, multiplication is commutative). For the present, we shall simply assume that, in general, the inverse of a square matrix *does* exist, and that its elements can be found by some procedure.

4.3. Linear simultaneous equations in matrix form

So far, all we have achieved is a number of what may appear to be rather sterile definitions. We can now begin to see how the concept of a matrix can be used. It is indeed interesting to note that the concept was first formulated, by Cayley, in relation to linear simultaneous equations.

A general set of N such equations in N variables x_i can be written

$$\sum_{j=1}^{N} a_{ij}x_j = b_i, \qquad i = 1, \ldots N \tag{4.38}$$

which can be presented in full as

$$\left.\begin{array}{l} a_{11}x_1 + a_{12}x_2 + a_{13}x_3 + \quad \ldots\ldots + a_{1N}x_N = b_1 \\ a_{21}x_1 + a_{22}x_2 + a_{23}x_3 + \quad \ldots\ldots + a_{2N}x_N = b_2 \\ \ldots\ldots\ldots\ldots\ldots\ldots\ldots\ldots\ldots\ldots \\ a_{N1}x_1 + a_{N2}x_2 + a_{N3}x_3 + \quad \ldots\ldots + a_{NN}x_N = b_N \end{array}\right\} \tag{4.39}$$

The set of coefficients on the left hand side form a matrix **a**:

$$\mathbf{a} = \begin{bmatrix} a_{11}a_{12}a_{13} & \ldots & a_{1N} \\ a_{21}a_{22}a_{23} & \ldots & a_{2N} \\ \cdots\cdots & \cdots & \cdots \\ a_{N1}a_{N2}a_{N3} & \ldots & a_{NN} \end{bmatrix} \tag{4.40}$$

and if we define vectors **x** and **b** as

$$\mathbf{x} = \begin{bmatrix} x_1 \\ x_2 \\ \vdots \\ x_N \end{bmatrix} \tag{4.41}$$

$$\mathbf{b} = \begin{bmatrix} b_1 \\ b_2 \\ \vdots \\ b_N \end{bmatrix} \tag{4.42}$$

then the system 4.39 can be written

$$\begin{bmatrix} a_{11}a_{12}a_{13} & \ldots & a_{1N} \\ a_{21}a_{22}a_{23} & \ldots & a_{2N} \\ \cdots\cdots & \cdots & \cdots \\ a_{N1}a_{N2}a_{N3} & \ldots & a_{NN} \end{bmatrix} \begin{bmatrix} x_1 \\ x_2 \\ \vdots \\ x_N \end{bmatrix} = \begin{bmatrix} b_1 \\ b_2 \\ \vdots \\ b_N \end{bmatrix} \tag{4.43}$$

or, in brief,

$$\mathbf{ax} = \mathbf{b} \tag{4.44}$$

Thus, at the very least, we are beginning to develop a powerful short-hand notation, which is often a sign of progress in mathematics. But the real achievement is that, if we know the inverse $\mathbf{a}^{-1}$ we can pre-multiply by this in equation 4.44 to give

$$\mathbf{a}^{-1}\mathbf{ax} = \mathbf{a}^{-1}\mathbf{b} \tag{4.45}$$

and so, since

$$\mathbf{a}^{-1}\mathbf{a} = \mathbf{I} \tag{4.46}$$

by definition, and

$$\mathbf{Ix} = \mathbf{x} \tag{4.47}$$

since pre- or post-multiplication by the identity matrix can easily be checked to be equivalent to multiplication by 1 in ordinary algebra, we have

$$\mathbf{x} = \mathbf{a}^{-1}\mathbf{b} \tag{4.47}$$

as the solution of the equations. We could re-introduce subscripts explicitly and write this in the form

$$x_i = \sum_j (a^{-1})_{ij} b_j \tag{4.48}$$

where $(a^{-1})_{ij}$ is the (i, j)th element of $\mathbf{a}^{-1}$.

We illustrate these concepts with the two-variable – two-equation system which we solved by other means in section 2.5 of Chapter 2. Equations 2.72 and 2.73 can be written in the form

$$\begin{bmatrix} 0{\cdot}8 & -0{\cdot}1 \\ -0{\cdot}3 & 0{\cdot}7 \end{bmatrix} \begin{bmatrix} x_1 \\ x_2 \end{bmatrix} = \begin{bmatrix} y_1 \\ y_2 \end{bmatrix} \tag{4.49}$$

or as

$$\mathbf{ax} = \mathbf{y} \tag{4.50}$$

where

$$\mathbf{a} = \begin{bmatrix} 0{\cdot}8 & -0{\cdot}1 \\ -0{\cdot}3 & 0{\cdot}7 \end{bmatrix} \tag{4.51}$$

$$\mathbf{x} = \begin{bmatrix} x_1 \\ x_2 \end{bmatrix} \tag{4.52}$$

$$\mathbf{y} = \begin{bmatrix} y_1 \\ y_2 \end{bmatrix} \tag{4.53}$$

The solution is

$$\mathbf{x} = \mathbf{a}^{-1}\mathbf{y} \tag{4.54}$$

The inverse in this case is

$$\mathbf{a}^{-1} = \begin{bmatrix} \frac{70}{53} & \frac{10}{53} \\ \frac{30}{53} & \frac{80}{53} \end{bmatrix} \tag{4.55}$$

so that

$$\begin{bmatrix} x_1 \\ x_2 \end{bmatrix} = \begin{bmatrix} \frac{70}{53} & \frac{10}{53} \\ \frac{30}{53} & \frac{80}{53} \end{bmatrix} \begin{bmatrix} y_1 \\ y_2 \end{bmatrix} \tag{4.56}$$

That is,

$$x_1 = \frac{70}{53} y_1 + \frac{10}{53} y_2 \tag{4.57}$$

$$x_2 = \frac{30}{53} y_1 + \frac{80}{53} y_2 \tag{4.58}$$

which agrees with the solutions given in section 2.5. We can check that $\mathbf{a}^{-1}$ as given in equation 4.55 is the inverse of **a** by direct multiplication:

$$\mathbf{aa}^{-1} = \begin{bmatrix} 0{\cdot}8 & -0{\cdot}1 \\ -0{\cdot}3 & 0{\cdot}7 \end{bmatrix} \begin{bmatrix} \frac{70}{53} & \frac{10}{53} \\ \frac{30}{53} & \frac{80}{53} \end{bmatrix} \tag{4.59}$$

$$= \begin{bmatrix} 1 & 0 \\ 0 & 1 \end{bmatrix} \tag{4.60}$$

which is the 2×2 identity matrix.

Although we have not given here the procedure for finding the inverse of a matrix, we should note that in any particular case, it can easily be found with the use of a standard programme on a computer. To give the usual method would take us beyond the scope of this book. However, the above discussion shows that one 'manual' method is available. If the simultaneous equations associated with a matrix **a**, as in equation 4.51 above, are solved, then the coefficients of the elements of the 'right hand side' vector (y_1 and y_2 in the above example) give the elements of the inverse matrix. This can be seen by comparing $\mathbf{a}^{-1}$ in equation 4.55 with the coefficients in equations 4.57 and 4.58.

4.4. Matrices and linear transformations

4.4.1. Rotations

Figure 4.1 shows two Cartesian co-ordinate systems, one with axes Ox and Oy, the other with axes OX and OY. The second has been rotated

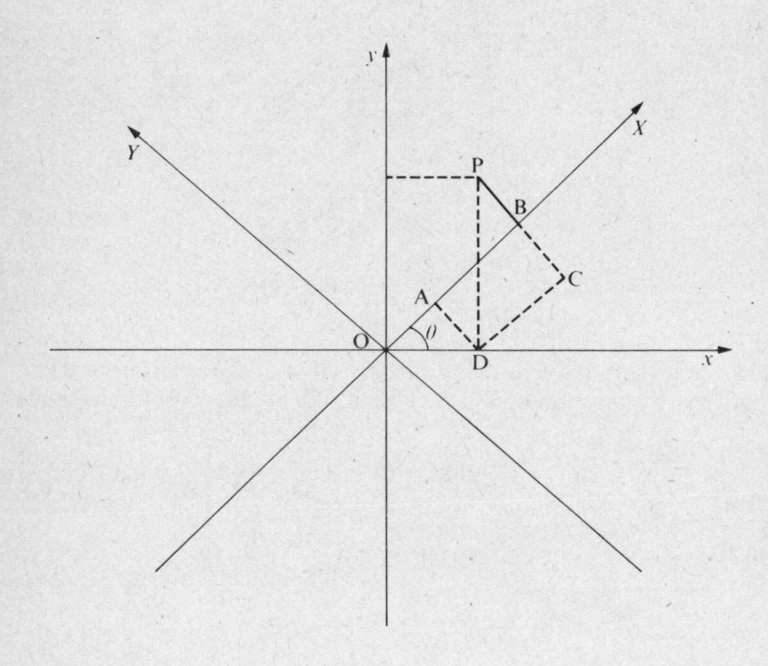

FIG. 4.1. Rotation of axes

through an angle θ from the first. Let any point P have co-ordinates (x, y) in system 1 and (X, Y) in system 2. What is the algebraic relationship between the two systems? Note that, in the second system

$$X = \text{OA} + \text{AB} \tag{4.61}$$

But that

$$\text{OA} = x \cos \theta \tag{4.62}$$

and

$$AB = CD = y \sin \theta \tag{4.63}$$

so that

$$X = x \cos \theta + y \sin \theta \tag{4.64}$$

Similarly,

$$Y = PC - BC \tag{4.65}$$

$$PC = y \cos \theta \tag{4.66}$$

$$BC = AD = x \sin \theta \tag{4.67}$$

and so

$$Y = -x \sin \theta + y \cos \theta \tag{4.68}$$

Equations 4.64 and 4.68 therefore specify the relation between the two co-ordinate systems. If we write

$$\mathbf{x} = \begin{bmatrix} x \\ y \end{bmatrix} \tag{4.69}$$

as the vector of co-ordinates in the first system, and

$$\mathbf{X} = \begin{bmatrix} X \\ Y \end{bmatrix} \tag{4.70}$$

as that for the second, and

$$\mathbf{T} = \begin{bmatrix} \cos \theta & \sin \theta \\ -\sin \theta & \cos \theta \end{bmatrix} \tag{4.71}$$

as the *matrix of the rotation,* then these equations can be written in matrix form as

$$\mathbf{X} = \mathbf{T}\mathbf{x} \tag{4.72}$$

This is an appropriate moment to indulge in a change of notation. The above description is fine for gaining a first understanding of the two-dimensional case, but does not easily extend to higher dimensions. Let $(x_1^{(1)}, x_2^{(1)})$ be the co-ordinates of P in the first system and $(x_1^{(2)}, x_2^{(2)})$ in the second. Then, **T** can be defined as before, and

$$\mathbf{x}^{(1)} = \begin{bmatrix} x_1^{(1)} \\ x_2^{(1)} \end{bmatrix} \tag{4.74}$$

and

$$\mathbf{x}^{(2)} = \begin{bmatrix} x_1^{(2)} \\ x_2^{(2)} \end{bmatrix} \tag{4.75}$$

The result can now be generalized to three and higher dimensions without difficulty. We state the corresponding results for three dimensions, though without proof. Let

$$\mathbf{x}^{(1)} = \begin{bmatrix} x_1^{(1)} \\ x_2^{(1)} \\ x_3^{(1)} \end{bmatrix} \tag{4.76}$$

be the co-ordinates of some point P, and let $\mathbf{x}^{(2)}$ be the co-ordinates of the same point in a second co-ordinate system obtained by rotating about the origin. Then the matrix of the transformation is

$$\mathbf{T} = \begin{bmatrix} m_{11} & m_{12} & m_{13} \\ m_{21} & m_{22} & m_{23} \\ m_{31} & m_{32} & m_{33} \end{bmatrix} \tag{4.77}$$

where m_{ij} is the cosine of the angle between the i-axis in the first system and the j-axis in the second. Then

$$\mathbf{x}^{(2)} = \mathbf{T}\mathbf{x}^{(1)} \tag{4.78}$$

The m_{ij} s are known as direction cosines. This transformation is of particular importance in photogrammetry (Thompson, 1966).

It can easily be checked that $\mathbf{T}$ in the two-dimensional case can be specified in the form

$$\mathbf{T} = \begin{bmatrix} m_{11} & m_{12} \\ m_{21} & m_{22} \end{bmatrix} \tag{4.78}$$

with the same formal definitions of the m_{ij} s.

4.4.2. Linear transformations of vectors and matrices

Rotations of the form 4.73 or 4.77 are special linear transformations in two and three dimensions respectively. If we also allowed a constant term to be added, so that

$$\mathbf{x}^{(2)} = \mathbf{T}\mathbf{x}^{(1)} + \mathbf{c} \tag{4.79}$$

say, where **c** is a vector of constants, then this would represent a shift in origin also. However, we are mostly interested in rotations and so we shall restrict ourselves to the case $\mathbf{c} = 0$.

We have obtained equations which give the effect of a rotation **T** on vectors. It is also of importance to see the affect of such a rotation on matrices. Let $\mathbf{a}^{(1)}$ be some matrix specified in relation to a co-ordinate system – say a 3×3 matrix for a three-dimensional co-ordinate system (though what follows will apply to an $n \times n$ matrix and an n dimensional co-ordinate system), and suppose it relates two vectors $\mathbf{y}^{(1)}$ and $\mathbf{x}^{(1)}$ specified in relation to system 1 by

$$\mathbf{y}^{(1)} = \mathbf{a}^{(1)}\mathbf{x}^{(1)} \tag{4.80}$$

In the new system, it will relate $\mathbf{y}^{(2)}$ and $\mathbf{x}^{(2)}$ by

$$\mathbf{y}^{(2)} = \mathbf{a}^{(2)}\mathbf{x}^{(2)} \tag{4.81}$$

where

$$\mathbf{y}^{(2)} = \mathbf{T}\mathbf{y}^{(1)} \tag{4.82}$$

and

$$\mathbf{x}^{(2)} = \mathbf{T}\mathbf{x}^{(1)} \tag{4.83}$$

Pre-multiply by $\mathbf{T}^{-1}$ (assuming that the inverse exists), to give

$$\mathbf{y}^{(1)} = \mathbf{T}^{-1}\mathbf{y}^{(2)} \tag{4.84}$$

and

$$\mathbf{x}^{(1)} = \mathbf{T}^{-1}\mathbf{x}^{(2)} \tag{4.85}$$

Substitute from equations 4.84 and 4.85 into equation 4.80:

$$\mathbf{T}^{-1}\mathbf{y}^{(2)} = \mathbf{a}^{(1)}\mathbf{T}^{-1}\mathbf{x}^{(2)} \tag{4.86}$$

Pre-multiply by **T**:

$$\mathbf{y}^{(2)} = \mathbf{T}\mathbf{a}^{(1)}\mathbf{T}^{-1}\mathbf{x}^{(2)} \tag{4.87}$$

and so, comparing equations 4.81 and 4.87 we see that

$$\mathbf{a}^{(2)} = \mathbf{T}\mathbf{a}^{(1)}\mathbf{T}^{-1} \tag{4.88}$$

which gives the transformation equation for a matrix under a linear transformation T.

4.4.3. Canonical forms, eigenvalues, and eigenvectors

It is often of some interest to find transformations **T** under which some given matrix takes a particularly convenient form. The most well known of such transformations is that which diagonalizes a matrix. That is, given some matrix $\mathbf{a}^{(1)}$, find **T** such that $\mathbf{a}^{(2)}$ given in equation 4.88 is diagonal (which means that all non-diagonal matrix elements are zero). To indicate how this is done, it is necessary, as a preliminary, to introduce the concepts of eigenvalues and eigenvectors of a matrix.

Let **a** be any $n \times n$ matrix. We can then ask whether there exists a scalar number λ, and a vector **x** such that

$$\mathbf{a}\mathbf{x} = \lambda\mathbf{x} \tag{4.89}$$

If they can be found, λ is known as an eigenvalue of **a** and **x** as the corresponding eigenvector. In general, an $n \times n$ matrix has n distinct eigenvalues, $\lambda_1, \lambda_2, \lambda_3, \ldots \lambda_n$, and n associated eigenvectors $\mathbf{x}_1, \mathbf{x}_2, \ldots \mathbf{x}_n$ say. It is beyond the scope of this book to give a procedure for calculating these. For our present purposes, we simply wish to note that if we form the matrix **T** whose columns are the eigenvectors of **a**:

$$\mathbf{T} = [\mathbf{x}_1 \mathbf{x}_2 \ldots \mathbf{x}_n] \tag{4.90}$$

then the matrix $\mathbf{TaT}^{-1}$ is diagonal, and the non-zero diagonal elements are the eigenvalues of **a**.

This is a useful result in relation to the technique of *principal components' analysis,* which has been much used by geographers and planners. Essentially, given a set of n variables describing some system as a vector **x**, principal components analysis involves finding the transformation **T** which diagonalizes the matrix of correlation coefficients **R** relating the variables of **x**. The principal components are then **y** given by

$$\mathbf{y} = \mathbf{T}\mathbf{x} \tag{4.91}$$

A more extensive discussion can easily be found elsewhere (see especially, Gould, 1967 and Rogers, 1972).

4.5. Examples of the use of matrices

4.5.1. The analysis of nodal structure

This illustration is not strictly concerned with matrix algebra as such, but since so much geographical data and information comes in matrix form, it offers an interesting way of interpreting such data. The technique to be discussed was invented by Nystuen and Dacey (1961), and is also reported by Haggett (1965). This account is given in relation to their own hypothetical data matrix for a twelve-zone spatial system as presented in Table 4.1.

Let us call the matrix $\{T_{ij}\}$, and then proceed as follows. M_i is the largest element in the ith row:

$$M_i = \operatorname*{Max}_j\{T_{ij}\} \tag{4.92}$$

using an obvious notation. Let J_i be the actual j value which gives rise to M_i. That is,

$$M_i = T_{iJ_i} \tag{4.93}$$

Let D_j be the column sum:

$$D_j = \sum_i T_{ij} \tag{4.94}$$

TABLE 4.1

*Example of an interaction matrix**

to	1	2	3	4	5	6	7	8	9	10	11	12
from												
1	0	75	15	20	28	2	3	2	1	20	1	0
2	*69*	0	45	50	58	12	20	3	6	35	4	2
3	5	*51*	0	12	40	0	6	1	3	15	0	1
4	19	*57*	140	0	30	7	6	2	11	18	5	1
5	7	40	*48*	26	0	7	10	2	37	39	12	6
6	1	6	1	1	10	0	*27*	1	3	4	2	0
7	2	16	3	3	13	31	0	3	18	8	3	1
8	0	4	0	1	3	3	6	0	12	*38*	4	0
9	2	28	3	6	43	4	16	12	0	*98*	13	1
10	7	40	10	8	40	5	17	34	*98*	0	35	12
11	1	8	2	1	18	0	6	5	12	*30*	0	15
12	0	2	0	0	7	0	1	0	1	6	*12*	0
Total/R	113	337	141	128	290	71	118	65	202	311	91	39
Rank	8	1	5	6	3	10	7	11	4	2	9	12

* Source Nystuen and Dacey (1961)

Then D_j is taken to measure the *rank order* of the zones, the centre with highest D_j being first, the next second, and so on.

A plot of nodal structure is then obtained as follows. If M_i is a flow to J_i which is a lower-order centre than i, then i is a terminal point. (That is, i is a terminal point if the largest outward flow is to a lower order centre). The terminal points (in this example nodes 2, 4, 7 and 10) should be plotted on a map.

Then a vector is plotted for each non-terminal zone i, joining i and J_i. This process will connect each zone directly or indirectly to a terminal zone. The result for this example is shown in Figure 4.2.

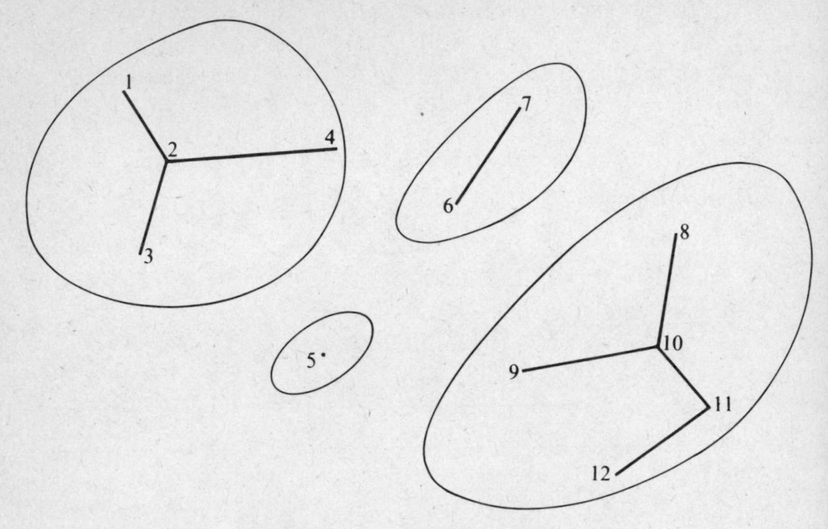

FIG. 4.2. Nodal structure analysis

The rules used by Nystuen and Dacey can be expressed formally as follows. Let $N(i, j)$ be 1 if i and j are to be connected, and zero otherwise. Then

$$N(i, j) = 1 \tag{4.95}$$

if and only if

$$\text{(i) } j = J_i \tag{4.96}$$

$$\text{and} \quad \text{(ii) } D_i > D_j \tag{4.97}$$

and (iii) i is not a terminal.

These rules produce as many connections as there are nodes. It is interesting to generalize them to create the possibility of producing more connections. For example, we could say that i is a terminal if, when $j = J_i$, *either* $D_i > D_j$ *or* $D_i > \text{A}$, where A is some constant. And then we could connect terminal i s to J_i s, and also set $N(i,j) = 1$ if $T_{ij} > \text{B}$, where B is some constant. This rule would allow terminals to be connected in suitable circumstances, while Nystuen's and Dacey's does not.

This form of analysis, of course, has a subjective basis only, but may offer a systematic way of pictorially representing the most important elements in a matrix and the associated nodal structure.

4.5.2. *An account-based population model*

In section 2.2.9 we defined $K_{ij}(t, t+T)$ to be the number of people who were resident in zone i at time t, but zone j at $t+T$, and $P_i(t)$ as the total population of zone i at time t and $P_i(t+T)$ as the total population of zone i at time $t+T$. $P_i(t)$ and $P_i(t+T)$ are, respectively, the row and column sums of the matrix $K_{ij}(t,t+T)$ as shown in Table 4.3.

TABLE 4.3

$K_{11}(t,t+T)$	$K_{12}(t,t+T)$	......	$K_{1N}(t,t+T)$	$P_1(t)$
$K_{21}(t,t+T)$	$K_{22}(t,t+T)$	......	$K_{2N}(t,t+T)$	$P_2(t)$
$\vdots$				$\vdots$
$K_{N1}(t,t+T)$	$K_{N2}(t,t+T)$	.....	$K_{NN}(t,t+T)$	$P_N(t)$
$P_1(t+T)$	$P_2(t+T)$	.....	$P_N(t+T)$	

A simple matrix model can be constructed as follows. Let **K** be the matrix $\{K_{ij}\}$. We can then form the matrix **R** of *rates* by dividing each element of **K** by the corresponding row sum. Such matrices can also sometimes be considered as Markov matrices (cf. Chapter 8). In full, this means that

$$\mathbf{R} = \begin{bmatrix} \dfrac{K_{11}(t,t+T)}{P_1(t)} & \dfrac{K_{12}(t,t+T)}{P_2(t)} \cdots\cdots & \dfrac{K_{1N}(t,t+T)}{P_1(t)} \\[2ex] \dfrac{K_{21}(t,t+T)}{P_2(t)} & \dfrac{K_{22}(t,t+T)}{P_2(t)} \cdots\cdots & \dfrac{K_{2N}(t,t+T)}{P_1(t)} \\[2ex] \dfrac{K_{N1}(t,t+T)}{P_N(t)} & \dfrac{K_{N2}(t,t+T)}{P_N(t)} \cdots\cdots & \dfrac{K_{NN}(t,t+T)}{P_N(t)} \end{bmatrix} \tag{4.98}$$

This kind of operation can be represented more conveniently by simply specifying what happens to the general term. Thus, **R** is the matrix whose (i,j)th term is

$$R_{ij} = K_{ij}(t,t+T)/P_i(t) \tag{4.99}$$

and henceforth we shall use this method. The next step is to form the matrix **G** which is the transpose of **R**:

$$G_{ij} = R_{ji} \tag{4.100}$$

Then, the reader can easily check (by specifying the matrices and vectors in full if necessary) that

$$\mathbf{P}(t+T) = \mathbf{G}\,\mathbf{P}(t) \tag{4.101}$$

In other words, the vector of $t+T$ populations is obtained by operating on the vector of old populations by the so-called *growth matrix* **G**. This can be used as a projection model if some independent way can be found of *forecasting* the rates which form the elements of **G**.

This idea is the basis of a number of spatial demographic models, though considerable refinement is needed – to take account of births, deaths, and migration, and age–sex disaggregation–before such models are realistic (see Rogers, 1966, Rees and Wilson, 1973, Wilson and Rees, 1974, for example).

We can give an idea of what is involved in age disaggregation by reinterpreting the i s and j s in the above example. Suppose they now represent age groups of population in a single region. Then, $K_{ij}(t, t+T)$ is the number of people in age group i at time t who have survived into age group j at time $t+T$. If we make the additional assumption that each age group interval is of length T – the same as the projection period – then survival is always from the next earlier age group. That is, the super-diagonal terms $K_{i\,i+1}(t, t+T)$ will be non-zero, but all others zero. 'Non-survival' in this case will include net out-migration as well as death.

Then, the matrices **R** and **G** can be formed as in equation 4.98 and 4.100, except that this time, most of the terms will be zero. For example,

$$\mathbf{G} = \begin{bmatrix} 0 & 0 & 0 \ldots\ldots 0 \\ G_{21} & 0 & 0 \ldots\ldots 0 \\ 0 & G_{32} & 0 \ldots\ldots 0 \\ 0 & 0 & G_{43} \ldots\ldots 0 \\ 0 \ldots\ldots & G_{NN-1} & G_{NN} \end{bmatrix} \tag{4.102}$$

Only the sub-diagonal terms will be non-zero, except for G_{NN} which represents the rate of survival within the last age group. The non-zero terms are given by

$$G_{ii+1} = K_{ii+1}(t, t+T)/P_i(t) \tag{4.103}$$

$$G_{NN} = K_{NN}(t, t+T)/P_N(t) \tag{4.104}$$

The model equation 4.101 then holds as before. The only problem is that $P_1(t+T)$ is zero since we have not allowed for births. This can be dealt with by adding some non-zero elements in the first row which correspond to birth rates for child-bearing age groups. **G** then takes the form

$$\mathbf{G} = \begin{bmatrix} 0 & 0 & 0 & b_\alpha \cdots & b_\beta & 0 \\ G_{12} & 0 & 0 & 0 & 0 & 0 \\ 0 & G_{23} & 0 & 0 \cdots & & \\ 0 & 0 & G_{34} & \cdots & & \\ 0 & 0 & & & G_{N-1\,N} & G_{NN} \end{bmatrix} \tag{4.105}$$

where α and β are the lower and upper limits of the child-bearing age groups. Thus the model, written out in full, is now

$$\begin{bmatrix} P_1(t+T) \\ P_2(t+T) \\ P_3(t+T) \\ P_4(t+T) \\ P_N(t+T) \end{bmatrix} = \begin{bmatrix} 0 & 0 & 0 & b_\alpha \ldots b_\beta & 0 \\ G_{12} & 0 & 0 & & 0 \\ 0 & G_{23} & 0 & & 0 \\ 0 & 0 & G_{34} & & 0 \\ 0 & 0 & & G_{N-1\,N} & G_{NN} \end{bmatrix} \begin{bmatrix} P_1(t) \\ P_2(t) \\ P_3(t) \\ P_4(t) \\ P_N(t) \end{bmatrix} \tag{4.106}$$

This, in essence, is the demographic model of Leslie (1945).

4.5.3. *The input–output model*

In section 2.2.9 also, we defined the main variables for describing the inter-industry transactions within an economy: Z_{mn} is the amount of the product of industry m used by industry n during some time period, Y_m is the amount of n consumed directly – the so-called 'final demand', and X_n is the total amount produced. For convenience, we have dropped the time labels. Then, we had the obvious accounting relationship

$$\sum_{n=1}^{N} Z_{mn} + Y_m = X_m \tag{4.107}$$

A model can be formed from this as follows: define

$$a_{mn} = \frac{Z_{mn}}{X_n} \tag{4.108}$$

as the amount of the product of industry m used to produce a unit of industry n. Then, replace Z_{mn} in equation 4.107 by $a_{mn}X_n$ (from 4.108) to give

$$\sum_n a_{mn}X_n + Y_m = X_m \quad (4.109)$$

In matrix notation, this can be written

$$\mathbf{a}\mathbf{X} + \mathbf{Y} = \mathbf{X} \quad (4.110)$$

or, more concisely,

$$(\mathbf{I} - \mathbf{a})\mathbf{X} = \mathbf{Y} \quad (4.111)$$

where **I** is the unit matrix. Thus, this represents a set of linear simultaneous equations in the elements of **X** which can be solved in the usual way to give

$$\mathbf{X} = (\mathbf{I} - \mathbf{a})^{-1}\mathbf{Y} \quad (4.112)$$

This is the usual statement of the input–output model. Given final demand, **Y**, and the input–output matrix **a** which represents the structure of the economy, **X** can be calculated. The model can, of course, be extended in all sorts of ways, and in particular, multi-regional versions can be developed (Leontief and Strout, 1963, Wilson, 1970).

A two-sector example of such an equation system was presented in section 2.5 of Chapter 2 and the same example in matrix form in section 4.3 [though note that the matrix **a** in section 4.3 is $(\mathbf{I} - \mathbf{a})$ above].

We can interpret this model further by using an extension (without proof) of a result given in Chapter 2. In equation 2.118, we showed that $1/(1-r)$, or $(1-r)^{-1}$, could be expanded as an infinite series as

$$(1-r)^{-1} = 1 + r + r^2 + r^3 + \ldots\ldots \quad (4.113)$$

This result can also be applied, under suitable conditions, to matrices, and so the right-hand side of equation 4.112 can be expanded as

$$\mathbf{X} = (\mathbf{I}-\mathbf{a})^{-1}\mathbf{Y} = (\mathbf{I} + \mathbf{a} + \mathbf{a}^2 + \mathbf{a}^3 + \ldots\ldots)\mathbf{Y} \quad (4.114)$$

$$= \mathbf{Y} + \mathbf{a}\mathbf{Y} + \mathbf{a}(\mathbf{a}\mathbf{Y}) + \mathbf{a}(\mathbf{a}^2\mathbf{Y}) + \ldots\ldots \quad (4.115)$$

So, we see from the expansion in equation 4.115 that total product **X** is made up of final demand **Y**, plus **a Y** which is the set of intermediate products needed to produce **Y**, plus **a(aY)**, which is the set of products generated by **a Y** and so on. Thus, by inverting the matrix, $(\mathbf{I}-\mathbf{a})$, and calculating total product as $(\mathbf{I}-\mathbf{a})^{-1}\mathbf{Y}$, we have avoided having to sum directly an infinite series.

Exercises

Section 4.2

1. If

$$\mathbf{A} = \begin{bmatrix} 1 & 2 & 1 \\ 2 & 1 & 1 \\ 0 & 0 & 2 \end{bmatrix}, \quad \mathbf{B} = \begin{bmatrix} 3 & 1 & 1 \\ 4 & 1 & 1 \\ 0 & 0 & 1 \end{bmatrix},$$

form **C** given by (*a*) **A** + **B**, (*b*) 2**A** − **B**, (*c*) **AB**, (*d*) **BA**

2. If

$$\mathbf{A} = \begin{bmatrix} 2 & 1 \\ 2 & 3 \\ 3 & 1 \end{bmatrix}, \quad \mathbf{B} = \begin{bmatrix} 1 & 2 & 3 & 4 \\ 3 & 3 & 2 & 1 \end{bmatrix}$$

form **C** = **AB** (Note that **BA** is undefined).

3. If

$$\mathbf{A} = \begin{bmatrix} 10 & 6 & 3 \\ 2 & 2 & 1 \\ 1 & 0 & 2 \end{bmatrix}, \quad \mathbf{X} = \begin{bmatrix} 2 \\ 1 \\ 3 \end{bmatrix}$$

form **Y** = **AX**

4. If

$$\mathbf{A} = \begin{bmatrix} a_{11} & a_{12} & a_{13} \\ a_{12} & a_{22} & a_{23} \\ a_{31} & a_{32} & a_{33} \end{bmatrix}, \quad \mathbf{X} = \begin{bmatrix} x_1 \\ x_2 \\ x_3 \end{bmatrix},$$

form **Y** = **AX**

Section 4.3

(5) If

$$\mathbf{A} = \begin{bmatrix} 0.8 & -0.1 \\ -0.3 & 0.7 \end{bmatrix}, \quad \mathbf{X} = \begin{bmatrix} x_1 \\ x_2 \end{bmatrix}$$

form **Y** = **AX**. By solving the equations **AX** = **Y** for the elements of **X**, find $\mathbf{A}^{-1}$.

(6) Repeat question (5) with

$$\mathbf{A} = \begin{bmatrix} 1 & 2 & 3 \\ 1 & 1 & 4 \\ 3 & 1 & 1 \end{bmatrix}, \quad \mathbf{X} = \begin{bmatrix} x_1 \\ x_2 \\ x_3 \end{bmatrix}$$

Find numerical values for **X** and $\mathbf{A}^{-1}$ if $\mathbf{Y} = \begin{bmatrix} 14 \\ 15 \\ 8 \end{bmatrix}$. Check that $\mathbf{AA}^{-1} = \mathbf{A}^{-1}\mathbf{A} = \mathbf{I}$.

Section 4.4

(7) A point P has co-ordinates (x,y) in system 1 and (X, Y) in system 2, which is formed by a rotation through θ from system 1. Write down the equations for X and Y in terms of x, y and θ. Solve these equations for x and y, and hence find the elements of T^{-1} where

$$T = \begin{bmatrix} \cos\theta & \sin\theta \\ -\sin\theta & \cos\theta \end{bmatrix}$$

Check that $\mathbf{TT}^{-1} = \mathbf{T}^{-1}\mathbf{T} = \mathbf{I}$

Section 4.5

(8) Find an example of an interaction matrix (migration, journey to work, etc.) and repeat Nystuen's and Dacey's analysis as described in section 4.5.1.

(9) If

$$\mathbf{K} = \begin{bmatrix} K_{11} & K_{12} & K_{13} \\ K_{21} & K_{22} & K_{23} \\ K_{31} & K_{32} & K_{33} \end{bmatrix}$$

write down the matrix **G** whose (i, j)th element is K_{ji}/K_{j*} (where $K_{j*} = \sum_{i=1}^{3} K_{ji}$). If

$$\mathbf{K} = \begin{bmatrix} 500 & 500 & 0 \\ 300 & 300 & 400 \\ 200 & 600 & 200 \end{bmatrix}$$

find **G**. Find $w(t+T)$ given by

$$\mathbf{w}(t+T) = \mathbf{G}\mathbf{w}(t) \text{ if } \mathbf{w}(t) = \begin{bmatrix} 1000 \\ 1000 \\ 1000 \end{bmatrix}$$

(10) If $\mathbf{K} = \begin{bmatrix} K_{11} & K_{12} \\ K_{21} & K_{22} \end{bmatrix} = \begin{bmatrix} 800 & 200 \\ 300 & 700 \end{bmatrix}$,

(a) form $\mathbf{w}(t) = \begin{bmatrix} w_1(t) \\ w_2(t) \end{bmatrix} = \begin{bmatrix} K_{1*} \\ K_{2*} \end{bmatrix}$

whose elements are the row sums of **K**,

(b) form $\mathbf{w}(t+T) = \begin{bmatrix} w_1(t+T) \\ w_2(t+T) \end{bmatrix} = \begin{bmatrix} K_{*1} \\ K_{*2} \end{bmatrix}$

whose elements are the column sums of **K**,
(c) divide each element of **K** by its row sum to form **L**
(d) transpose **L** to form **G**
(e) show that $\mathbf{G}\mathbf{w}(t) = \mathbf{w}(t+T)$.

(11) If $\mathbf{G} = \begin{bmatrix} 0 & 0.5 & 0.5 & 0 \\ 0.9 & 0 & 0 & 0 \\ 0 & 0.9 & 0 & 0 \\ 0 & 0 & 0.7 & 0.2 \end{bmatrix}$ and $\mathbf{w}(t) = \begin{bmatrix} 1000 \\ 1000 \\ 1000 \\ 1000 \end{bmatrix}$

find $\mathbf{w}(t+T)$ given by $\mathbf{w}(t+T) = \mathbf{G}\mathbf{w}(t)$. Let $\mathbf{L} = \mathbf{G}'$ be the transpose of **G**. Form the matrix **K** whose (i, j)th element is $L_{ij}w_i(t)$ and eheck that it forms an accounting matrix.

(12) Given an input–output matrix

$$\mathbf{A} = \begin{bmatrix} 0.2 & 0.3 & 0.1 \\ 0.1 & 0.2 & 0.3 \\ 0.3 & 0.2 & 0.1 \end{bmatrix} \text{ and final demand } \mathbf{y} = \begin{bmatrix} 10 \\ 120 \\ 20 \end{bmatrix}$$

find $\mathbf{x}$ if $(\mathbf{I}-\mathbf{A})\mathbf{x} = \mathbf{y}$. Using $\mathbf{y}$ in the form $\begin{bmatrix} y_1 \\ y_2 \\ y_3 \end{bmatrix}$ solve the equation again and hence obtain the elements of $(\mathbf{I}-\mathbf{A})^{-1}$.

5. Calculus

5.1. Introduction

The aim of this chapter is to provide a basic understanding of calculus. There will be less illustration with geographical and planning examples than was the case with some earlier chapters; most examples will in fact come in later chapters. Here, we are simply providing the main tools.

There are two main branches, differential and integral calculus, the nature of which we can identify by reference to Figure 5.1. A function $y = f(x)$ is shown; P and Q are two points on the curve, and we have also shown the chord, the straight line, connecting P and Q. We earlier defined the gradient of a straight line. Differential calculus is concerned with the *gradient of a curve at a point*. As the point Q approaches P, the chord PQ becomes, in the limit, the *tangent* to the curve at P. The techniques of the differential calculus will allow us to calculate this gradient for a wide variety of functions.

Another quantity we may be interested in, and which at first sight seems very difficult to calculate is the *area under a curve*, say ABQP in Figure 5.1. The calculation of such areas is the subject matter of integral

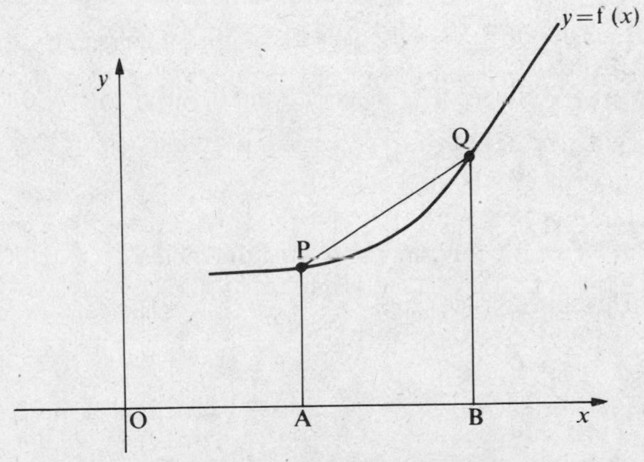

FIG. 5.1. Gradient of a chord

calculus. We will see that the two branches of the calculus are in fact closely related to each other.

5.2. The basic concepts of differential calculus

5.2.1. The gradient as a limit

Suppose the curve in Figure 5.1 is $y = x^2 + 2$ and the point P is the point (2, 6). What is the gradient of the curve at P? Let the abscissa of the point Q be 2 + h, where h is some positive number, or increment. Then the ordinate, the y-value, is $(2 + h)^2 + 2$, which is $h^2 + 4h + 6$. This is, Q is the point $(2 + h, 6 + 4h + h^2)$. These quantities are shown on Figure 5.2.

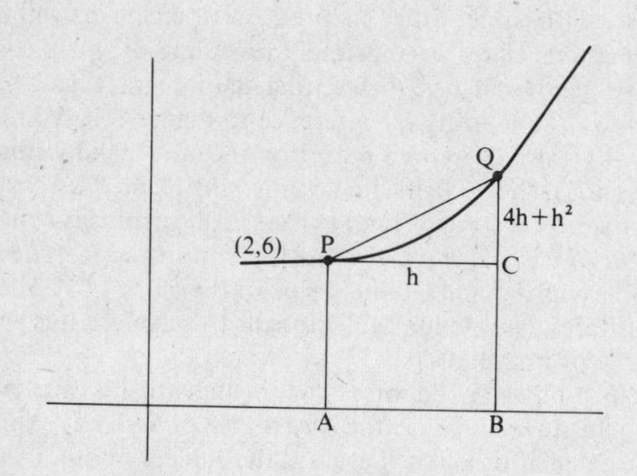

FIG. 5.2. Towards the gradient at a point

Note that the height QC is $4h + h^2$. We know that the gradient of the chord PQ is QC/PC, so

$$\text{gradient of PQ} = \frac{QC}{PC} = \frac{4h + h^2}{h} = 4 + h \qquad (5.1)$$

We can now find the gradient at P by letting Q tend towards P, which we write $Q \rightarrow P$. This is equivalent to letting the increment h tend to zero: $h \rightarrow 0$. This process is written $\underset{h \rightarrow 0}{\text{Limit}}$, or $\underset{h \rightarrow 0}{\text{Lt}}$. Hence,

$$\text{gradient at P} = \underset{h \rightarrow 0}{\text{Lt}} (4 + h) = 4 \qquad (5.2)$$

Thus, by taking a limit, we have found the gradient at $y = x^2 + 2$ at the point (2, 6), and it is 4.

We can obviously obtain a more general result by taking P as any point (x, y). Then Q is the point $(x + h, (x + h)^2 + 2)$, that is

$(x + h, x^2 + 2xh + h^2 + 2)$. The gradient of the chord PQ, by a similar argument to that above, is $(2xh + h^2)/h$, which is $2x + h$. So

$$\text{gradient at P} = \underset{h \to 0}{\text{Lt}} (2x + h) = 2x \tag{5.3}$$

We can put $x = 2$ and confirm the particular result for the point (2, 6).

5.2.2. Notation and main concepts

Consider again points P, Q on a curve $y = f(x)$ as in Figure 5.3. The increment in the x direction, which we previously called h, is written δx.

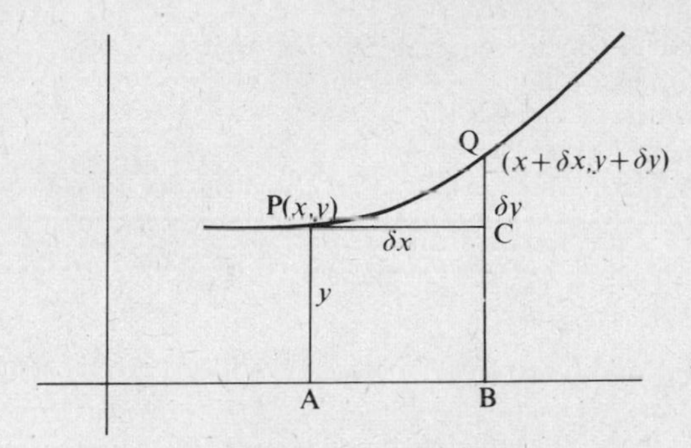

FIG. 5.3. Increments

It is all one quantity—certainly never to be thought of as $\delta \times x$. 'δ' is a symbol meaning the 'increment in x'. Similarly, the corresponding increment in y is written as δy. Such small increments are sometimes known as 'infinitesimals'. Then

$$\text{gradient of chord PQ} = \frac{QC}{PC} = \frac{\delta y}{\delta x} \tag{5.4}$$

The notation which is used for the gradient at a point is dy/dx. At this stage, this should be considered to be one symbol—not a quantity dy divided by dx. In the limit, we have the result

$$\frac{dy}{dx} = \underset{\delta x \to 0}{\text{Lt}} \frac{\delta y}{\delta x} \tag{5.5}$$

We can now repeat our earlier example in the proper notation. If $y = x^2 + 2$

$$\frac{dy}{dx} = \underset{\delta x \to 0}{\text{Lt}} \frac{\delta y}{\delta x} \tag{5.6}$$

$$= \underset{\delta x \to 0}{\mathrm{Lt}} \frac{(x + \delta x)^2 + 2 - x^2 - 2}{\delta x} \tag{5.7}$$

$$= \underset{\delta x \to 0}{\mathrm{Lt}} \frac{2x \cdot \delta x + (\delta x)^2}{\delta x} \tag{5.8}$$

$$= \underset{\delta x \to 0}{\mathrm{Lt}} (2x + \delta x) \tag{5.9}$$

$$= 2x \tag{5.10}$$

In particular, for the point (2, 6) which we started with, $\mathrm{d}y/\mathrm{d}x = 4$ when $x = 2$.

The procedure which we have described may be written for a general function as follows. If

$$y = \mathrm{f}(x) \tag{5.11}$$

then

$$\frac{\mathrm{d}y}{\mathrm{d}x} = \underset{\delta x \to 0}{\mathrm{Lt}} \left[\frac{\mathrm{f}(x + \delta x) - \mathrm{f}(x)}{\delta x} \right] \tag{5.12}$$

since $$\delta y = \mathrm{f}(x + \delta x) - \mathrm{f}(x) \tag{5.13}$$

The gradient, $\mathrm{d}y/\mathrm{d}x$, is known as the 'derivative of y with respect to x'. The process of finding the derivative is known as 'differentiation'. We have already noted that δx and δy are called increments; $\mathrm{d}y$ and $\mathrm{d}x$ are called differentials. If $\mathrm{d}x$ is considered finite and taken to be the same as some δx, the corresponding $\mathrm{d}y$ is shown on Figure 5.4.

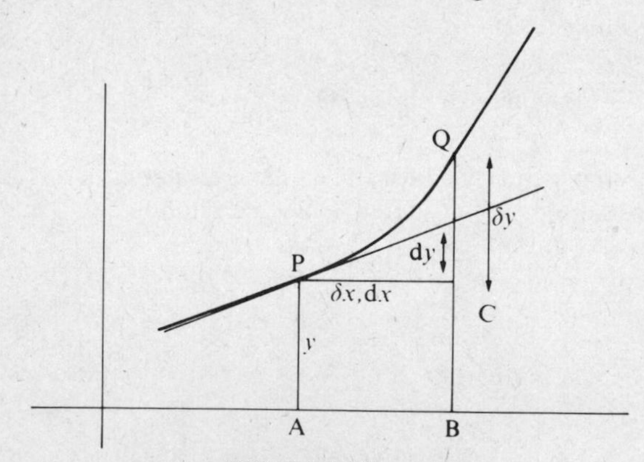

FIG. 5.4. Differentials

Other notations are sometimes used for the gradient of $y = \mathrm{f}(x)$ at

(x, y). Perhaps the most common alternative to dy/dx is $f'(x)$; a less common alternative is y'. Thus if

$$y = f(x) \tag{5.14}$$

we sometimes say that we differentiate this equation to give

$$\frac{dy}{dx} = f'(x) \tag{5.15}$$

or

$$y' = f'(x) \tag{5.16}$$

Equation 5.15 shows that the differentials defined in Figure 5.4 are related by

$$dy = f'(x)dx \tag{5.17}$$

though, more importantly in practice, since $\delta y \simeq dy$ for small δx, we have the approximate relationship

$$\delta y \simeq f'(x)\delta x \tag{5.18}$$

5.2.3. The sign of the derivative, and a preliminary note on maxima and minima

We know from our earlier discussion of the geometry of the straight line that if a tangent is upward sloping, the gradient is positive at that point. This means that for a positive δx, $f(x + \delta x) > f(x)$. Conversely, for a backward sloping tangent, the gradient is negative, and $f(x + \delta x) < f(x)$ for positive δx. In the first case the function $f(x)$ is increasing at x; in the second case it is decreasing.

If the gradient is actually zero, at that point the function is neither increasing nor decreasing, and is known as a *stationary* point. We can therefore identify stationary points by solving the equation.

$$\frac{dy}{dx} = 0 \tag{5.19}$$

Particular cases of stationary points will turn out to be the maximum and minimum values of functions, and so an important role of differential calculus will be to identify such points by solving equation 5.19.

5.3. Derivatives of standard functions

So far, we have only differentiated $y = x^2 + 2$ specifically. We can use the methods outlined above to differentiate each of the standard functions of Chapter 3. For example, if

$$y = x^n \tag{5.20}$$

$$\frac{dy}{dx} = \operatorname*{Lt}_{\delta x \to 0} \frac{(x + \delta x)^n - x^n}{\delta x} \tag{5.21}$$

It can be shown that [cf. equation 3.156 in Chapter 3]:

$$(x + \delta x)^n = x^n + nx^{n-1} \cdot \delta x + 0(\delta x^2) \tag{5.22}$$

where $0(\delta x^2)$ represents a series of terms in $(\delta x)^2$ or higher powers of δx. Thus

$$\frac{(x + \delta x)^n - x^n}{\delta x} = nx^{n-1} + 0(\delta x) \tag{5.23}$$

Since

$$\operatorname*{Lt}_{\delta x \to 0} 0(\delta x) = 0 \tag{5.24}$$

We see that

$$\frac{dy}{dx} = nx^{n-1} \tag{5.25}$$

TABLE 5.1

Derivatives of standard functions

$y\,[= f(x)]$	$f'(x) = \frac{dy}{dx}$
x^n	nx^{n-1}
$\log_e x$	$1/x$
e^x	e^x
$\sin x$	$\cos x$
$\cos x$	$-\sin x$
$\tan x$	$\sec^2 x$
$\cosh x$	$\sinh x$
$\sinh x$	$\cosh x$
$\tanh x$	$\operatorname{sech}^2 x$
$\sin^{-1} x$	$1/\sqrt{(1 - x^2)}$
$\cos^{-1} x$	$-1/\sqrt{(1 - x^2)}$
$\tan^{-1} x$	$1/(1 + x^2)$
$\sinh^{-1} x$	$1/\sqrt{(1 + x^2)}$
$\cosh^{-1} x$	$1/\sqrt{(x^2 - 1)}$
$\tanh^{-1} x$	$1/(1 - x^2)$

The explicit calculation of the derivatives of other standard functions is beyond the scope of this book, as limits of the form 5.12 are in general difficult to handle. However, since our readers will be mainly concerned with the meaning of the results, and the results themselves, we simply state the derivatives of these functions in Table 5.1 above. We will show in section 5.4 below how to use these results to obtain the derivatives of a wide range of other functions.

Table 5.1 gives us much new information about our standard functions, of course. For example, e^x turns out to have the important characteristic that its derivative at a point is equal to the value of the function at the point.

5.4. Rules for differentiating non-standard functions

5.4.1. Introduction

We saw in Chapter 3, section 3.5, that a wide range of functions can be generated from the standard functions. Although it is necessary to learn by heart (rather like multiplication tables!) the derivatives of standard functions, it is neither necessary nor practical to do so for others. We consider in turn a range of cases which will enable us to differentiate almost anything given a knowledge of the standard functions and these rules.

5.4.2. Scalar multiples

If

$$y = \mathrm{f}(x) = \mathrm{kg}(x) \tag{5.26}$$

where k is a constant, then

$$\frac{\mathrm{d}y}{\mathrm{d}x} = \mathrm{f}'(x) = \mathrm{kg}'(x) \tag{5.27}$$

In other words, the derivative is the constant, times the derivative of the function. For example, if

$$y = 2x^n \tag{5.28}$$

since the derivative of x^n is known as nx^{n-1}, we can deduce from 5.27 that

$$\frac{\mathrm{d}y}{\mathrm{d}x} = 2nx^{n-1} \tag{5.29}$$

5.4.3. The sum of two functions

If

$$y = f(x) + g(x) \tag{5.30}$$

$$\frac{dy}{dx} = f'(x) + g'(x) \tag{5.31}$$

That is, the derivative of the sum is the sum of the derivatives. For example, if

$$y = x^3 + \sin x \tag{5.32}$$

$$\frac{dy}{dx} = 3x^2 + \cos x \tag{5.33}$$

A similar result holds for the difference of two functions.

5.4.4. Function of a function

If

$$y = f(z) \tag{5.34}$$

and

$$z = g(x) \tag{5.35}$$

then

$$\frac{dy}{dx} = \frac{dy}{dz}\frac{dz}{dx} \tag{5.36}$$

or, in an alternative notation,

$$\frac{dy}{dx} = f'(z)\, g'(x) \tag{5.37}$$

For example, if

$$y = (x^2 - 3x + 7)^5 \tag{5.38}$$

put

$$z = x^2 - 3x + 7 \tag{5.39}$$

Then

$$y = z^5 \tag{5.40}$$

$$\frac{dy}{dx} = \frac{dy}{dz}\frac{dz}{dx} \tag{5.41}$$

Differentiating the expression 5.40 with respect to z gives

$$\frac{dy}{dz} = 5z^4 \tag{5.42}$$

and the expression 5.39 with respect to x gives

$$\frac{dz}{dx} = 2x - 3 \tag{5.43}$$

so

$$\frac{dy}{dx} = 5z^4(2x - 3) \tag{5.44}$$

Hence, substituting again for z from equation 5.39

$$\frac{dy}{dx} = 5(x^2 - 3x + 7)^4(2x - 3) \tag{5.45}$$

Another example would be if

$$y = e^{x^2} \tag{5.46}$$

Put

$$z = x^2 \tag{5.47}$$

so

$$y = e^z \tag{5.48}$$

Then

$$\frac{dy}{dx} = \frac{dy}{dz}\frac{dz}{dx} \tag{5.49}$$

$$= e^z \cdot 2x \tag{5.50}$$

$$= 2xe^{x^2} \tag{5.51}$$

5.4.5. Differentiation of a product

If

$$y = u(x)v(x) \tag{5.52}$$

where u and v are each functions of x, then

$$\frac{dy}{dx} = u(x)v'(x) + u'(x)v(x) \tag{5.53}$$

For example, if

$$y = x^2 \sin x \tag{5.54}$$

take

$$u(x) = x^2 \tag{5.55}$$

$$v(x) = \sin x \tag{5.56}$$

and substitute in equation 5.53

$$\frac{dy}{dx} = x^2 \cos x + 2x \sin x \tag{5.57}$$

or, if

$$y = \sin x \cdot \log x \tag{5.58}$$

$$\frac{dy}{dx} = \frac{\sin x}{x} + \cos x \cdot \log x \tag{5.59}$$

5.4.6. Differentiation of a quotient

Suppose now,

$$y = u(x)/v(x) \tag{5.60}$$

$$\frac{dy}{dx} = \frac{u'(x)v(x) - u(x)v'(x)}{v(x)^2} \tag{5.61}$$

For example, if

$$y = \sin x/\cos x \tag{5.62}$$

$$\frac{dy}{dx} = \frac{\cos x \cdot \cos x - \sin x(-\sin x)}{(\cos x)^2} \tag{5.63}$$

$$= \frac{\cos^2 x + \sin^2 x}{\cos^2 x} \tag{5.64}$$

Since

$$\cos^2 x + \sin^2 x = 1 \tag{5.65}$$

$$\frac{dy}{dx} = \frac{1}{\cos^2 x} = \sec^2 x \tag{5.66}$$

(which gives a 'proof' for the derivative of tan x in terms of known derivatives of sin x and cos x).

5.4.7. Differentiation of a parametrically specified function

It is sometimes convenient to express a functional relationship between y and x through a common parameter, t say. Let

$$y = y(t) \quad (5.67)$$

$$x = x(t) \quad (5.68)$$

Then

$$\frac{dy}{dx} = \frac{dy}{dt}\frac{dt}{dx} = \frac{dy}{dt}\bigg/\frac{dx}{dt} \quad (5.69)$$

Thus, if

$$y = \cos t \quad (5.70)$$

$$x = \sin t \quad (5.71)$$

$$\frac{dy}{dx} = -\sin t/\cos t = -\tan t \quad (5.72)$$

We could check this by noting that, since

$$\cos^2 t + \sin^2 t = 1 \quad (5.73)$$

$$y^2 + x^2 = 1 \quad (5.74)$$

so

$$y = (1 - x^2)^{\frac{1}{2}} \quad (5.75)$$

We can differentiate this by the 'function of a function' rule:

$$\frac{dy}{dx} = \frac{1}{2}(1 - x^2)^{-\frac{1}{2}}(-2x) \quad (5.76)$$

$$= -x(1 - x^2)^{-\frac{1}{2}} \quad (5.77)$$

which does agree with expression 5.72 if we substitute for x from 5.71. When alternative methods are available like this, it is often a useful check to investigate in this way.

5.4.8. The 'rules' in combination

Sometimes, the rules may have to be applied in combination. For example, if

$$y = x^2 e^{x^2} \quad (5.78)$$

then it is necessary to treat e^{x^2} on a 'function of a function' basis, and then $x^2 e^{x^2}$ as a product. For example, put

$$u = x^2 \tag{5.79}$$

$$v = e^{x^2} \tag{5.80}$$

$$z = x^2 \tag{5.81}$$

$$v = e^z \tag{5.82}$$

$$\frac{dv}{dx} = \frac{dv}{dz}\frac{dz}{dx} = e^z \cdot 2x = 2xe^{x^2} \tag{5.83}$$

$$\frac{dy}{dx} = u\frac{dv}{dx} + v\frac{du}{dx} = x^2 \cdot 2x \cdot e^{x^2} + e^{x^2} \cdot 2x \tag{5.84}$$

$$= 2xe^{x^2}(x^2 + 1) \tag{5.85}$$

In this way, most functions which are likely to occur in practice can be differentiated.

5.4.9. Differentiation in practice

The inexperienced reader will only be able to learn how to differentiate a wide range of functions fluently by attempting the exercises provided and more, and will probably need the advice of a skilled teacher. Initially, the rules should be systematically applied, as in the above examples. Eventually, however, fluency should mean that many of these steps can be accomplished 'in the head' and derivatives of quite complicated functions can be written down accurately with only a line or two of working.

5.5. Second and higher order derivatives

The derivative is itself a function of x, and so has a derivative also. This is known as the second derivative of $f(x)$ and is denoted by d^2y/dx^2, $f''(x)$ or y''. For example, if

$$y = x^n \tag{5.86}$$

$$\frac{dy}{dx} = nx^{n-1} \tag{5.87}$$

and then differentiating again with respect to x,

$$\frac{d^2y}{dx^2} = n(n-1)x^{n-2} \tag{5.88}$$

Clearly, we can go on indefinitely and define higher order derivatives: d^3y/dx^3 is the derivative of d^2y/dx^2 – in the above example

$$\frac{d^3y}{dx^3} = n(n-1)(n-2)x^{n-3} \tag{5.89}$$

and so on. The nth derivative of $y = f(x)$ is denoted by d^ny/dx^n or $f^{(n)}(x)$. The first important application of these concepts is in relation to the identification of maxima and minima in the next section.

5.6. Maxima, minima, and points of inflexion

A number of different kinds of stationary points (that is, points at which the gradient is zero) are exhibited in Figure 5.5. The situation in (a)

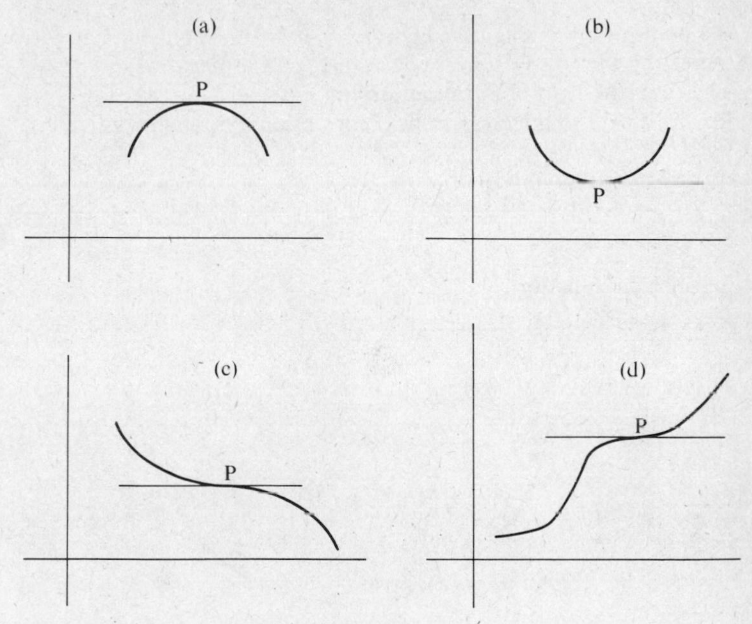

FIG. 5.5. Stationary points

clearly represents a maximum, and that in (b) a minimum; (c) and (d) are stationary points known as points of inflexion. Since these are all solutions of

$$\frac{dy}{dx} = 0 \tag{5.90}$$

it is important to be able to distinguish between them.

An examination of Figure 5.5. (a) shows that as we progress towards P from the left, and then beyond, the gradient is at all times decreasing. That is, the rate of change of the gradient itself is negative. This is formally expressed in terms of the second derivative. Thus, the conditions for a maximum are

$$\frac{dy}{dx} = 0; \qquad \frac{d^2y}{dx^2} < 0 \tag{5.91}$$

By a similar argument, the conditions for a minimum are

$$\frac{dy}{dx} = 0; \qquad \frac{d^2y}{dx^2} > 0 \tag{5.92}$$

At a point of inflexion, as in Figure 5.5, in (c) the gradient is negative but increasing towards P from the left, but remains negative and decreasing from the right of P. In fact, in this case, and the (d) case, $d^2y/dx^2 = 0$, and so the conditions for a stationary point of inflexion are

$$\frac{dy}{dx} = 0, \qquad \frac{d^2y}{dx^2} = 0 \tag{5.93}$$

Consider the function

$$y = x^2 - 2x + 3 \tag{5.94}$$

which is plotted in Figure 5.6.

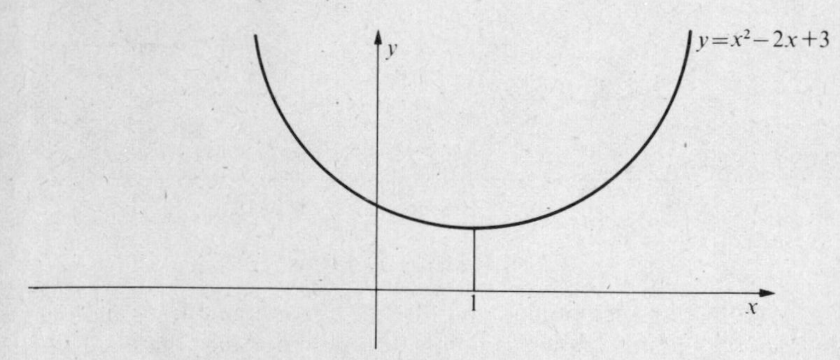

FIG. 5.6. Graph of $y = x^2 - 2x + 3$

$$\frac{dy}{dx} = 2x - 2 \tag{5.95}$$

so

$$\frac{dy}{dx} = 0 \tag{5.96}$$

when

$$2x - 2 = 0, \quad \text{or} \quad x = 1 \tag{5.97}$$

Thus, this is the one stationary point. We can differentiate 5.95 again:

$$\frac{d^2y}{dx^2} = 2 > 0 \tag{5.98}$$

which confirms, as is clear from the Figure 5.6, that $x = 1$ is a minimum.

As a second example, consider

$$y = x^3 \tag{5.99}$$

plotted in Figure 5.7.

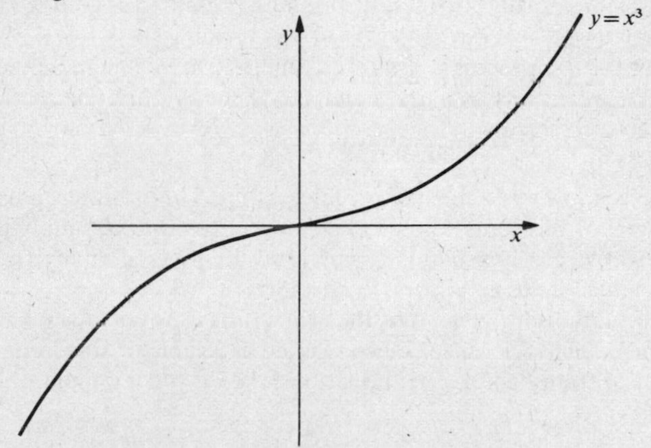

FIG. 5.7. Graph of $y = x^3$

In this case,

$$\frac{dy}{dx} = 3x^2 \tag{5.100}$$

and so

$$\frac{dy}{dx} = 0 \text{ when } x = 0 \tag{5.101}$$

Also

$$\frac{d^2y}{dx^2} = 6x \tag{5.102}$$

and is zero when $x = 0$, confirming that $x = 0$ is a point of inflexion as shown.

When we introduced a wide range of functions in Chapter 3, we also indicated how to sketch the graph of any particular function. With the tools near at hand, we can clearly improve our curve sketching ability by finding the location of stationary points (and their type) in any particular case.

5.7. The basic concepts of integral calculus

5.7.1. Integral calculus as the inverse of differential calculus

Integration can be defined as the reverse process of differentiation. Thus, if nx^{n-1} is the derivative of x^n, we can define x^n to be the integral of nx^{n-1}. (More usually, this result would be presented as $(1/n)x^n$ being the integral of x^{n-1}, or even $(x^{n+1})/(n+1)$ being the integral of x^n). The sign for this process is $\int \mathrm{d}x$, the function to be integrated being placed between the two parts, $\int$ and $\mathrm{d}x$, of the symbol. Thus,

$$\int nx^{n-1}\,\mathrm{d}x = x^n + c \qquad (5.103)$$

where c is known as a constant of integration. This is always present in an integral of this form, known as an *indefinite integral*, simply because the derivative of a constant is zero, and so the integral of any function is determined up to an arbitrary constant.

These definitions mean that the table of derivatives of standard functions, Table 5.1, can also be regarded as a table of standard integrals if it is read from right to left. Equation 5.103 is one example. Further examples are:

$$\int \cos\,\mathrm{d}x = \sin x + \mathrm{c} \qquad (5.104)$$

$$\int \sin x\,\mathrm{d}x = -\cos x + \mathrm{c} \qquad (5.105)$$

$$\int \mathrm{e}^x\,\mathrm{d}x = \mathrm{e}^x + \mathrm{c} \qquad (5.106)$$

$$\int \frac{1}{x}\,\mathrm{d}x = \log_{\mathrm{e}} x + \mathrm{c} \qquad (5.107)$$

The last of these examples is important in relation to the integrals of power functions. Equation 5.103 does not hold for $n = 0$; equation 5.107 gives the correct relationship in this case.

5.7.2. Integrals as areas under curves

For many applications, as we shall see later, we are more interested in integrals which measure an area under a curve. Consider the curve $y = \mathrm{f}(x)$ in Fig. 5.8. Suppose we want to find the area under the curve, shown in the figure as ABCD, where OA $=$a, OD $=$ b. Let P and Q be any two points on the curve, and complete the rectangle UVQR. Let P be the point (x, y) and Q the point $(x + \delta x, y + \delta y)$. Then ABPU is the

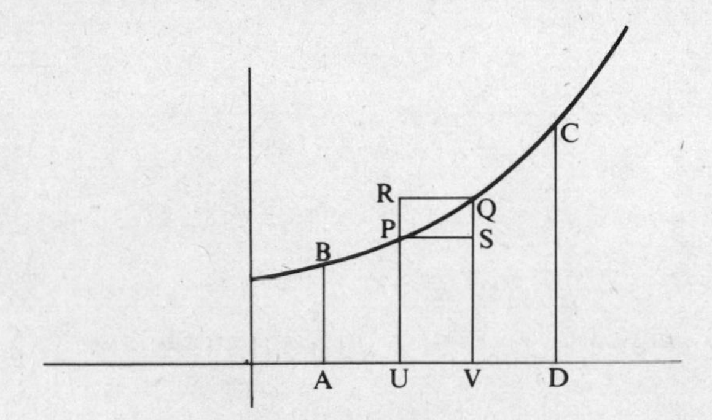

FIG. 5.8. The area under a curve

area under the curve from $x = a$ up to x, ABQV that from $x = a$ up to $x + \delta x$. The increment in area, say δA, is UPQV. Note that

$$\text{UVSP} \leqslant \delta \text{A} \leqslant \text{UVQR} \tag{5.108}$$

That is,

$$y\,\delta x \leqslant \delta \text{A} \leqslant (y + \delta y)\delta x \tag{5.109}$$

We can divide through by the positive number, δx, to give

$$y \leqslant \frac{\delta \text{A}}{\delta x} \leqslant y + \delta y \tag{5.110}$$

We now let $\delta x \to 0$; then, $\delta y \to 0$ and so lower and upper bounds of $\delta \text{A}/\delta x$ tend to y. Further

$$\underset{\delta x \to 0}{\text{Lt}} \frac{\delta \text{A}}{\delta x} = \frac{\text{dA}}{\text{d}x} \tag{5.111}$$

by definition, and so we have

$$\frac{\text{dA}}{\text{d}x} = y \tag{5.112}$$

Thus, since integration is the reverse of differentiation, we can write

$$\text{A} = \int y\text{d}x = \int \text{f}(x)\text{d}x \tag{5.113}$$

It is convenient to develop the notation a stage further in relation to a particular example. Suppose the curve is

$$y = f(x) = x^2 + 2 \quad (5.114)$$

Since, if

$$z = \tfrac{1}{3}x^3 + 2x \quad (5.115)$$

$$\frac{dz}{dx} = x^2 + 2 \quad (5.116)$$

then

$$\int f(x)dx = \tfrac{1}{3}x^3 + 2x + c \quad (5.117)$$

where c is the constant of integration. So A, the area from $x = a$ up to x is given by

$$A = \int f(x)dx = \tfrac{1}{3}x^3 + 2x + c \quad (5.118)$$

But $A = 0$ when $x = a$, so

$$0 = \tfrac{1}{3}a^3 + 2a + c \quad (5.119)$$

Thus

$$c = -\tfrac{1}{3}a^3 - 2a \quad (5.120)$$

and

$$A = \tfrac{1}{3}x^3 + 2x - \tfrac{1}{3}a^3 - 2a \quad (5.121)$$

Hence, the area up to $x = b$ can be obtained by putting $x = b$:

$$A = \tfrac{1}{3}b^3 + 2b - \tfrac{1}{3}a^3 - 2a \quad (5.122)$$

A concise notation for this is.

$$A = [\tfrac{1}{3}x^3 + 2x]_a^b \quad (5.123)$$

This means: take the expression in square brackets, substitute b, then substitute a and subtract. a and b are the *lower and upper limits of integration*, and such an integral is known as a *definite integral* and is written in general for $y = f(x)$,

$$A = \int_a^b f(x)dx \quad (5.124)$$

Thus, for our example, the calculation can be concisely expressed as follows:

$$A = \int_a^b (x^2 + 2)dx \quad (5.125)$$

$$= [\tfrac{1}{3}x^3 + 2x]_a^b \qquad (5.126)$$

$$= \tfrac{1}{3}b^3 + 2b - \tfrac{1}{3}a^3 - 2a \qquad (5.127)$$

5.7.3. An interpretation of the integral notation

This kind of integration can be seen as the limiting process of a summation. Figure 5.9 is a representation of Figure 5.8 with the area divided

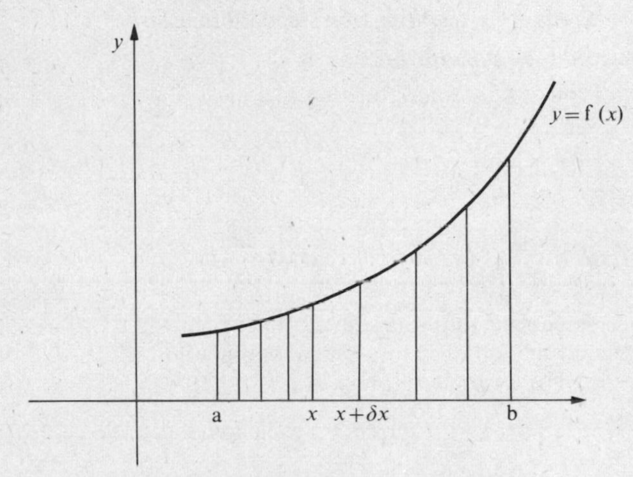

FIG. 5.9. The integral as a sum

into strips. The area of a 'typical strip' is approximately (i.e. taking a rectangular approximation) $f(x)\delta x$, so,

$$A \simeq \sum_{x=a}^{b} f(x)\delta x \qquad (5.128)$$

(Formulae of this type are used in numerical analysis to obtain areas under curves in an approximate manner).
Now, as $\delta x \to 0$, A tends to the area under the curve. Thus

$$A = \underset{\delta x \to 0}{\text{Lt}} \sum_{x=a}^{b} f(x)\delta x = \int_a^b f(x)\,dx \qquad (5.129)$$

The integral sign $\int$ is in fact an elongated s, which stands for the sum of a large number of infinitesimal areas $f(x)dx$.

5.7.4. Integration of standard functions

As indicated earlier, the integrals of standard functions can be obtained by reading Table 5.1 from right to left, and possibly making minor adjustments with multiplicative constants, as we did earlier to obtain the integral of x^n. Integration in practice relies very much on 'spotting' the correct solution. A guess can always be tested by differentiating again, and making any adjustments to the original guess that are required. In the next section however, we describe a number of rules which can be used to transform non-standard functions into combinations of more recognizable standard functions.

5.8. Two methods for integrating non-standard functions

5.8.1. Integration by substitution

Suppose we replace x by some function of z:

$$x = x(z) \tag{5.130}$$

Then

$$\int f(x)dx = \int f[x(z)] \frac{dx}{dz} dz \tag{5.131}$$

The trick in this case is to find a substitution which will transform $f(x)$, a non-standard function of x, into a standard function of z, as $f[x(z)]\ dx/dz$. For example, suppose

$$I = \int (\cos\theta + 1)^3 \sin\theta d\theta \tag{5.132}$$

Put

$$\cos\theta + 1 = z \tag{5.133}$$

[in this case defining $\theta(z)$ implicitly]

Differentiate to give

$$-\sin\theta\ d\theta = dz \tag{5.134}$$

so

$$\frac{d\theta}{dz} = -1/\sin\theta \tag{5.135}$$

Then, using equation 5.131,

$$I = -\int z^3 \cdot dz \tag{5.136}$$

(since the $\sin\theta$ s cancel), so

$$I = -\frac{z^4}{4} + c \tag{5.137}$$

giving

$$I = -\frac{(\cos\theta + 1)^4}{4} + c \qquad (5.138)$$

in terms of θ. Note that, in effect, $\sin\theta d\theta$ (in 5.132) is 'replaced' by dz, and this is a clue to the usefulness of the transformation in this case. Unfortunately, there are few standard rules for integration by substitution: only experience and practice can generate the necessary skills.

5.8.2. Integration by parts

Integration by parts is based on the formula for differentiating a product. If

$$f(x) = u(x)v(x) \qquad (5.139)$$

then

$$\frac{df(x)}{dx} = \frac{d}{dx}(uv) = u\frac{dv}{dx} + v\frac{du}{dx} \qquad (5.140)$$

We now integrate this with respect to x:
Note that

$$\int \frac{d}{dx}(uv)dx = uv \qquad (5.141)$$

so

$$uv = \int u\frac{dv}{dx}dx + \int v\frac{du}{dx}dx \qquad (5.142)$$

which is written

$$\int u\frac{dv}{dx}dx = uv - \int v\frac{du}{dx}dx \qquad (5.143)$$

Thus, the trick in this case is to define u and v such that, although $u(dv/dx)$ is not standard function, $v(du/dx)$ is. An interesting case is the following. Suppose

$$I = \int \log_e x \, dx \qquad (5.144)$$

There is no standard integral for $\log_e x$.
Take

$$u = \log_e x \qquad (5.145)$$

$$\frac{dv}{dx} = 1 \tag{5.146}$$

in the above formula, and note that

$$\frac{du}{dx} = \frac{1}{x} \tag{5.147}$$

$$v = x \tag{5.148}$$

Thus

$$\int 1 \cdot \log_e x \cdot dx = x \log_e x - \int x \cdot \frac{1}{x} dx \tag{5.149}$$

Note that

$$\int x \cdot \frac{1}{x} dx = \int 1 \cdot dx = x \tag{5.150}$$

so

$$\int \log_e x \, dx = x \log_e x - x + c \tag{5.151}$$

Such results can always be checked by differentiating again:

$$\frac{d}{dx}(x \log x - x + c) = \log_e x + x \frac{1}{x} - 1 + 0 \tag{5.152}$$

$$= \log_e x \tag{5.153}$$

showing that our answer is correct. Another more important example connected to urban densities and resulting populations will turn up later.

5.9. Calculus of functions of several variables

5.9.1. Partial differentiation

We saw in Chapter 3 that functions which are of interest to us in modelling geographical systems are, typically, functions of several independent variables. We now explore how to apply the concepts of the calculus to such functions, beginning with differentiation.

The concept of differentiation can be applied as follows: assume all the variables but one are fixed—that is, for these purposes, treat them as constants. Then it is possible to differentiate *partially* with respect to the remaining variable. To indicate that the differentiation is partial, a symbol '∂' is used instead of 'd' in dy/dx, to form such *partial derivatives* as $\partial y/\partial x$. Suppose, for example that y is a function of three variables x_1, x_2 and x_3, and

$$y = ax_1^2 + bx_2^2 + cx_3^2 \tag{5.154}$$

Then, according to the above definitions

$$\frac{\partial y}{\partial x_1} = 2ax_1 \tag{5.155}$$

since the term bx_2^2 and cx_3^2 are treated as constants for partial differentiation with respect to x_1. By the same argument

$$\frac{\partial y}{\partial x_2} = 2bx_2 \tag{5.156}$$

$$\frac{\partial y}{\partial x_3} = 2\,cx_3 \tag{5.157}$$

The other variables would not necessarily be eliminated from the partial derivatives, as in this example. For instance, if

$$y = ax_1^2x_2 + bx_2^2x_1 \tag{5.158}$$

$$\frac{\partial y}{\partial x_1} = 2ax_1x_2 + 6x_2^2 \tag{5.159}$$

$$\frac{\partial y}{\partial x_2} = ax_1^2 + 2bx_2x_1 \tag{5.160}$$

Two results from single variable differential calculus can be extended to the multi-dimensional case. First, we had for increments and differentials:

$$\delta y \approx f'(x)\delta(x) \tag{5.161}$$

$$dy = f'(x)dx \tag{5.162}$$

These equations now become, if y is a function of n variables $x_1, x_2, x_3, \ldots x_n$:

$$y = f(x_1, x_2, x_3, \ldots x_n) \tag{5.163}$$

$$\delta y \approx \sum_i \frac{\partial f}{\partial x_i}\delta x_i \tag{5.164}$$

$$dy = \sum_i \frac{\partial f}{\partial x_i}dx_i \tag{5.165}$$

In other words, individual increments and differentials, say $(\partial f/\partial x_i)\delta x_i$ for one i, obtained assuming all other variables remain fixed, can be summed to give total increments and differentials.

Second, we saw in the single independent variable case that stationary points, and in particular maxima and minima, can be obtained as the solution of

$$\frac{dy}{dx} = 0 \tag{5.166}$$

In the multi-dimensional case, say for the function in equation 5.163, the stationary points are given by the solutions of the simultaneous equations

$$\frac{\partial f}{\partial x_i} = 0, \quad i = 1, 2, \ldots\ldots n \tag{5.167}$$

or, using the alternative notation,

$$\frac{\partial y}{\partial x_i} = 0, \quad i = 1, 2, \ldots\ldots n \tag{5.168}$$

An important special case which arises in the multi-variable example is to find maxima and minima *under constraints* – that is further relations on the independent variables $x_1, x_2, \ldots x_n$. We shall tackle this problem in Chapter 9.

5.9.2. Estimation of a 'best fitting' straight line

We can illustrate the multi-dimensional optimization problem of the previous section with an example which draws several threads of argu-

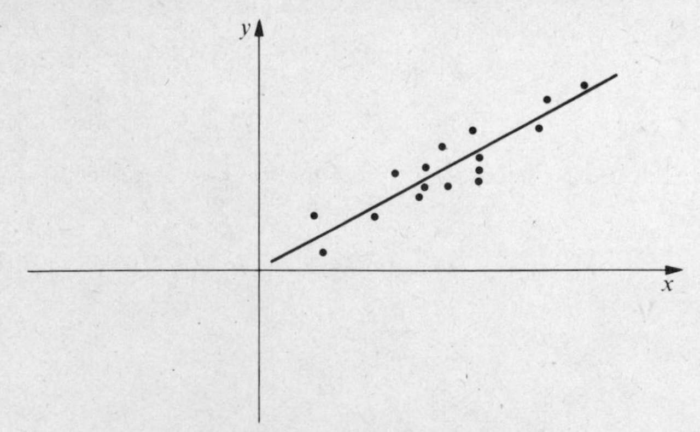

FIG. 5.10. A 'best-fitting' straight line

ment together and which provides a very useful result for geographers and planners. Suppose we have some observations of two quantities y and x which we think are related linearly. Such a situation is depicted in Figure 5.10. Let there be N points in all, and let (x_i, y_i) be the Cartesian co-ordinates of the ith point. We can hypothesize that the line shown on the figure is

$$y = \mathrm{a}x + \mathrm{b} \tag{5.169}$$

The problem is to find a 'best' estimate of a and b. For each x_i, we can obtain an estimate of the corresponding y, which we call $\hat{y}_i$, from equation 5.169 as

$$\hat{y}_i = \mathrm{a}x_i + \mathrm{b} \tag{5.170}$$

The error in the estimate is $\hat{y}_i - y_i$. As a measure of 'goodness of fit', we define the sum-of-squares S(a, b), a function of a and b, as

$$\mathrm{S(a, b)} = \sum_{i=1}^{N} (\hat{y}_i - y_i)^2 \tag{5.171}$$

We can now obtain a 'best' line by finding the a and b which minimize S(a, b). Such an a and b can be found by solving (in the manner of equation 5.168)

$$\frac{\partial \mathrm{S}}{\partial \mathrm{a}} = 0 \tag{5.172}$$

and

$$\frac{\partial \mathrm{S}}{\partial \mathrm{b}} = 0 \tag{5.173}$$

We can substitute from equation 5.170 for each $\hat{y}_i$ into equation 5.171 to obtain

$$\mathrm{S(a, b)} = \sum_{i=1}^{N} (\mathrm{a}x_i + \mathrm{b} - y_i)^2 \tag{5.174}$$

so that,

$$\begin{aligned} \mathrm{S(a, b)} &= \sum_{i=1}^{N} (\mathrm{a}^2 x_i^2 + 2\mathrm{ab}x_i + \mathrm{b}^2 - 2\mathrm{b}y_i + y_i^2 - 2\mathrm{a}x_i y_i) \\ &= \mathrm{a}^2 \sum_{i=1}^{N} x_i^2 + 2\mathrm{ab} \sum_{i=1}^{N} x_i + N\mathrm{b}^2 - 2\mathrm{b} \sum_{i=1}^{N} y_i + \sum_{i=1}^{N} y_i^2 \\ &\quad - 2\mathrm{a} \sum_{i=1}^{N} x_i y_i \end{aligned} \tag{5.175}$$

We can obtain an expression in a more convenient form if we define $\bar{x}$ to be the average of the x_i s, $\bar{y}$ to be the average of the y_i s, $\overline{x^2}$ to be the average of the x_i^2 s, and $\overline{xy}$ to be the average of $x_i y_i$. That is

$$\bar{x} = \frac{1}{N}\sum_{i=1}^{N} x_i \tag{5.176}$$

$$\bar{y} = \frac{1}{N}\sum_{i=1}^{N} y_i \tag{5.177}$$

$$\overline{x^2} = \frac{1}{N}\sum_{i=1}^{N} x_i^2 \tag{5.178}$$

$$\overline{xy} = \frac{1}{N}\sum_{i=1}^{N} x_i y_i \tag{5.179}$$

We can then substitute for the summed terms, using these expressions, to give

$$S(a, b) = N\overline{x^2}\cdot a^2 + N\bar{x}\cdot 2ab + Nb^2 - N\bar{y}\cdot 2b + \sum_{i=1}^{N} y_i^2 - N\overline{xy}\cdot 2a \tag{5.180}$$

Now,

$$\frac{\partial S}{\partial a} = 2N\overline{x^2}\cdot a + 2N\bar{x}\cdot b - 2N\overline{xy} \tag{5.181}$$

and

$$\frac{\partial S}{\partial b} = 2N\bar{x}\cdot a + 2N\cdot b - 2N\bar{y} \tag{5.182}$$

Thus, we can form the equations 5.172 and 5.173 using these values of $\partial S/\partial a$, $\partial S/\partial b$ and cancel the $2N$ factors and re-arrange to give linear simultaneous equations in a and b:

$$\overline{x^2}\cdot a + \bar{x}\cdot b = \overline{xy} \tag{5.183}$$

$$\bar{x}\cdot a + b = \bar{y} \tag{5.184}$$

Multiply the second of these by $\bar{x}$ and subtract to give

$$a = \frac{\overline{xy} - \bar{x}\,\bar{y}}{\overline{x^2} - (\bar{x})^2} \tag{5.185}$$

Substitute in equation 5.184:

$$b = \bar{y} - \bar{x} \cdot a$$

$$= \frac{\bar{y}\,(\overline{xy} - \bar{x}\,\bar{y}) - \bar{x}\,(\overline{xy} - \bar{x}\,\bar{y})}{\overline{x^2} - (\bar{x})^2}$$

$$= \frac{(\bar{y} - \bar{x}\,)(\overline{xy} - \bar{x}\,\bar{y})}{\overline{x^2} - (\bar{x})^2} \qquad (5.186)$$

This result forms the basis of regression analysis. It has been presented here mainly to illustrate the method of optimizing a function of two variables. It should only be used in statistical analysis if certain conditions are satisfied, and the reader should consult a more advanced statistical text to inform himself of these. .

5.9.3. Weber's problem

A second example of optimization of a function of two variables is provided by Weber's location problem outlined in sections 1.3.3. and 2.2.4. The problem is to locate optimally at a point P a firm which uses inputs which have to be transported from points A, B and C (see Figure 2.7). Although the problem was formulated as a net profit maximization problem it can easily be seen, that since prices are assumed to be fixed, this is equivalent to minimizing

$$S = c_A q_A r_1 + c_B q_B r_2 + c_C q_C r_3 \qquad (5.187)$$

To simplify the notation, let

$$t_1 = c_A q_A,\ t_2 = c_B q_B,\ t_3 = c_C q_C \qquad (5.188)$$

and let (x_1, y_1) be the co-ordinates of A, (x_2, y_2) be the co-ordinates of B, (x_3, y_3) be the co-ordinates of C, and (x, y) be the co-ordinates of P in a Cartesian co-ordinate system. Then, the distances r_1, r_2 and r_3 are given by

$$r_i = [(x - x_i)^2 + (y - y_i)^2]^{\frac{1}{2}} \qquad (5.189)$$

(cf. section 3.3.4.) and S can be written

$$S = \sum_{i=1}^{3} t_i [(x - x_i)^2 + (y - y_i)^2]^{\frac{1}{2}} \qquad (5.190)$$

Our task is to find the values of x and y which minimize S. In formulating the problem in this way, and in the rest of this section, we have followed Scott (1971). x and y must be the solutions of

$$\frac{\partial S}{\partial x} = 0 \tag{5.191}$$

$$\frac{\partial S}{\partial y} = 0 \tag{5.192}$$

That is,

$$\sum_{i=1}^{3} t_i \frac{x - x_i}{[(x - x_i)^2 + (y - y_i)^2]^{\frac{1}{2}}} = 0 \tag{5.193}$$

$$\sum_{i=1}^{3} t_i \frac{y - y_i}{[(x - x_i)^2 + (y - y_i)^2]^{\frac{1}{2}}} = 0 \tag{5.194}$$

These are non-linear equations in x and y, and there is no immediately obvious solution procedure. However, Scott presents an iterative solution technique developed by Kuhn and Kuenne and Cooper. If we use r_i as defined in equation 5.189 and index n, in brackets, to label iterations, they show that the solution can be obtained to any degree of accuracy from

$$x(n+1) = \frac{t_i x_i}{r_i(n)} \bigg/ \sum_i \frac{t_i}{r_i(n)} \tag{5.195}$$

and

$$y(n+1) = \frac{t_i y_i}{r_i(n)} \bigg/ \sum_i \frac{t_i}{r_i(n)} \tag{5.196}$$

Any starting values, $x(1)$ and $y(1)$ can be chosen for x and y [and hence for $r(1)$] but it has been shown that convergence is particularly rapid if the centre of gravity of the points A, B and C is taken. That is:

$$x(1) = \sum_{i=1}^{3} t_i x_i \bigg/ \sum_{i=1}^{3} t_i \tag{5.197}$$

and

$$y(1) = \sum_{i=1}^{3} t_i y_i \bigg/ \sum_{i=1}^{3} t_i \tag{5.198}$$

Then, starting with these values, we set $n = 1, 2, 3, \ldots.$ in turn in equations 5.195 and 5.196 and a solution can be obtained to any desired degree of accuracy.

5.9.4. Multiple integration

The concept in the integral calculus which corresponds to partial differentiation in differential calculus is the *multiple integral*. We will consider this as a sequence of integrals as follows. Suppose y is a function of three variables

$$y = f(x_1, x_2, x_3) \tag{5.199}$$

We can integrate with respect to one of these variables, say x_1, keeping the others constant, to give

$$I_1(x_2, x_3) = \int f(x_1, x_2, x_3) dx_1 \tag{5.200}$$

The answer, which we have called I_1 is a function of x_2 and x_3 as shown. We could now integrate $I_1(x_2, x_3)$ with respect to x_2 keeping x_3 constant to form I_3, which will be a function of x_3:

$$I_2(x_3) = \int I_1(x_2, x_3) dx_2 \tag{5.201}$$

However, we can substitute the expression 5.200 for I_1 into 5.201:

$$I_2(x_3) = \int\int f(x_1, x_2, x_3) dx_1\, dx_2 \tag{5.202}$$

This is known as a *double integral*. We could integrate again, with respect to x_3, now, to give

$$I_3 = \int\int\int f(x_1, x_2, x_3) dx_1\, dx_2\, dx_3 \tag{5.203}$$

which is a triple integral. Such integrals, as we can see from the method of construction used here, are evaluated from the inside outwards. In 5.203 for example, $\int f(x_1, x_2, x_3) dx_1$ is evaluated first, and so on.

These notions can be generalized to deal with any number of variables. Their use will be explained when the occasion arises, as a detailed account would be out of balance with the treatment of other topics in this chapter.

5.10. Series expansions and the mean-value theorem

5.10.1. Taylor's theorem

It is sometimes the case that we have extensive knowledge of the value of a function $f(x)$ at a point x and its derivatives $f'(x)$, $f''(x)$, $f'''(x)$, ... $f^{(n)}(x)$. Taylor's theorem gives an expression for $f(x + h)$, for some small increment h in terms of $f(x)$ and its derivatives.

Assume that $f(x + h)$ can be expanded as a power series in h:

$$f(x + h) = a_0 + a_1 h + a_2 h^2 + \ldots . \tag{5.204}$$

$$= \sum_{n=0}^{\infty} a_n h^n \tag{5.205}$$

We can find the coefficients a_n as follows.
First, put $h = 0$ in equation 5.174:

$$f(x) = a_0 \tag{5.206}$$

which gives a_0. Then, differentiate equation 204 with respect to h:

$$f'(x+h) = a_1 + 2a_2h + 3a_3h^2 + \ldots. \tag{5.207}$$

Again, put $h = 0$:

$$f'(x) = a_1 \tag{5.208}$$

which gives a_1. Differentiate again:

$$f''(x+h) = 2a_2 + 3 \cdot 2 \cdot a_3h + 4 \cdot 3 \cdot a_4h^2 + \ldots \tag{5.209}$$

and put $h = 0$:

$$f''(x) = 2a_2 \tag{5.210}$$

and so on. In general,

$$a_n = \frac{f^{(n)}(x)}{n!} \tag{5.211}$$

and so the series expansion is

$$f(x+h) = f(x) + hf'(x) + \frac{h^2}{2!}f''(x) + \frac{h^3}{3!}f'''(x) + \ldots\ldots\ldots \tag{5.212}$$

5.10.2. Maclaurin's theorem

Taylor's theorem is sometimes used in the form (5.212), but to generate series expansion of functions, it is more useful in the form of Maclaurin's theorem. This is an expansion of $f(x)$ about $x = 0$, and is thus obtained from (5.212) by setting $x = 0$, and $h = x$:

$$f(x) = f(0) + xf'(0) + \frac{x^2}{2!}f''(0) + \frac{x^3}{3!}f'''(0) + \ldots. \tag{5.213}$$

For example, if

$$f(x) = e^x \tag{5.214}$$

$$f'(x) = e^x \tag{5.215}$$

so
$$f'(0) = 1 \tag{5.216}$$

and indeed, in this case,

$$f^{(n)}(0) = 1 \tag{5.217}$$

for all n. Thus, if we substitute in (5.213):

$$e^x = 1 + x + \frac{x^2}{2!} + \frac{x^3}{3!} + \ldots\ldots \tag{5.218}$$

which is the well-known series expansion for e^x which we presented in section 3.6 of Chapter 3. The other series expansions presented there can be proved using Maclaurin's theorem provided a convenient formula can be obtained for $f^{(n)}(x)$, or at least $f^{(n)}(0)$.

5.10.3. Series expansion by the differentiation and integration of known series

We can also note briefly that another method of generating series expansions is to differentiate or integrate known series. For example, equation 3.163 in Chapter 3 gives $\sin x$ as

$$\sin x = x - \frac{x^3}{2!} + \frac{x^5}{5!} \ldots\ldots\ldots \tag{5.219}$$

We can differentiate to obtain a series expansion for $\cos x$:

$$\cos x = 1 - 3.\frac{x^2}{3!} + 5.\frac{x^4}{2!} \ldots\ldots\ldots \tag{5.220}$$

$$= 1 - \frac{x^2}{2!} + \frac{x^4}{4!} \ldots\ldots \tag{5.221}$$

5.10.4. The differential and integral mean-value theorems

Consider the plot of $y = f(x)$ in Figs. 5.11. Let P be the point whose abscissa is x, Q that whose abscissa is $x + a$. Then, at some intermediate point R, the gradient of the curve will be equal to the gradient of the chord PQ. Suppose the abscissa of R is $x + \theta a$ where $0 \leqslant \theta \leqslant 1$; then this statement can be expressed as

$$\frac{f(x+a) - f(x)}{a} = f'(x + \theta a) \tag{5.222}$$

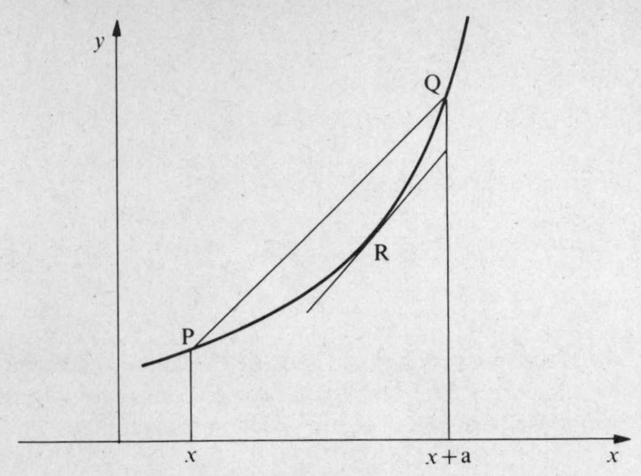

FIG. 5.11. The basis of the differential mean-value theorem

Since the left-hand side is the mean increment in the function from x to $x + \mathrm{a}$, this is known as the mean value theorem. Note that equation 5.222 can be re-arranged in the form

$$\mathrm{f}(x + \mathrm{a}) = \mathrm{f}(x) + \mathrm{af}'(x + \theta \mathrm{a}) \qquad (5.223)$$

which shows it to be a 'cousin' of Taylor's theorem – the series being taken to two terms only.

From equation 5.223 we can also obtain a second useful form of the mean value theorem – the *integral* mean value theorem. Let the function $\mathrm{f}(x)$ be defined by

$$\mathrm{f}(x) = \int_0^x \phi(x')\mathrm{d}x' \qquad (5.224)$$

Then

$$\mathrm{f}(x + \mathrm{a}) = \int_0^{x+\mathrm{a}} \phi(x')\mathrm{d}x' \qquad (5.225)$$

$$\mathrm{f}'(x) = \phi(x) \qquad (5.226)$$

and

$$\mathrm{f}'(x + \theta\mathrm{a}) = \phi(x + \theta\mathrm{a}) \qquad (5.227)$$

Then, from equation 5.223

$$\int_0^{x+\mathrm{a}} (x')\mathrm{d}x' = \int_0^x \phi(x')\mathrm{d}x' + \mathrm{a}\phi(x + \theta\mathrm{a}) \qquad (5.228)$$

or, since

$$\int_{\mathrm{a}}^{x+\mathrm{a}} \phi(x')\mathrm{d}x' = \int_0^{x+\mathrm{a}} \phi(x')\mathrm{d}x' - \int_0^x \phi(x')\mathrm{d}x' \qquad (5.229)$$

we have

$$\int_{a}^{x+a} \phi(x')\,dx' = a\phi(x + \theta a) \qquad (5.230)$$

In graphical terms (cf. Figure 5.12), we are saying that the area under

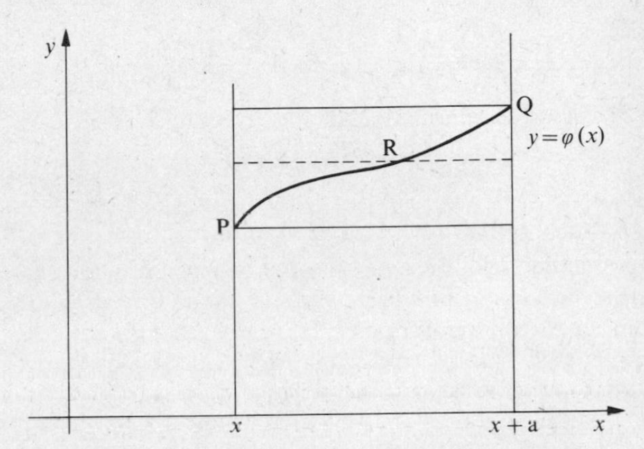

FIG. 5.12. The basis of the integral mean-value theorem

the curve, which is the definite integral $\int_{x}^{x+a} \phi(x')dx'$, must be equal to the area of a rectangle of width 'a', and with a height which is equal to the height at some point R on the curve between P and Q.

This theorem sometimes allows us to approximate, and to set limits to the value of an integral when we cannot work it out exactly. The expression $\phi(x + \theta a)$ must lie between the extreme maximum and extreme minimum values of $\phi(x)$ between P and Q.
That is

$$\min_{x,\ x+a} \{\phi(x)\} \leqslant \phi(x + a) \leqslant \max_{x,\ x+a} \{\phi(x)\} \qquad (5.231)$$

and consequently

$$a \cdot \min_{x,\ x+a} \{\phi(x)\} \leqslant \int_{a}^{x+a} \phi(x')\,dx' \leqslant a \cdot \max_{x,\ x+a} \{\phi(x)\} \qquad (5.232)$$

As an example, consider the function

$$\phi(x) = e^{-x^2} \qquad (5.233)$$

and let us obtain limits for its integral over the range $x = 2$ to 2·5.

The integral mean value theorem gives

$$0{\cdot}5 \cdot \min_{2,\ 2\cdot 5} \{e^{-x^2}\} \leqslant \int_{2}^{2\cdot 5} e^{-x^2}\,dx \leqslant 0{\cdot}5 \cdot \max_{2,\ 2\cdot 5} \{e^{-x^2}\} \qquad (5.234)$$

Since e^{-x^2} is monotonic,

$$\min_{2,\,2\cdot5} \{e^{-x^2}\} = e^{-2\cdot5^2} = 0\cdot0019 \qquad (5.235)$$

and

$$\max_{2,\,2\cdot5} \{e^{-x^2}\} = e^{-2\cdot0^2} = 0\cdot0183 \qquad (5.236)$$

so, substituting in the relation 5.222

$$0\cdot0009 \leqslant \int_2^{2\cdot5} e^{-x^2}\,dx \leqslant 0\cdot0091 \qquad (5.237)$$

5.10.5. A word of caution on series expansions

We have presented various results on series expansions above, in each case giving some form of proof. However, it should be emphasized that we have not been fully rigorous and the results only hold if certain conditions are satisfied – for example, that the successive derivatives, $f^{(n)}(x)$ exist and that the series is convergent. Thus, the results should be used with some caution and more advanced texts consulted if there is any doubt.

5.11. Some examples

5.11.1. Introduction

We noted at the outset of this chapter that there would be relatively few examples within it, as most of the applications will come later chapters, particularly 6, 8 and 9. However, here, we present a number of simple examples which show how a knowledge of both branches of calculus is needed for some quite elementary tasks. In some cases, this also gives us the opportunity of exhibiting alternative modelling procedures for some examples which have been presented in Chapter 3.

5.11.2. Population models

Suppose we consider a particularly simple single region version of the population model given by equation 3.218 in Chapter 3:

$$P(t+\delta t) = [1 + a(t, t+\delta t)]\,P(t) \qquad (5.238)$$

where $a(t, t+\delta t)$ represents net additions (births plus net in-migration minus deaths) in the period t to $t+\delta t$. (We now use δt instead of T for convenience in taking the limit as $\delta t \to 0$ later).

We can now apply Taylor's theorem, up to the first term (or, more precisely, the mean value theorem, though this gives the same result in the limit) to $a(t, t+\delta t)$ to expand it around $\delta t = 0$, considering it as a function of δt:

$$a(t, t+\delta t) = a(t, t) + \frac{\partial a}{\partial \delta t} \cdot \delta t + 0[(\delta t)^2] \qquad (5.239)$$

where $\partial a/\partial \delta t$ is the derivative of $a(t, t+\delta t)$ with respect to δt evaluated at $\delta t = 0$. Call this quantity $\alpha(t)$ – the 'instantaneous addition rate'! Note also that $a(t, t)$ must be zero, since there can be no change in zero time. Thus

$$a(t, t+\delta t) = \alpha(t)\delta t + 0[(\delta t)^2] \qquad (5.240)$$

We can substitute back into equation 5.238 to give

$$P(t+\delta t) = P(t) + \alpha(t)\delta t\, P(t) + 0[(\delta t)^2] \qquad (5.241)$$

This can be rearranged as

$$\frac{P(t+\delta t) - P(t)}{\delta t} = \alpha(t)P(t) + 0(\delta t) \qquad (5.242)$$

Now let $\delta t \to 0$, so that

$$\frac{\mathrm{d}P(t)}{\mathrm{d}t} = \alpha(t)P(t) \qquad (5.243)$$

This is an example of a simple *differential* equation (of which we shall see many more examples in Chapter 6). Thus, we have now built a model of population within which time is a *continuous* variable, and the model turns out to be a differential equation. Much more complicated examples than this are possible, of course. For ecological examples, see Watt (1968), and for a more complicated (and, hopefully, more realistic) human example, see Wilson (1972).

The equation 5.243 turns out to have a very simple solution if $a(t)$ can be taken as a constant – independent of t. Then,

$$\frac{\mathrm{d}P}{\mathrm{d}t} = \alpha P \qquad (5.244)$$

and the solution is

$$P(t) = \mathrm{e}^{at} \qquad (5.245)$$

since

$$\frac{\mathrm{d}P}{\mathrm{d}t} = \alpha \mathrm{e}^{at} = \alpha P \qquad (5.246)$$

in this case. If $\alpha > 0$, this would represent exponential growth, if $\alpha < 0$, exponential decline.

5.11.3. The intervening opportunities model

In section 3.8.7 of Chapter 3, we presented a number of spatial interaction models – all of them, so called 'gravity models'. An alternative form of spatial interaction model is the intervening opportunities model. The basic concepts of the gravity model related to attraction between 'masses', say O_i and D_j of equation 3.268 at some distance c_{ij}. This intervening opportunities model relies on a hypothesis of there being a fixed probability L, that any given opportunity is accepted.

Suppose T_{ij}, O_i, D_j and c_{ij} are the main variables, defined as in section 3.8.7. Let U_{ij} be the probability that a traveller who has reached j from i, to assess its D_j opportunities moves on from there. Thus U_{ij} is the probability of moving beyond the zone immediately preceeding j in rank order from i, $U_{ij'}$, say, times the probability of not being satisfied at j, which is $1 - LD_j$.
Thus

$$U_{ij} = U_{ij'}(1 - LD_j) \tag{5.247}$$

This can be written

$$\frac{U_{ij} - U_{ij}'}{U_{ij}'} = -LD_j \tag{5.48}$$

Let A_j be the total number of opportunities passed up to and including zone j. Then,

$$D_j = A_j - A_j' \tag{5.249}$$

and equation 5.248 can be written

$$\frac{U_{ij} - U_{ij'}}{U_{ij'}} = L(A_j - A_{j'}) \tag{5.250}$$

or

$$\frac{\delta U}{U} = -L\delta A \tag{5.251}$$

if we use δ to denote an increment, and drop subscript labels for the time being. Then

$$\frac{\delta U}{\delta A} = -LU \tag{5.252}$$

If we let $\delta A \to 0$ (implicitly assuming that the zones take on a very small size), this becomes the differential equation

$$\frac{dU}{dA} = -LU \tag{5.253}$$

and we know from the previous example that a solution is

$$U = e^{-LA} + C \tag{5.254}$$

where C is a constant. That is, if we re-introduce zone labels,

$$U_{ij} = e^{-LA_j} + C_i$$

For the 'previous zone', we would have

$$U_{ij'} = e^{-LA_{j'}} + C_i \tag{5.256}$$

and so we can subtract to give

$$U_{ij'} - U_{ij} = e^{-LA_{j'}} - e^{-LA_j} \tag{5.257}$$

as the probability that one tripper from i stops in j. But since O_i trippers leave i altogether, the expected number of trips from i to j is T_{ij} given by

$$T_{ij} = O_i(U_{ij'} - U_{ij}) \tag{5.258}$$

so, using equation 5.257,

$$T_{ij} = O_i(e^{-LA_{j'}} - e^{-LA_j}) \tag{5.259}$$

which is the usual statement of the intervening opportunities model as invented by Schneider (Chicago Transportation Study, 1960). A more detailed and rigorous presentation of the above argument is given in Wilson (1970).

5.11.4. Error propagation in photogrammetry

We present the bare bones of this situation only, relying on Hallert (1960) to whom the reader is referred for more details and discussion. We derive the result in a slightly different way than Hallert, relying on our equation 5.164 for the differential in a function given increments in several independent variables. Suppose

$$y = f(x_1, x_2, x_3 \ldots \ldots x_n) \tag{5.260}$$

where $x_1, x_2, \ldots \ldots x_n$ are n photogrammetric measurements, and y is some dimension which is required and is a known function of these. The question is: given measurement errors on $x_1, x_2, \ldots x_n$, what is the associated error on y? Equation 5.164 says that, for increments δx_1, $\delta x_2, \ldots \delta x_n$.

$$\delta y = \sum_i \frac{\partial f}{\partial x_i} \delta x_i \qquad (5.261)$$

The $\partial f/\partial x_i$s will, in most photogrammetric cases, be known. Hallert then shows that, given such a relationship which can be taken as approximately linear in the increments δx_i the associated measurement error is

$$M = \left[\left(\frac{\partial f}{\partial x_1} m_1\right)^2 + \left(\frac{\partial f}{\partial x_2} m_2\right)^2 + \ldots + \left(\frac{\partial f}{\partial x_n} m_n\right)^2\right]^{1/2} \qquad (5.262)$$

where m_1 is the standard measurement error in a particular case on the measured variable.

5.11.5. Consumers' surplus

We now consider some examples of simple integration. This example is taken from economics. A typical demand curve is shown in Figure 5.13,

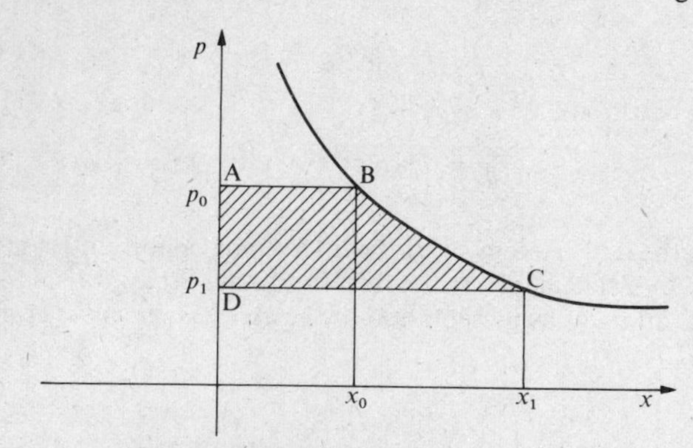

FIG. 5.13. Change in consumers' surplus

relating demand x to price p and showing two points (x_0, p_0) and (x_1, p_1). The change in consumers' surplus for a price reduction from p_0 to p_1 is defined as the area ABCD.

The axes in Figure 5.13 are those conventionally used in economics. This means that the area we are interested in is that between a curve and the vertical axis. The reader who feels uncomfortable with this should use Figure 5.14 in which the axes have been reversed.

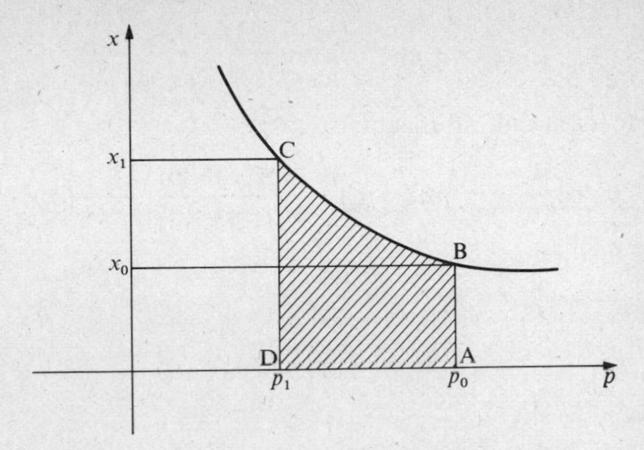

FIG. 5.14. Consumers' surplus, with axes reversed

If we denote the area of interest as ΔC.S., standing for 'change in consumers' surplus', our definition of a definite integral tells us that

$$\Delta\text{C.S.} = \int_{p_1}^{p_0} x(p)\,\mathrm{d}p \tag{5.263}$$

where

$$x = x(p) \tag{5.264}$$

is the demand function of x against p.

Suppose we take this function as the backward sloping straight line

$$x = -\mathrm{m}p + \mathrm{c},\ \mathrm{m} > 0 \tag{5.265}$$

Then,

$$\Delta\text{C.S.} = \int_{p_1}^{p_0} (-\mathrm{m}p + \mathrm{c})\,\mathrm{d}p \tag{5.266}$$

$$= \left[-\frac{\mathrm{m}p^2}{2} + \mathrm{c}p\right]_{p_1}^{p_0} \tag{5.267}$$

$$= -\frac{\mathrm{m}}{2}(p_0^2 - p_1^2) + \mathrm{c}\,(p_0 - p_1) \tag{5.268}$$

However, we know that the line passes through the points (x_0 p_0) and (x_1, p_1) and so we can use the methods of section 3.3.5 of Chapter 3 to find m and c. Using equations 3.54 and 3.55, we get

$$-\mathrm{m} = \frac{x_0 - x_1}{p_0 - p_1} \tag{5.269}$$

$$c = \frac{p_0x_1 - p_1x_0}{p_0 - p_1} \tag{5.270}$$

Substitute for m and c in equation (5.268)

$$\Delta\text{C.S.} = \tfrac{1}{2}\frac{(x_0 - x_1)}{(p_0 - p_1)}(p_0^2 - p_1^2) + \frac{(p_0x_1 - p_1x_0)}{p_0 - p_1}(p_0 - p_1) \tag{5.271}$$

Note that

$$p_0^2 - p_1^2 = (p_0 - p_1)(p_0 + p_1) \tag{5.272}$$

so this simplifies to

$$\Delta\text{C.S.} = \tfrac{1}{2}(x_0 - x_1)(p_0 + p_1) + p_0x_1 - p_1x_0 \tag{5.273}$$

$$= \tfrac{1}{2}(x_0 + x_1)(p_0 - p_1) \tag{5.274}$$

which is a well-known result.
Note that, in Figure 5.14, if BC is taken as a straight line, ΔC.S. is the

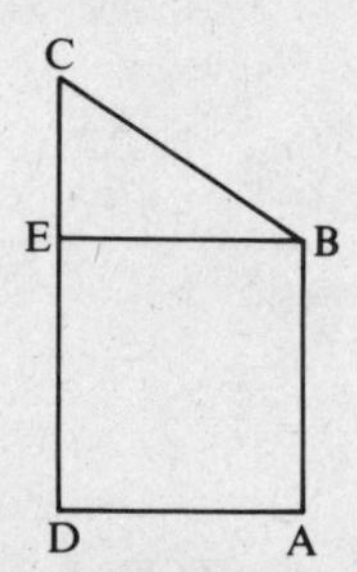

FIG. 5.15. Consumers' surplus as a trapezium

area of the trapezium shown in Figure 5.15.
Since

$$DA = p_0 - p_1 \tag{5.275}$$

$$AB = x_0 \tag{5.276}$$

$$CE = x_1 - x_0 \tag{5.277}$$

$$ABCD = ABED + BCE \tag{5.278}$$

$$= x_0(p_0 - p_1) + \tfrac{1}{2}(x_1 - x_0)(p_0 - p_1) \tag{5.279}$$

$$= \tfrac{1}{2}(x_0 + x_1)(p_0 - p_1) \tag{5.280}$$

This confirms our earlier result for the straight line demand curve in a much simpler way. However, the general result can and must be used for other demand functions, and in any case, the 'long way round' has given us the opportunity to illustrate integration methods linked with the co-ordinate geometry of the straight line.

The formula 5.274 has been commonly applied in the transport planning field, where x-variable is taken as T_{ij}, the number of trips from i to j and c_{ij} is the cost of travel from i to j. (cf. Tressider et al, 1968, Wilson and Kirwan, 1969). If a new road, for example, reduces c_{ij} from c_{ij}^0 to c_{ij}^1, and trips increase from T_{ij}^0 to T_{ij}^1, then

$$\Delta\text{C.S.} = \tfrac{1}{2}(T_{ij}^0 + T_{ij}^1)(c_{ij}^0 - c_{ij}^1) \qquad (5.281)$$

is taken as a measure of the social benefit and used as the basis of cost – benefit analysis.

5.11.6. Total population given a density distribution

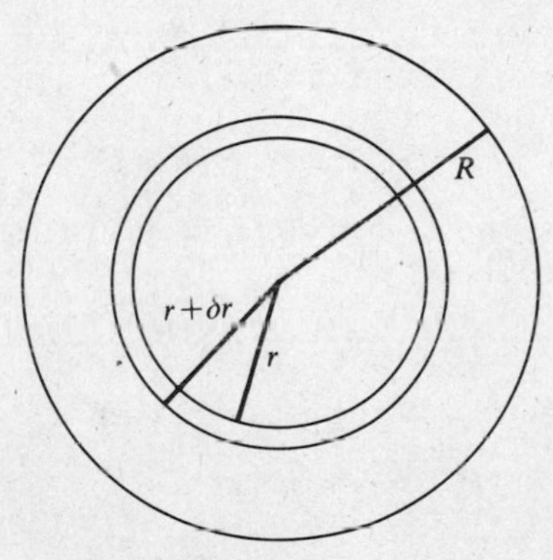

FIG. 5.16. Circular city of radius R

Consider a circular city of radius R as shown in Fig. 5.16. Form an annulus of inner radius r, outer radius $r + \delta r$ as shown. Define $P(r_1, r_2)$ to be the population in an annulus of inner radius r_1, outer radius r_2. The area of the thin annulus between r and $r + \delta r$ is approximately $2\pi r\delta r$, and so if the density of population at distance r from the centre is $d(r)$, the population in the annulus is

$$P(r, r + \delta r) = 2\pi r\delta r \cdot d(r) \qquad (5.282)$$

If δr is infinitesimal, we now see that we can find the population in a finite annulus by evaluating the definite integral

$$\text{P}(r_1, r_2) = \int_{r_1}^{r_2} 2\pi r d(r)\text{d}r \qquad (5.283)$$

The population of the whole city is

$$P(0, R) = \int_0^R 2\pi r d(r) \mathrm{d}r \tag{5.284}$$

Suppose

$$d(r) = \begin{cases} \frac{1}{r}, r > \mathrm{a} & (5.285) \\ 0, r < \mathrm{a} & (5.286) \end{cases}$$

This implies that no people live in the inner a miles. Then the population is

$$P(\mathrm{a}, R) = \int_\mathrm{a}^R 2\pi r \cdot \frac{1}{r} \mathrm{d}r \tag{5.287}$$

$$= \int_\mathrm{a}^R 2\pi \cdot \mathrm{d}r \tag{5.288}$$

$$= [2\pi r]_\mathrm{a}^R \tag{5.289}$$

so

$$P(\mathrm{a}, R) = 2\pi(R - \mathrm{a}) \tag{5.290}$$

A more interesting case is when $d(r)$ is taken as a negative exponential function (Clark, 1951):

$$d(r) = d_0 \mathrm{e}^{-\mathrm{a}r} \tag{5.291}$$

say, where d_0 is the density at the centre. Then

$$P(0, R) = \int_0^R 2\pi r \cdot d_0 \mathrm{e}^{-\mathrm{a}r} \mathrm{d}r \tag{5.292}$$

$$= 2\pi d_0 \int_0^R r\mathrm{e}^{-\mathrm{a}r} \mathrm{d}r \tag{5.293}$$

$r\mathrm{e}^{-\mathrm{a}r}$ is not one of our standard functions. In this case we can use the method of integration by parts, using the formula 5.143 with

$$\mathrm{u} = r \tag{5.294}$$

$$\mathrm{v} = -\frac{\mathrm{e}^{-\mathrm{a}r}}{\mathrm{a}} \tag{5.295}$$

so

$$\frac{\mathrm{du}}{\mathrm{d}r} = 1 \tag{5.296}$$

$$\frac{\mathrm{dv}}{\mathrm{d}r} = \mathrm{e}^{-\mathrm{a}r} \tag{5.297}$$

Hence

$$\int re^{-ar}\, dr = -\frac{re^{-ar}}{a} - \int (1)\left(-\frac{e^{-ar}}{r}\right) dr \tag{5.298}$$

$$= -re^{-ar}/a + \int \frac{e^{-ar}}{a}\, dr \tag{5.299}$$

$$= -\frac{re^{-ar}}{a} - \frac{e^{-ar}}{a^2} + C \tag{5.300}$$

where C is the constant of integration. We can now use this result in the definite integral 5.293 to give

$$P(0, R) = 2\pi d_0 \left[-\frac{re^{-ar}}{a} - \frac{e^{-ar}}{a^2}\right]_0^R \tag{5.301}$$

$$= 2\pi d_0 \left[-R\frac{e^{-aR}}{a} - \frac{e^{-aR}}{a^2} + \frac{1}{a^2}\right] \tag{5.302}$$

$$= \frac{2\pi d_0}{a}\left[\frac{1}{a} - Re^{-aR} - \frac{e^{-aR}}{a}\right] \tag{5.303}$$

If the city is considered to have *infinite* radius, then we can let $R \to \infty$, and since $e^{-aR} \to 0$,

$$P(0, \infty) = \frac{2\pi d_0}{a^2} \tag{5.304}$$

5.11.7. A continuous space model of spatial interaction

In section 3.8.7. of Chapter 3, we presented a spatial interaction model of T_{ij} for a system of N zone $i = 1, 2, 3, \ldots$ N. The number zones and hence, implicitly, the size of the zones, determine the level of spatial resolution – the greater the number the finer the level of resolution. Angel and Hyman (1972) have taken this procedure to its limit and built a continuous space interaction model analogous to that given by equations 3.364, 3.369 and 3.370 of Chapter 3. They base their model on a polar co-ordinate system, and a trip *density* function $T(r_1, \theta_1, r_2, \theta_2)$ which is the number of trips from a unit area at (r_1, θ_1) to a unit area at (r_2, θ_2), as shown in Figure 5.17. They use polar co-ordinates to take advantage of circular symmetry in some applications of the model.

In Figure 5.17 we have also shown the points (r, θ) and $(r + \delta r, \theta + \delta\theta)$, together with the increment in area generated by these increments. It

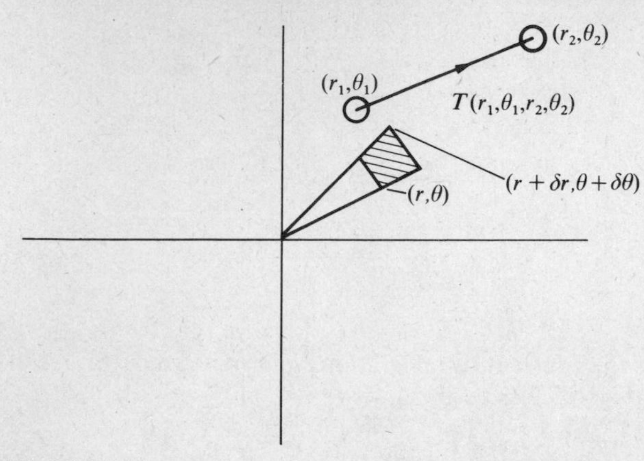

FIG. 5.17. Polar co-ordinates for a continuous space interaction model

can easily be seen that this area is approximately $r\delta r\delta\theta$. Thus, we take the number of trips from a small area around the point (r, θ) to be $T(r, \theta, r', \theta')r\delta r\delta\theta$ and we obtain trip bundles by integrating over both variables. It can now be seen, for example, that the equations which are equivalent to equation 3.263 and 3.264 for the continuous model are

$$\int_0^\infty \int_0^{2\pi} T(r_1,\theta_1,r_2,\theta_2)r_2\mathrm{d}\theta_2\mathrm{d}r_2 = 0(r_1,\theta_1) \tag{5.305}$$

where $0(r_1,\theta_1)$ is the given density of trip productions at (r_1,θ_1), and

$$\int_0^\infty \int_0^{2\pi} T(r_1,\theta_1,r_2,\theta_2)r_1\mathrm{d}\theta_1\mathrm{d}r_1 = D(r_2,\theta_2) \tag{5.306}$$

where $D(r_2,\theta_2)$ is the given density of trip attractions at each (r_2,θ_2). It is then shown in the paper cited that the continuous model can be written.

$$T(r_1,\theta_1,r_2,\theta_2) = A(r_1,\theta_1)B(r_2,\theta_2)0(r_1,\theta_1)D(r_2,\theta_2)\mathrm{e}^{-\mu c(r_1,\theta_1,r_2,\theta_2)} \tag{5.307}$$

where $c(r_1,\theta_1,r_2,\theta_2)$ is the cost of travel from (r_1,θ_1) to (r_2,θ_2) and

$$\frac{1}{A(r_1,\theta_1)} = \int_0^\infty \int_0^{2\pi} B(r_2,\theta_2)D(r_2,\theta_2)\mathrm{e}^{-\mu c(r_1,\theta_1,r_2,\theta_2)}r_2\mathrm{d}\theta_2\mathrm{d}r_2 \tag{5.308}$$

$$\frac{1}{B(r_2,\theta_2)} = \int_0^\infty \int_0^{2\pi} A(r_1,\theta_1)0(r_1,\theta_1)\mathrm{e}^{-\mu c(r_1,\theta_1,r_2,\theta_2)}r_1\mathrm{d}\theta_1\mathrm{d}r_1 \tag{5.309}$$

These are equivalent to equations 3.268, 3.273 and 3.274 of Chapter 3. The reader is referred to the paper by Angel and Hyman for further details and applications, but the above should suffice to illustrate briefly the concepts of multiple integrals introduced in this chapter, and of polar co-ordinates, introduced in Chapter 3.

Exercises

Section 5.2

(1) Find, from first principles, the gradients of the following curves at the points mentioned: (a) $y = x^2 + 2x$ at $x = 3$; (b) $y = x^3 + x^2$ at $x = 2$.

Sections 5.3 and 5.4

(2) Differentiate the following functions with respect to x:
(a) x^n, (b) $x^{\frac{1}{2}}$, (c) x^{-n}, (d) $x^{-\frac{1}{3}}$, (e) $x^{-\frac{1}{2}} + x^{\frac{1}{2}}$,
(f) $x^2 - x$, (g) $2x^2 + 3x^3$, (h) $x^2 \sin x$, (i) $x^2 e^{-x}$,
(j) $3x^n \sin 2x$, (k) $e^x / \sin x$, (l) $(\sin x)/x^2$, (m) $e^{\sin x}$,
(n) e^{-ax^2}, (o) $(x + \sin x)^2$

Section 5.5

(3) Find the *second* derivatives with respect to x of the functions listed in question 2.

Section 5.6

(4) Find the stationary points (and identify the type of each), of the following functions, and sketch the curves:

(a) $y = x^2 - 3x + 2$, (b) $y = x^3 + x^2 - x + 6$,
(c) $y = -4 + 6x - x^2$, (d) $y = \sin x, 0 < x < \pi$,
(e) $y = x \sin x, 0 < x < \pi$

(5) Return to Chapter 3, Exercise 8, and see if you can improve your sketches with a knowledge of stationary points.

Sections 5.7 and 5.8

(6) Integrate the following functions with respect to x: (a) x^n,
(b) x^{-n}, (c) $1/x^2$, (d) $1/x^{\frac{1}{2}}$, (e) $x^{\frac{1}{3}}$, (f) $1/x$, (g) $3x^{-2n}$,
(h) $4 \sin x$, (i) $4 \sin 4x$, (j) $(\cos 2x)/2$, (k) e^{3x},
(e) $(x^2 + 3x)^5 (2x + 3)$, (m) $(\sin\theta + 1)^6 \cos\theta$, (n) xe^{-x^2},
(o) $\log 3x$, (p) xe^x, (q) $x^2 e^x$.

(7) Evaluate the following definite integrals:
(a) $\int_0^1 x^3 dx$, (b) $\int_2^3 x^{\frac{1}{2}} dx$, (c) $\int_0^{\pi/2} \sin x \, dx$,
(d) $\int_0^{\pi/2} \cos x \, dx$, (e) $\int_a^b (x^2 + 3) dx$, (f) $\int_{p_1}^{p_2} p dp$.

Section 5.9

(8) If $u = x_1^{\alpha_1} x_2^{\alpha_2} x_3^{\alpha_3} \sin x_4$, find $\frac{\partial u}{\partial x_1}, \frac{\partial u}{\partial x_2}, \frac{\partial u}{\partial x_3}, \frac{\partial u}{\partial x_4}, \frac{\partial^2 u}{\partial x_1^2}, \frac{\partial^2 u}{\partial x_1 x_2}, \frac{\partial^2 u}{\partial x_3 x_4}$

(9) If $S = -\sum_i \sum_j \log T_{ij}!$, and $\log T_{ij}! = T_{ij} \log T_{ij} - T_{ij}$, find $\frac{\partial S}{\partial T_{ij}}$.

Section 5.11

10) The density of population in a uniform circular town of radius R is constant and equal to d. Find the population of an annulus of inner radius r and outer radius $r + \delta r$ and hence show that the population of the town is $\pi R^2 d$ by integrating over r.

(11) A demand function is $x = \frac{100}{p^2}$, $1 \leqslant p \leqslant 10$. At price 5, therefore, 4 units are purchased, and at price 2, 25 units are purchased. Calculate the change in consumers' surplus consequent on this price reduction. What would the change in consumers' surplus have been if the demand curve was a straight line joining the two points mentioned: (5,4) and (2,25)?

Part 2

6. Differential and difference equations

6.1. Definitions

At the end of the last chapter some examples of differential equations were introduced as among the uses of calculus. As we look at geographical problems more closely, we come across more and more examples of equations which express relationships between *differences* or between *differentials*. In analysing any continuous surface or function, such as the topography of a landform, the density of traffic along a road, or the variation of temperature at a point, we are constantly dealing with the *differences* of value (elevation, density and so on) between neighbouring points or times, or with the rates of change of value – that is with *differentials*. Because differentials are originally obtained by a limiting process from differences (section 5.2), there is a close relationship between differential equations which relate derivatives, and difference equations which contain finite differences, and they are treated together in this chapter. The choice between them depends in practice on whether we wish to use the formal methods of calculus on differentials; or to deal numerically or on a computer with a finite number of points and with the differences between values at these points. Where convenient we can approximate to a differential equation by using a difference equation, (for example where the differential equation has no formal solution) and *vice-versa*.

6.2. Ordinary differential equations

6.2.1. Introduction

Differential equations which involve only two variables, say y and x; or x and t, are called ordinary differential equations, to distinguish them from partial differential equations which contain three or more variables and consequently partial derivatives. In this and following sections, we will consistently use x as the independent variable; and y as the dependent.

The simplest differential equations are those containing a differential and a function of x, the independent variable. For example:

$$\frac{dy}{dx} = 2x \qquad (6.1)$$

$$\frac{dy}{dx} = e^x \tag{6.2}$$

These equations can be integrated directly, using the methods of sections 5.7 and 5.8. Only slightly more complex are equations like

$$\frac{dy}{dx} = y \tag{6.3}$$

$$\frac{dy}{dx} = \sqrt{(1-y^2)} \tag{6.4}$$

This type of equations contains dy/dx and y, but not the independent variable x. In solving this type of equation, and in handling differential equations generally, it is usually permissible to treat dy/dx exactly like $\delta y/\delta x$.

This means that one can usually treat dy and dx *as if* they were separate entities which can be multiplied and divided. There are a few cases where this method of *separation of variables* breaks down, for which more advanced texts should be consulted.

Thus we can proceed from equation 6.4 as follows:

$$\frac{dy}{\sqrt{(1-y^2)}} = dx \tag{6.5}$$

which we can integrate to give

$$\int \frac{1}{\sqrt{(1-y^2)}}\,dy = \int 1\,dx \tag{6.6}$$

Because these are indefinite integrals, we must introduce a constant c, so that noting the derivative of $1/\sqrt{(1-x^2)}$ in Table 5.1 of Chapter 5:

$$\sin^{-1} y = x + c \tag{6.7}$$

Our general solution for y then becomes

$$y = \sin(x + c) \tag{6.8}$$

for an arbitrary constant, c.

6.2.2. Solution by inspection

In the rest of this chapter, we give an account of methods for the solution of differential and difference equations, and illustrate these with geographical examples. At first, some readers may find this difficult relative to some earlier chapters. It should be borne in mind, however, that a minimum gain in understanding can be achieved by knowing the basic idea behind, say, a differential equation in a geographical example, and then a stated solution can be checked by inspection. For example, if the equation given in 6.4 arises in the geographical literature, then direct differentiation and substitution of $y = \sin(x + c)$ will confirm that this is a solution.

6.2.3. First-order linear differential equations

The next most complicated type of equation is one containing dy/dx, y and x. If y and its derivatives appear in their simple form (that is, not squared, $\log y$ or whatever) then the equation is said to be a *linear* differential equation, even though it contains complex functions of x. If the equation contains no higher order derivatives than dy/dx (that is, not d^2y/dx^2 and the like) then the equation is a *first-order* differential equation. A general linear first-order differential equation can be expressed as

$$\frac{dy}{dx} + Py = Q \tag{6.9}$$

where P and Q are any functions of x.

Equations which are not exactly in this form can often be manipulated into this form, for example.

$$x^2\frac{dy}{dx} - x^3y = x^3 \tag{6.10}$$

can be converted, by dividing through by x^2, to

$$\frac{dy}{dx} - xy = x \tag{6.11}$$

which is in the standard form of equation 6.9 where

$$P = -x;\ Q = x \tag{6.12}$$

We obtain a solution to equation 6.9 by looking at the expression

$$z = y.e^{\int P.dx} \tag{6.13}$$

By applying the rule for differentiating a product given in section 5.4.5,

$$\frac{dz}{dx} = y\frac{d}{dx}(e^{\int P.dx}) + \frac{dy}{dx}e^{\int P.dx} \tag{6.14}$$

By applying the rule for differentiating the function of a function (section 5.4.4),

$$\frac{d}{dx}(e^{\int Pdx}) = \frac{d}{dx}\left(\int Pdx\right)e^{\int Pdx} = P.e^{\int P.dx} \tag{6.15}$$

Substituting back into equation 6.14,

$$\frac{dz}{dx} = Py\,e^{\int Pdx} + \frac{dy}{dx}e^{\int Pdx}$$
$$= e^{\int P\,dx}\left(\frac{dy}{dx} + Py\right) \tag{6.16}$$

We observe therefore, that if the left hand side of equation 6.9 were multiplied by $e^{\int Pdx}$, we could integrate it as a whole. We now proceed to do this as follows:

$$e^{\int Pdx}\left(\frac{dy}{dx} + Py\right) = Q\,e^{\int Pdx} \tag{6.17}$$

Integrating with respect to x,

$$z = ye^{\int Pdx} = \int Qe^{\int Pdx}dx + A, \tag{6.18}$$

where A is an arbitrary constant of integration. Assuming we can integrate the expression on the right-hand side, we now have a formal solution:

$$y = e^{-\int Pdx}\int Qe^{\int Pdx}dx + Ae^{-\int Pdx} \tag{6.19}$$

This rather formidable expression is often quite manageable in practise. As an illustration, we can repeat the steps for equation 6.11 above:

$$\frac{dy}{dx} - xy = x \tag{6.20}$$

where $P = -x$

$$\int Pdx = \int -x\,dx = -\tfrac{1}{2}x^2 \tag{6.21}$$

We therefore wish to multiply both sides of equation 6.11 by

$$e^{\int Pdx} = e^{-\frac{1}{2}x^2} \tag{6.22}$$

We then have

$$e^{-\frac{1}{2}x^2}\left(\frac{dy}{dx} - xy\right) = xe^{-\frac{1}{2}x^2} \tag{6.23}$$

The left-hand side should now be the derivative of $y\,e^{-\frac{1}{2}x^2}$, which we can confirm by direct differentiation. Integrating equation 6.14 we therefore obtain

$$ye^{-\frac{1}{2}x^2} = \int xe^{-\frac{1}{2}x^2}\,dx \qquad (6.24)$$

The right-hand side of this equation can be integrated to

$$\int xe^{-\frac{1}{2}x^2}\,dx = -e^{-\frac{1}{2}x^2} + A. \qquad (6.25)$$

where A is an arbitrary constant, as can be confirmed by differentiation (using function of a function; section 5.4.4). We then have equation 6.15 in the form

$$ye^{-\frac{1}{2}x^2} = -e^{-\frac{1}{2}x^2} + A, \qquad (6.26)$$

and multiplying both sides by $e^{\frac{1}{2}x^2}$,

$$y = -1 + Ae^{\frac{1}{2}x^2} \qquad (6.27)$$

as the general solution to the differential equation 6.11, containing one arbitrary constant A.

As in integration, it is often helpful to check the solution of a differential equation at this stage by substitution in the original equation. From equation 6.27 we derive:

$$\frac{dy}{dx} = O + Axe^{\frac{1}{2}x^2} \qquad (6.28)$$

$$xy = \quad x + Axe^{\frac{1}{2}x^2} \qquad (6.29)$$

Subtracting to substitute in the left-hand side of equation 6.11, we have

$$\frac{dy}{dx} - xy = Axe^{\frac{1}{2}x^2} - (-x + Axe^{\frac{1}{2}x^2}) = +x \qquad (6.30)$$

confirming that 6.18 is a solution to equation 6.11. The appearance of one arbitrary constant in the solution arises through integrating *once* to get rid of the first-order derivative dy/dx. There is a general rule which we state here without proof:

'An ordinary differential equation of nth order has a unique general solution with n arbitrary constants.'

Conversely: 'A solution with n independent arbitrary constants is the unique general solution.'

Thus we may be sure that equation 6.18 is the *only* complete solution to equation 6.11.

6.2.4. Particular solutions

For a particular problem, we need an appropriate *particular solution* which fits the problem. The solution of equation 6.18 represents a family of curves, each one corresponding to a value of the constant A, as shown in Fig. 6.1.

Only one of these curves will satisfy the conditions of our particular problem, which might specify, for example, that $y = 0$ when $x = 0$. Substituting this 'initial condition' in the solution (equation 6.18),

$$0 = Ae^0 - 1, \tag{6.31}$$

and since $e^\circ = 1$, the arbitrary constant

$$A = 1. \tag{6.32}$$

The particular solution is then

$$y = e^{\frac{1}{2}x^2} - 1, \tag{6.33}$$

corresponding to the A = 1 curve in Figure 6.1.

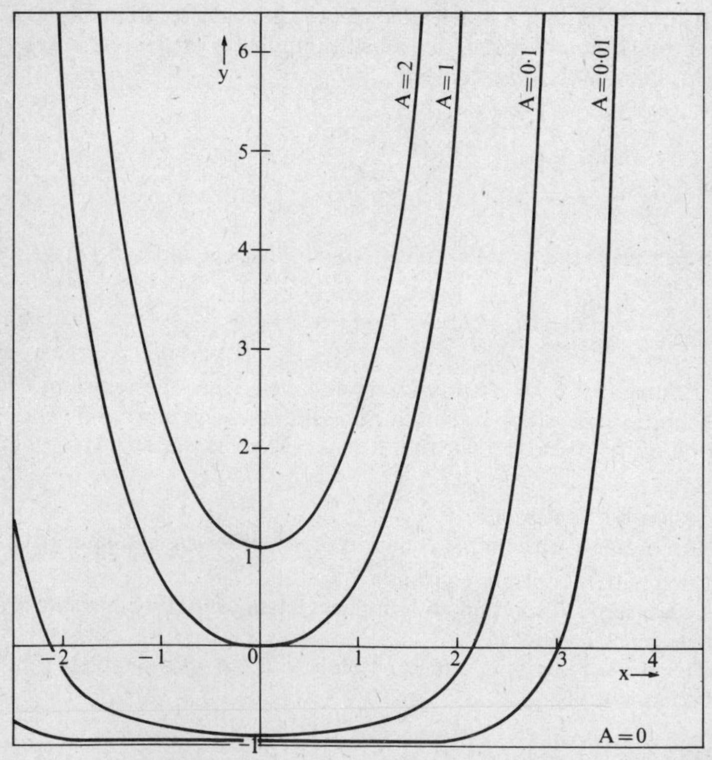

FIG. 6.1. The family of curves $y = Ae^{1/2x^2} - 1$

6.3. Ordinary linear differential equations

6.3.1. *Introduction*

A second set of differential equations which can be solved with relative ease is the set with higher differentials, but with each appearing as a linear term, multiplied by some function of x. That is

$$\frac{d^n y}{dx^n} + a_1 \frac{d^{n-1}y}{dx^{n-1}dy} + a_2 \frac{d^{n-2}y}{dx^{n-2}} + \ldots + a_n \cdot y = F(x) \qquad (6.34)$$

where $a_1, a_2 \ldots a_n$ and $F(x)$ are in general all functions of x. We will build up to this general case in two stages, first treating $a_1, a_2 \ldots a_n$ as constants and $F(x) = 0$, and secondly allowing $F(x)$ to be non-zero.

6.3.2. *Constant coefficients; $F(x) = 0$*

We can begin by considering d/dx as an *operator* acting on y.

$$\frac{d^2y}{dx^2} = \frac{d}{dx}\left(\frac{d}{dx}y\right) \qquad (6.35)$$

is the same operator acting twice in succession on y. Let us write this operator as D. Then equation 6.34, with $F(x) = 0$ can be written

$$\phi(D)y = (D^n + a_1 D^{n-1} + a_2 D^{n-2} + \ldots + a_n)y = 0 \qquad (6.36)$$

where $\phi(D)$ represents the function enclosed in brackets. In a rather similar way to the approach in section 6.1.2, we multiply through by e^{-px}, where p is a positive constant, and integrate with respect to x from 0 to ∞. We can 'integrate by parts' (section 5.8.2) to give

$$\begin{aligned}\int_0^\infty e^{-px} D \cdot y \, dx &= [e^{-px} \cdot y]_0^\infty + p\int_0^\infty e^{-px} \cdot y \, dx \\ &= -y_0 + p\int_0^\infty e^{-px} \cdot \; y_1 \end{aligned} \qquad (6.37)$$

where y_0 is the value of y at $x = 0$.
Similarly,

$$\begin{aligned}\int_0^\infty e^{-px} D^2 y dx &= [e^{-px} Dy]_0 + p\int_0^\infty e^{-px} Dy dx \\ &= -y_1 + p(-y_0 + p\int_0^\infty e^{-px} y dx) \\ & -y_1 - py_0 + p^2\int_0^\infty e^{-px} y dx \end{aligned} \qquad (6.38)$$

where y_1 is the value of Dy at $x = 0$. Proceeding further in this way, we get

$$\int_0^\infty e^{-px} D^r y \cdot dx = -(p^{r-1}y_0 + p^{r-2}y_1 \quad \ldots + py_{r-2} + y_{r-1}) + p^r \int_0^\infty e^{-px} y \cdot dx \qquad (6.39)$$

where $y_2 \ldots y_{r-2}y_{r-1}$ are similarly defined and $r \leqslant n$.
Substituting these values into equation 6.36 we obtain:

$$\begin{aligned}\phi(p)\int_0^\infty e^{-px} y dx = {} & (p^{n-1}y_0 + p^{n-2}y_1 + \ldots + py_{n-2} + y_{n-1}) \\ & + a_1(p^{n-2}y_0 + p^{n-3}y_1 + \ldots + py_{n-3} + y_{n-2}) \\ & \ldots\ldots\ldots\ldots\ldots\ldots\ldots\ldots \\ & + a_{n-2}(py_0 + y_1) \\ & a_{n-1}y_0 \end{aligned} \qquad (6.40)$$

This is called the 'subsidiary equation' and brings us much closer to a solution. Because it looks rather difficult, it may be worth illustrating what is involved by following through a simple example. Let us take the differential equation:

$$\frac{d^2y}{dx^2} + k^2 y = 0 \qquad (6.41)$$

This equation arises commonly in connection with many simple wave motions, including those found in river meanders (Ferguson, 1973). In the 'D' notation this becomes:

$$\phi(D)y = (D^2 + k^2)y = 0 \qquad (6.42)$$

where $\phi(D) = D^2 + k^2$.
Multiplying through by e^{-px} and integrating, we substitute

$$\int_0^\infty e^{-px} D^2 y = -y_1 - py_0 + p^2 \int_0^\infty e^{-px} y dx, \qquad (6.38)$$

so that

$$\int_0^\infty e^{-px} \phi(D) y dx = (p^2 + k^2)\int_0^\infty e^{-px} y dx - y_1 - py_0 \qquad (6.43)$$

or

$$\int_0^\infty e^{-px} y dx = \frac{y_1 + py_0}{p^2 + k^2} \qquad (6.44)$$

At this stage, we need a set of standard forms, comparable to standard integrals, for finding the functions, y which correspond to particular functions on the right-hand side. The expression on the left-hand side is called the 'Laplace Transform' of y, and Table 6.1 is a list of some of the simpler cases.

TABLE 6.1

*Standard Laplace Transforms**

$\int_0^\infty e^{-px} y\,dx$	y
$1/p$	1
$1/p^n$	$\dfrac{x^{n-1}}{(n-1)!}$ when n is a positive integer
$1/p-a$	e^{ax}, $p > a$
$a/p^2 + a^2$	$\sin ax$
$p/p^2 + a^2$	$\cos ax$
$a/p^2 - a^2$	$\sinh ax$, $p > \lvert a \rvert$
$p/p^2 - a^2$	$\cosh ax$, $p > \lvert a \rvert$
$p/(p^2 + a^2)^2$	$\dfrac{x}{2a} \sin ax$
$1/(p^2 + a^2)^2$	$\dfrac{1}{2a^3}$ $(\sin ax - ax \cos ax)$

* This table is extracted with permission from Carslaw and Jaeger (1941), *Operational Methods in Applied Mathematics*, which contains a much fuller table

Referring our example to this table, the right-hand side can be decomposed as

$$\frac{y_1}{k}\left(\frac{k}{p^2 + k^2}\right) + y_0\left(\frac{p}{p^2 + k^2}\right) \tag{6.45}$$

The values of y corresponding to these Laplace transforms can be read from Table 6.1. as

$$y = \frac{y_1}{k} \cdot \sin kx + y_0 \cos kx \tag{6.46}$$

This can be seen to be a general solution to the second-order equation 6.41, because it contains two arbitrary constants (y_0 and y_1).

6.3.3. Constant coefficients: $F(x) \neq 0$

If we follow the argument of the previous section through, but retain a non-zero value of $F(x)$, then it can be seen that the subsidiary equation 6.40 is modified by the addition, on the right-hand side, of an extra term, i.e.

$$+ \int_0^\infty e^{-px} F(x)\,dx \tag{6.47}$$

which can be evaluated using the Laplace transform Table (6.1) in reverse, going from $y = F(x)$ to its transform.

Thus, for example, if instead of equation 6.41, we had to solve

$$\frac{d^2y}{dx^2} + k^2 y = x \tag{6.48}$$

then equation (6.40) would be followed by

$$(p^2 + k^2)\int_0^\infty e^{-px} y\,dx - y_1 - py_0 = \int_0^\infty e^{-px} x\,dx \tag{6.49}$$

From the second line in Table 6.1, the right-hand side is seen to be $1/p^2$ (remember that $0! = 1$), so that:

$$\int_0^\infty e^{-px} y\,dx = \frac{y_1 + py_0}{p^2 + k^2} + \frac{1}{p^2(p^2 + k^2)} \tag{6.50}$$

The second term on the right-hand side can be re-arranged as

$$\frac{1}{k^2}\left(\frac{1}{p^2} - \frac{1}{p^2 + k^2}\right) \tag{6.51}$$

so that:

$$\int_0^\infty e^{-px} y\,dx = \frac{y_1 - 1/k^2}{k} \cdot \frac{k}{p^2 + k^2} + y_0 \frac{p}{p^2 + k^2} + \frac{1}{k^2}\frac{1}{p^2} \tag{6.52}$$

Referring again to the table of transforms, we obtain the solution:

$$y = \frac{1}{k}\left(y_1 - \frac{1}{k^2}\right) \sin kx + y_0 \cos kx + \frac{1}{k^2} x \tag{6.53}$$

6.3.4. Non-constant coefficients; Solution in series

When the coefficients of the linear differential equation are not constant, we can still solve some simple equations as a series, by the method of Frobenius. Solutions are most easily obtained when the coefficients are polynomials in x. This method is best shown by an example.

$$(u^2 + x^2)\frac{d^2y}{dx^2} + 2x\frac{dy}{dx} + Ay = 0 \tag{6.54}$$

where u and A are positive constants. Let us look for a solution as an infinite series:

$$y = b_0 + b_1 x + b_2 x^2 + \ldots + b_r x^r + \ldots \tag{6.55}$$

where the b's are constants.
Differentiating, we obtain

$$\frac{dy}{dx} = b_1 + 2b_2x + 3b_3x^2 + \ldots + rb_rx^{r-1} + \ldots \tag{6.56}$$

$$\frac{d^2y}{dx^2} = 2b_2 + 6b_3x + \ldots + r(r-1)b_rx^{r-2} + \ldots \tag{6.57}$$

Substituting in equation 6.54

$$\begin{aligned}[2b_2u^2 + 6b_3u^2x + 12b_4u^2x^2 + \ldots + (r+2)(r+1)b_r + 2u^{2r}x + \ldots \\ + 2b_2x^2 + \ldots + r(r-1)b_rx^r + \ldots] \\ + (2b_1x + 4b_2x^2 + \ldots + 2rb_rx^r + \ldots) \\ + (Ab_0 + Ab_1x + Ab_2x^2 + \ldots + Ab_rx^r + \ldots) \\ = 0 \text{ (for all values of } x)\end{aligned} \tag{6.58}$$

Collecting together first constant terms, and then terms in x, x^2 and so on, we get

$$\left.\begin{aligned} 2b_2u^2 + Ab_0 &= 0 \\ 6b_3u^2 + (A+2)b_1 &= 0 \\ 12b_4u^2 + (A+6)b_2 &= 0 \\ \ldots\ldots\ldots\ldots\ldots\ldots \\ (r+2)(r+1)b_{r+2}u^2 + [A + r(r+1)]\,b_r &= 0 \end{aligned}\right\} \tag{6.59}$$

therefore

$$\begin{aligned} b_2 &= -\frac{1}{u^2}\frac{A}{2}b_0 \\ b_3 &= -\frac{1}{u^2}\frac{A+2}{6}b_1 \\ b_4 &= -\frac{1}{u^2}\frac{A+6}{12}b_2 = \frac{1}{u^4}\frac{A(A+6)}{4!}b_0 \\ b_5 &= \frac{1}{u^2}\frac{A+12}{20}b_3 = \frac{1}{u^4}\frac{(A+2)(A+12)}{5!}b_1 \end{aligned} \tag{6.60}$$

etc

so that

$$y = b_0\left[1 - \frac{A}{2!}\left(\frac{x}{u}\right)^2 + \frac{A(A+6)}{4!}\left(\frac{x}{u}\right)^4 - \ldots + (-1)^m \frac{A(A+6)\,[A+(2m-2)(2m-1)]}{(2m)!}\left(\frac{x}{u}\right)2m + \ldots\right]$$

$$+ b_1 u\left[\frac{x}{u} - \frac{A+2}{3!}\left(\frac{x}{u}\right)^3 + \frac{(A+2)(A+12)}{5!}\left(\frac{x}{u}\right)^5 - \ldots + (-1)^j\frac{(A+2)(A+12)\ldots[A+(2j-1)2j]}{(2j+1)!}\left(\frac{x}{u}\right)^{2j+1} + \ldots\right] \quad (6.61)$$

which is the general solution with arbitrary constants b_0 and b_1, convergent over the interval $0 \leqslant x/u < 1$

The first series in this solution is illustrated in Fig. 6.2. A relationship between successive terms, like that between the b_r s in equation 6.59 is called a recurrence relation, and we will return to this in section 6.5.

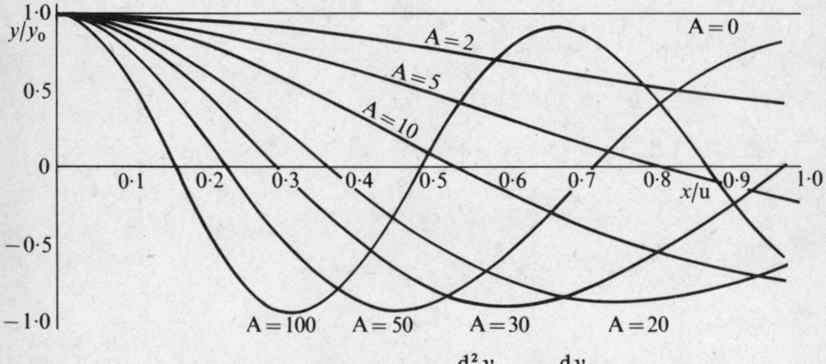

FIG. 6.2. Solutions of $(u^2 + x^2)\dfrac{d^2y}{dx^2} + 2x\dfrac{dy}{dx} + Ay = 0$

6.3.5. *Iterative solutions*

In some cases, including the last example for $x/u > 1$, the series solution is not convergent, and it may be possible to rearrange the differential equation so that successive terms are obtained by integration. As an example, for an equation which arises as a subsidiary equation to partial differential equations of continuity (sections 6.6.2 and 6.6.9 below)

$$\frac{dQ}{dx} = ky \quad (6.62)$$

where k is a constant and Q is a linear function of y, dy/dx and higher differentials. Integrating both sides in the interval $0 \rightarrow x$;

$$Q\left(y, \frac{dy}{dx}, \ldots\right) = +k\int_0^x y\,dx' \qquad (6.63)$$

If we now make a first approximation to y on the right-hand side, then equation 6.63 gives an equation of order one less for a second approximation to y. This process can be repeated, or iterated, to give any desired degree of accuracy.

This method is particularly suitable when Q takes the simple form:

$$Q = -f(x)\frac{dy}{dx} \qquad (6.64)$$

which occurs commonly in hillslope models (Kirkby, 1971). In this case we can formally write $y_0, y_1 \ldots$ as the successive approximations to y. Then:

$$\frac{d}{dx}\left[f(x)\frac{dy_{n+1}}{dx}\right] = -k\int_0^x y_n\,dx' \qquad (6.65)$$

For the boundary condition $y = y_0$ at $x = 0$; the first approximation can simply be $y = y_0$ (a constant).
Then

$$\frac{d}{dx}\, f(x)\frac{dy_1}{dx} = -ky_0\int_0^x 1\,dx'$$
$$= ky_0 x \qquad (6.66)$$

so that

$$y_1 = y_0 - ky_0\int_0^x \frac{x'}{f(x')}\,dx' \qquad (6.67)$$

applying the boundary condition to remove the arbitrary constant.
If we define

$$I(z) = \int_0^x \frac{\int_0^{x'} z\,dx''}{f(x')}\,dx' \qquad (6.68)$$

Then

$$\frac{y_{n+1}}{y_0} = 1 - kI\left(\frac{y_n}{y_0}\right) \qquad (6.69)$$

or

$$\frac{y_n}{y_0} = 1 - kI(1) + k^2I^2(1) - k^3I^3(1)\ldots \qquad (6.70)$$

where

$$I^2(z) = I[I(z)] \text{ etc} \qquad (6.71)$$

The example above, equation 6.54, is of this type:

$$(u^2 + x^2)\frac{d^2y}{dx^2} + 2x\frac{dy}{dx} + Ay = 0 \tag{6.54}$$

which can be rewritten as

$$\frac{d}{dx}\left[(u^2 + x^2)\frac{dy}{dx}\right] = -Ay \tag{6.72}$$

For this example,

$$f(x) = u^2 + x^2 \tag{6.73}$$

and

$$I(1) = \int_0^x \frac{\int_0^{x'} 1\,dx''}{u^2 + x'^2}\,dx'$$

$$= \int_0^x \frac{x'}{u^2 + x'^2}\,dx'$$

$$= \tfrac{1}{2}\log(1 + x^2/u^2) \tag{6.74}$$

$$I^2(1) = I[I(1)] = \int_0^x \frac{\int_0^x \frac{1}{2}\log(1 + x'^2/u^2)\,dx'}{u^2 + x'^2}\,dx'$$

$$= \frac{1}{8}\log^2(1 + x^2/u^2) - \tfrac{1}{2}\log(1 + x^2/u^2) + \tfrac{1}{2}(\tan^{-1}x/u)^2 \tag{6.75}$$

Thus the first approximations to a solution which is much more strongly convergent for large x/u are:

$$y_1 = 1 - k \cdot \tfrac{1}{2}\log(1 + x^2/u^2) \tag{6.76}$$

$$y_2 = 1 - k \cdot \tfrac{1}{2}\log(1 + x^2/u^2) + k^2\left[\frac{1}{8}\log^2(1 + x^2/u^2) - \tfrac{1}{2}\log(1 + x^2/u^2) + \tfrac{1}{2}(\tan^{-1}x/u)^2\right] \tag{6.77}$$

In many cases, this degree of approximation is adequate.

6.4. Finite differences

6.4.1. Introduction

Most of our data is only available at a finite number of points, so that we cannot fully justify the limiting process involved in differentiation as the step length δx tends to zero. In other cases we may, although possessing complete data from a mathematical function, wish to perform operations on the function by computation. This arises, for example, in connection with differential equations which cannot be solved analytically. In this case we also have to resort to a finite step length. We then have to make use of *finite* difference between the values of y at successive values of x, which we will assume to be evenly spaced, at intervals of h. We can then tabulate a function, y against x;

TABLE 6.2

Tabulation of a function in terms of finite differences

x	y				
$x_0 - 4h$	y_{-4}				
		$\Delta y_{-3\frac{1}{2}}$			
$x_0 - 3h$	y_{-3}		$\Delta^2 y_{-3}$		
		$\Delta y_{-2\frac{1}{2}}$		$\Delta^3 y_{-2\frac{1}{2}}$	
$x_0 - 2h$	y_{-2}		$\Delta^2 y_{-2}$		
		$\Delta y_{-1\frac{1}{2}}$		$\Delta^3 y_{-1\frac{1}{2}}$	
$x_0 - h$	y_{-1}		$\Delta^2 y_{-1}$		
		$\Delta y_{-\frac{1}{2}}$		$\Delta^3 y_{-\frac{1}{2}}$	etc
x_0	y_0		$\Delta^2 y_0$		
		$\Delta y_{\frac{1}{2}}$		$\Delta^3 y_{\frac{1}{2}}$	
$x_0 + h$	y_1		$\Delta^2 y_1$		
		$\Delta y_{1\frac{1}{2}}$		$\Delta^3 y_{1\frac{1}{2}}$	
$x_0 + 2h$	y_2		$\Delta^2 y_2$		
		$\Delta y_{2\frac{1}{2}}$		$\Delta^3 y_{2\frac{1}{2}}$	
$x_0 + 3h$	y_3		$\Delta^2 y_3$		
		$\Delta y_{3\frac{1}{2}}$			
$x_0 + 4h$	y_4				

The notation in this diagram is self-explanatory: y_{-2}, for example is the value of y at $x_0 - 2h$; y_{+3} is the value of y at $x_0 + 3h$, and so on. The first difference for example, is

$$\Delta y_{-2\frac{1}{2}} = y_{-2} - y_{-3}. \tag{6.78}$$

Because this difference is centred between $x_0 - 2h$ and $x_0 - 3h$, it is given the suffix, $-2\frac{1}{2}$.
Similarly the second difference is defined by

$$\Delta^2 y_{-2} = \Delta y_{-1\frac{1}{2}} - \Delta y_{-2\frac{1}{2}} \tag{6.79}$$

TABLE 6.3

Numerical example of a function tabulated by finite differences

	x	y	Δ	Δ^2	Δ^3	
$x_0 - 4h$	1·6	·2041				
			·0263			
$x_0 - 3h$	1·7	·2304		−·0014		
			·0249		0	
$x_0 - 2h$	1·8	·2553		−·0014		
			·0235		+·0001	
$x_0 - h$	1·9	·2788		−·0013		
			·0222		+·0003	
x_0	2·0	·3010		−·0010		
			·0212		0	etc
$x_0 + h$	2·1	·3222		−·0010		
			·0202		+·0001	
$x_0 + 2h$	2·2	·3424		−·0009		
			·0193		+·0001	
$x_0 + 3h$	2·3	·3617		−·0008		
			·0185			
$x_0 + 4h$	2·4	·3802				

and is positioned opposite $x_0 - 2h$.
A numerical example, for the function $y = \log_e x$, is tabulated above at intervals of h = 0·1, and for $x_0 = 2{\cdot}0$:
Thus, for example $y_{-2} = {\cdot}2553$, $\Delta y_{-2\frac{1}{2}} = {\cdot}0249$, $\Delta^2 y_{-1} = -{\cdot}0013$. Because the higher differences tend to magnify irregularities in the initial data, this table initially provides a basis for revealing, and if desired, smoothing these irregularities.

6.4.2. Differentiation and integration using finite differences

If it can be assumed that our tabulated data is smooth and continuous enough to allow differentiation and integration, then we make calculations from the table values. If there is some doubt about the smoothness of the data, it should be noted that integration tends to smooth out irregularities, and differentiation to accentuate them, which means that differentiation is likely to be the less reliable operation. Thus second derivatives are likely to be less regular than first derivatives, and so on.

The expressions used for computing integrals and derivatives are obtained by expressing successive values in the table using Taylor's theorem (section 5.10.1). Thus to obtain the value of dy/dx at $x = x_0$, we write down the Taylor series expansions:

$$y_1 = y_0 + hy'_0 + \frac{1}{2!}h^2y''_0 + \frac{1}{3!}h^3y'''_0 + \frac{1}{4!}h^4y^{iv}_0 + \frac{1}{5!}h^5y^{v}_0 + \ldots \tag{6.80}$$

$$y_{-1} = y_0 - hy'_0 + \frac{1}{2!}h^2y''_0 - \frac{1}{3!}h^3y'''_0 + \frac{1}{4!}h^4y^{iv}_0 - \frac{1}{5!}h^5y^{v}_0 + \ldots \tag{6.81}$$

$$y_2 = y_0 + 2hy'_0 + \frac{1}{2!}(2h)^2y''_0 + \frac{1}{3}(2h)^3y'''_0 + \frac{1}{4!}(2h)^4y^{iv}_0 + + \frac{1}{5}(2h)^5y^{v}_0 + \ldots \tag{6.82}$$

$$y_{-2} = y_0 - 2hy'_0 + \frac{1}{2!}(2h)^2y''_0 - \frac{1}{3!}(2h)^3y'''_0 + \frac{1}{4!}(2h)^4y^{iv}_0 - \frac{1}{5!}(2h)^5y^{v}_0 + \ldots \tag{6.83}$$

where $y'_0, y''_0, y'''_0, y^{iv}_0, y^{v}_0$ denote the values of

$$\frac{dy}{dx}, \frac{d^2y}{dx^2}, \frac{d^3y}{dx^3}, \frac{d^4y}{dx^4}, \frac{d^5y}{dx^5} \text{ at } x = x_0.$$

Subtracting the first pair of equations (6.80 – 6.81)

$$y_1 - y_{-1} = 2hy'_0 + \frac{1}{3}h^3y'''_0 + \frac{1}{60}h^5y^v_0 + \ldots \qquad (6.84)$$

Subtracting the second pair (6.82 – 6.83)

$$y_2 - y_{-2} = 4hy'_0 + \frac{8}{3}h^3y'''_0 + \frac{32}{60}h^5y^v_0 + \ldots \qquad (6.85)$$

Multiply equation 6.84 by 8, and subtract equation 6.85 from it:

$$8(y_1 - y_{-1}) - (y_2 - y_{-2}) = 12hy'_0 - \frac{24}{60}h^5y^v_0 - \ldots \qquad (6.86)$$

This expression gives an approximation for $y'_0 = dy/dx$ at $x = x_0$, as required, with a final error term. It is more convenient however to express the left-hand side in terms of differences as follows.
Referring back to Table 6.2 it can be seen that

$$\left.\begin{aligned} \Delta y_{\frac{1}{2}} &= y_1 - y_0 \\ \Delta y_{-\frac{1}{2}} &= y_0 - y_{-1} \end{aligned}\right\} \qquad (6.87)$$

Adding

$$\Delta y_{\frac{1}{2}} + \Delta y_{-\frac{1}{2}} = y_1 - y_{-1} \qquad (6.88)$$

Similarly

$$\Delta^3 y_{\frac{1}{2}} + \Delta^3 y_{-\frac{1}{2}} = \Delta^2 y_1 - \Delta^2 y_{-1} \qquad (6.89)$$

Expanding the right-hand side of this expression gives,

$$\begin{aligned} \Delta^2 y_1 &= \Delta y_{1\frac{1}{2}} - \Delta y_{\frac{1}{2}} \\ &= (y_2 - y_1) - (y_1 - y_{-0}) \end{aligned} \qquad (6.90)$$

so that

$$\begin{aligned} &\Delta^3 y_{\frac{1}{2}} + \Delta^3 y_{-\frac{1}{2}} \\ &= \Delta^2 y_1 - \Delta^2 y_{-1} = [(y_2 - y_1) - (y_1 - y_{-0})] - [(y_1 - y_{-1}) - (y_1 - y_{-2})] \\ &= y_2 - 2y_1 + 2y_{-1} - y_{-2} = (y_2 - y_{-2}) - 2(y_1 - y_{-1}) \end{aligned} \qquad (6.91)$$

Substituting equations 6.91 into the left-hand side of equation 6.86 gives:

$$\begin{aligned} &8(y_1 - y_{-1}) - (y_2 - y_{-2}) \\ &= 8(y_1 - y_{-1}) - [(\Delta^3 y_{\frac{1}{2}} + \Delta^3 y_{-\frac{1}{2}}) + 2(y_1 - y_{-1})] \\ &= 6(y_1 - y_{-1}) - (\Delta^3 y_{\frac{1}{2}} + \Delta^3 y_{-\frac{1}{2}}) \end{aligned} \qquad (6.92)$$

Now substituting equation (6.88) into this expression, it becomes:

$$= 6(\Delta y_{\frac{1}{2}} + \Delta y_{-\frac{1}{2}}) - (\Delta^3 y_{\frac{1}{2}} + \Delta^3 y_{-\frac{1}{2}}) \quad (6.93)$$

A natural notation for this expression is

$$[6\Delta - \Delta^3]\,(y_{\frac{1}{2}} + y_{-\frac{1}{2}}) \quad (6.94)$$

the Δ operators acting on the *y*s..
Returning finally to equation 6.86 we obtain

$$[6\Delta - \Delta^3]\,(y_{\frac{1}{2}} + y_{-\frac{1}{2}}) = 12\text{h}y'_0 - \frac{24}{60}\text{h}^5 y^{\text{v}}_0 \quad (6.95)$$

or

$$y'_1 = \frac{1}{2\text{h}}\left[\Delta - \frac{1}{6}\Delta^3\right](y_{\frac{1}{2}} + y_{-\frac{1}{2}}) + \frac{1}{30}\text{h}^5 y^{\text{v}}_0 \quad (6.96)$$

the last term giving a measure of the residual error. Although this expression is long enough for most purposes, additional terms can be added in higher differences. Table 6.4 lists this and other tabulation series which may be obtained in a similar way.

In a comparable way to the derivatives, finite difference formulae for integration can be built up using successively higher order differences. The first expression for an integral in Table 6.4 integrates over single steps in Δx, and the first term is equivalent to the area obtained by joining successive data points with straight lines, the so-called 'trapezoidal rule'. The second expression integrates over double increments, producing a more rapidly convergent expression. The first two terms of this expression are equivalent to fitting a second degree equation to three successive data points, which is 'Simpson's rule.'

As examples, let us take the function in Table 6.3 and calculate (a) $\frac{dy}{dx}$ at $x = 1{\cdot}95$ and (b) $\int_{1\cdot8}^{2\cdot2} y.dx$

TABLE 6.4

Derivatives and integrals using finite differences

Derivatives

$$\frac{dy}{dx} \text{ at } x = x_0 : y'_0 = \frac{1}{2\text{h}}\left[\Delta - \frac{1}{6}\Delta^3 + \frac{1}{30}\Delta^5 - \ldots\right](y_{\frac{1}{2}} + y_{-\frac{1}{2}})$$

$$\frac{dy}{dx} \text{ at } x = x_0 + \tfrac{1}{2}\text{h} : y'_{\frac{1}{2}} = \frac{1}{\text{h}}\left[\Delta - \frac{1}{24}\Delta^3 + \frac{3}{640}\Delta^5 - \ldots\right](y_{\frac{1}{2}})$$

The latter series converges more rapidly and should therefore be used when it is possible to calculate the derivative at the interval mid-point.

$$\frac{d^2y}{dx^2} \text{ at } x = x_0 : y''_0 = \frac{1}{h^2}\left[\Delta^2 - \frac{1}{12}\Delta^4 + \frac{1}{90}\Delta^6 \ldots\right](y_0)$$

Integrals

$$\int_{x_0}^{x_0+h} y.dx = \frac{h}{2}\left[1 - \frac{1}{12}\Delta^2 + \frac{11}{720}\Delta^4 - \ldots\right](y_0 + y_1)$$

$$\int_{x_0-h}^{x_0+h} y.dx = 2h\left[1 + \frac{1}{6}\Delta^2 - \frac{1}{180}\Delta^4 + \frac{1}{1512}\Delta^6\right](y_0)$$

Again the second series is preferable when it can be used. The first two terms of this series correspond to 'Simpson's rule.'

(a) $x_0 = 1{\cdot}9$; $h = 0{\cdot}1$;

$$y'_{\frac{1}{2}} = \frac{1}{0{\cdot}1}\left[\Delta - \frac{1}{24}\Delta^3\right] y_{1\cdot95}$$

$$= \frac{1}{0{\cdot}1}\left[0{\cdot}222 - \frac{1}{24} \times 0{\cdot}0003\right]$$

$$- \; 0{\cdot}222$$

(b) Using the last series of Table 6.4 twice;

$$\int_{1\cdot8}^{2\cdot2} y\text{d}x = \int_{1\cdot8}^{2\cdot0} y\text{d}x + \int_{2\cdot0}^{2\cdot2} y\text{d}x$$

$$= 2 \times 0{\cdot}1\left[0{\cdot}2788 + \frac{1}{6} \times -0{\cdot}0013\right] +$$

$$2 \times 0{\cdot}1\left[0{\cdot}3010 + \frac{1}{6} \times -0{\cdot}0010\right]$$

$$= 0{\cdot}2\left[(0{\cdot}2788 - 0{\cdot}0002) + (0{\cdot}3010 - 0{\cdot}0002)\right]$$

$$= 0{\cdot}1159$$

6.5 Difference equations

6.5.1 Relation to differential equations

In solving a differential equation numerically, we can either use the series of Table 6.4 directly, or we can substitute them into the differential equation. For example, let us substitute for $\mathrm{d}y/\mathrm{d}x$ and $\mathrm{d}^2y/\mathrm{d}x^2$ at $x = x_0$ in the differential equation 6.54:

$$(\text{u}^2 + x^2)\frac{\text{d}^2y}{\text{d}x^2} + 2x\frac{\text{d}y}{\text{d}x} + \text{A}y = 0$$

$$(u^2 + x^2)\frac{1}{h_3^2}\Delta^2 y_0 + 2x\frac{1}{2h}(\Delta y_{\frac{1}{2}} + \Delta y_{-\frac{1}{2}}) + Ay = 0 \qquad (6.97)$$

ignoring third and higher differences.

This is an example of a *difference equation,* which contains *differences* rather than *differentials*, but difference equations are not most simply treated using the Δ-notation. Instead, the Δ s are removed, replacing $\Delta y_{\frac{1}{2}}$ by $(y_1 - y_0)$; $y_{-\frac{1}{2}}$ by $(y_0 - y_{-1})$; and $\Delta^2 y_0$ by $(y_1 - 2y_0 + y_{-1})$, so that equation 6.97 is rewritten;

$$\frac{u^2 + x^2}{h^2}(y_1 - 2y_0 + y_{-1}) + \frac{2x}{2h}(y_1 - y_{-1}) + Ay = 0 \qquad (6.98)$$

or

$$(u^2 + x^2 + xh)\,y_1 + [Ah^2 - 2(u^2 + x^2)]\,y_0 + (u^2 + x^2 - xh)\,y_{-1} = 0 \qquad (6.99)$$

and the terms we have ignored in Δ^3 give an estimate of the error.

6.5.2 Linear difference equations with constant coefficients

Not only may difference and differential equations be used as approximations for one another, but very similar methods are used to solve both. To illustrate this similarity, we will consider the set of linear difference equations with constant coefficients;

$$y_{m+n} + a_1 y_{m+n-1} + \ldots + a_{m-1} y_{n+1} + a_m y_n = F(x_n) \qquad (6.100)$$

where the a_is are constant and $F(x_n)$ is a function of x_n; and x_n is $x_0 + nh$, thus measuring x in relation to increments from x_0. We will look for a general solution in the form $y_n = y(x_0 + nh)$. A direct comparison may be drawn with equation 6.34 and the methods of sections 6.2.2 and 6.2.3 above. The operators E, E^r are defined as

$$Ey_n = y_{n+1}, E^r y_n = y_{n+r}; \qquad (6.101)$$

and the equation is solved by multiplying through by $1/p^{n+1}$ and summing for $n = 0$ to ∞. This step is the analogue of the Laplace transform, and has the same properties, namely:

$$\sum_0^\infty \frac{1}{p^{n+1}} E^r y_n = \frac{1}{p}y_1 + \frac{1}{p^2}y_2 + \frac{1}{p^3}y_3 + \ldots$$

$$= -y_0 + p\left[\frac{1}{p}y_0 + \frac{1}{p^2}y_1 + \frac{1}{p^3}y_2 + \frac{1}{p^4}y_3 + \ldots\right]$$

$$= -y_0 + p \sum_0^\infty \frac{1}{p^{n+1}}y_n \tag{6.102}$$

Similarly

$$\frac{1}{p^{n+1}} E^r y_n = -(p^{r-1}y_0 + p^{r-2}y_1 + \ldots + py_{r-2} + y_{r-1})$$

$$+ p^r \sum_0^\infty \frac{1}{p^{n+1}} y_n \tag{6.103}$$

Equations 6.102 and 6.103 are the exact analogues of equations 6.37 and 6.39, and substitution into equation 6.102 yields:

$$\phi(p) \sum_0^\infty \frac{1}{p^{n+1}} y_n = (p^{m-1}y_0 + p^{m-2}y_1 + \ldots + py_{m+2} + y_{m-1})$$

$$+ a_1 (p^{m-2}y_0 + p^{m-3}y_1 + \ldots + py_{m+3} + y_{m-2})$$

$$\cdot \quad \cdot \quad \cdot \quad \cdot \quad \cdot \quad \cdot \quad \cdot \quad \cdot \quad \cdot$$

$$+ a_{m-2}(py_0 + y_{-1})$$

$$+ a_{m-1}y_0$$

$$+ \Gamma(x_n) \tag{6.104}$$

where $\phi(p) = p^m + a_1 p^{m-1} + \ldots + a_{m-1} p + a_m$ (6.105)

This method allows us, as with the differential equations, to calculate an expression in p for $\sum_0^\infty \frac{1}{p^{n+1}} y_n$, and this can be inverted to a value for y_n by using the standard function table below:

As an example, let us solve the equation:

$$y_{n+2} - 2y_{n+1} + y_n = 1 \tag{6.106}$$

In the E notation, this can be written

$$(E^2 - 2E + 1)y_n = 0$$

TABLE 6.5

Laplace transform analogues for difference equations

$\sum_0^\infty p^{-(r+1)} y_r$	y_n
$\frac{1}{p-1}$	1
$\frac{1}{(p-1)}^{r+1}$	$\frac{n!}{r!(n-r)!}$
$\frac{1}{p-1}$	$a^n \; p > a$
$\frac{a}{p^2+a^2}$	$a^n \sin(n\frac{\pi}{2})$
$\frac{p}{p^2+a^2}$	$a^n \cos(n\frac{\pi}{2})$
$\frac{1}{p}e^{a/p}$	$\frac{a^n}{n!}$
$\frac{a}{p^2-a^2}$	$\frac{1}{2}a^n[1^n-(-1)^n]$
$\frac{p}{p^2-a^2}$	$\frac{1}{2}a^n[1^n+(-1)^n]$

Multiplying through by $1/p^{n+1}$ and summing from $n = 0$ to ∞, we obtain, using equation (6.104):

$$(p^2 - 2p + 1) \sum_{n=0}^{\infty} \frac{1}{p^{n+1}} y_n = (py_0 + y_1) - 2y_0 + \sum_{n=0}^{\infty} \frac{1}{p^{n+1}} .1 \qquad (6.108)$$

The last term on the right can be read from Table 6.5 giving:

$$(p-1)^2 \sum_{n=0}^{\infty} \frac{1}{p^{n+1}} y_n = py_0 + (y_1 - 2y_0) + \frac{1}{(p-1)} \qquad (6.109)$$

so that

$$\sum_{n=0}^{\infty} \frac{1}{p^{n+1}} y_n = \frac{py_0 + (y_1 - 2y_0)}{(p-1)^2} + \frac{1}{(p-1)^3} \qquad (6.110)$$

Expanding the first term on the left-hand side gives

$$\sum_{n=0}^{\infty} \frac{1}{p^{n+1}} y_n = \frac{y_0}{p-1} + \frac{(y_1 - y_0)}{(p-1)^2} + \frac{1}{(p-1)^3} \tag{6.111}$$

Referring again to Table 6.5, we obtain the general solution:

$$y_n = y_0 + n(y_1 - y_0) + \tfrac{1}{2}n(n-1) \tag{6.112}$$

This solution has two arbitrary constants, y_0 and y_1, and by analogy with the second-order differential equation, this must be the *unique* general solution.

6.6 Some examples of ordinary differential and difference equations

6.6.1 Hillslope profile development with a steadily down-cutting river

If a hillslope profile, represented by the curve OPQ in Figure 6.3, reaches

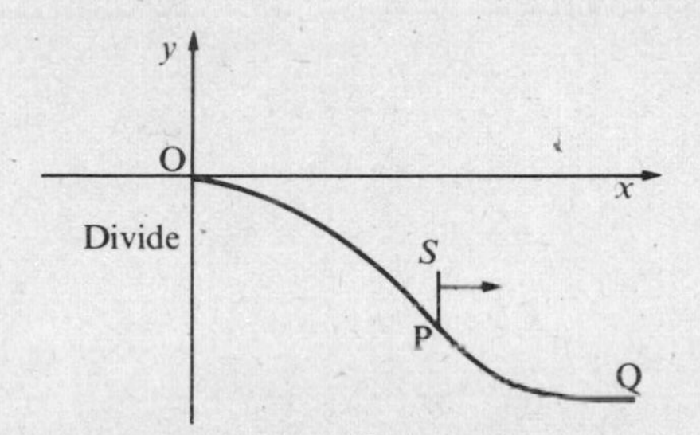

FIG. 6.3. Hillslope profile

equilibrium with a steady rate of river downcutting T then the whole profile is being lowered at rate T. Thus the rate of transport S of debris past the position P (co-ordinates x, y measured from the divide) is given by

$$S = Tx \tag{6.113}$$

Empirically, debris transport rates, can be expressed in the form

$$S = f(x)\left(-\frac{dy}{dx}\right) \tag{6.114}$$

where the function $f(x)$ represents the role of overland flow, so that a combination of creep and wash, say, might give

$$f(x) = \alpha(u^2 + x^2) \tag{6.115}$$

Combination of equations 6.113 and 6.114 gives the very simple differential equation

$$Tx = -\mathrm{f}(x)\frac{\mathrm{d}y}{\mathrm{d}x} \tag{6.116}$$

which can be integrated in principle, for a given f(x), as:

$$y = -T\int \frac{x}{\mathrm{f}(x)}\mathrm{d}x \tag{6.117}$$

For the value of f(x) in equation 6.115, this gives:

$$y = \frac{-\mathrm{T}}{\alpha}\int \frac{x}{\mathrm{u}^2 + x^2}\,\mathrm{d}x \tag{6.118}$$

This form can be converted to a standard integral by substitution of

$$z = \mathrm{u}^2 + x^2 \tag{6.119}$$

$$\therefore \mathrm{d}z = 2x.\mathrm{d}x \tag{6.120}$$

so that

$$y = \frac{-T}{2\alpha}\int \frac{\mathrm{d}z}{z}$$

$$= \frac{-T}{2\alpha}\log z + \mathrm{c}$$

$$= \frac{-T}{2\alpha}\log(\mathrm{u}^2 + x^2) + \mathrm{c} \tag{6.121}$$

Since when $x = 0, y = 0$ at the divide, we have

$$0 = \frac{-T}{2\alpha}\log \mathrm{u}^2 + \mathrm{c} \tag{6.122}$$

so that

$$y = \frac{-T}{2\alpha}[\log(\mathrm{u}^2 + x^2) - \log \mathrm{u}^2]$$

$$= \frac{-T}{2\alpha}\log\left(1 + \frac{x^2}{\mathrm{u}^2}\right), \tag{6.123}$$

the particular solution required in this case, which is illustrated in Figure 6.4.

6.6.2 Freezing of water in a lake or in a uniform soil from above

If freezing conditions persist, an increasing thickness of water will freeze from the surface downwards (Casagrande, 1931). Given a sufficient

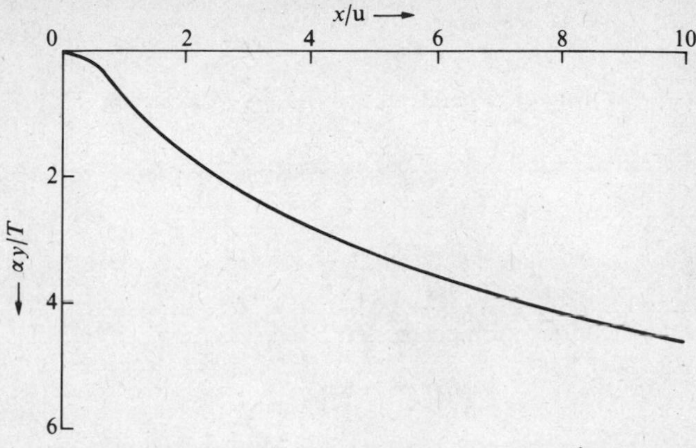

FIG. 6.4. Particular solution of $Tx = -f(x)\frac{dy}{dx}$

moisture supply, then the rate of advance of the freezing front depends on the rate at which heat can be carried away to the surface (figure 6.5). If it may be assumed that the ice or frozen soil has a constant ice content

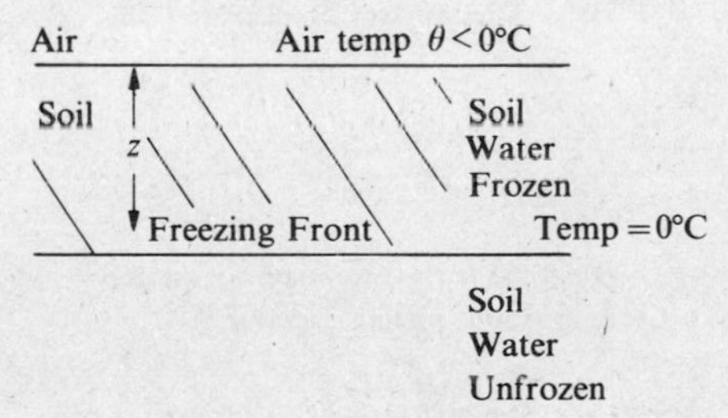

FIG. 6.5. Freezing of water in soil

and constant thermal characteristics, the rate of heat flow to the surface;

$$Q = K\frac{0-\theta}{z} \tag{6.124}$$

where K is a thermal constant. This quantity of heat can freeze water at a rate

$$\frac{dz}{dt} = \frac{Q}{L}, \tag{6.125}$$

where L is the latent heat involved in freezing water to ice. Eliminating Q from equations 6.124 and 6.125, we obtain the differential equation

$$L\frac{dz}{dt} = -K\frac{\theta}{z} \tag{6.126}$$

which can be solved by separation of variables (see section 6.2.1. above):

$$z\,dz = -\frac{K\theta}{L}\,dt \tag{6.127}$$

$$\int z\,dz = -\frac{K}{L}\int \theta\,dt \tag{6.128}$$

Performing the integration on the left-hand side.

$$\tfrac{1}{2}z^2 = -\frac{K}{L}\int \theta\,dt + c \tag{6.129}$$

If the thickness is zero at time $t = 0$, then

$$z^2 = \frac{2K}{L}\int_0^t -\theta\,dt' \tag{6.130}$$

The right-hand side is the accumulated temperature over time, that is the number of day-degrees below freezing point H.
Thus

$$z = \sqrt{\left(\frac{2K}{L}H\right)} \tag{6.131}$$

is the ice thickness developed during a freezing period which has, at that time, accumulated H day-degrees below 0° C.

6.6.3 Length probabilities for a queue in a steady state

For a simple queueing model, it is assumed that arrival (increasing the queue-length by 1) and service (reducing queue-length by 1) are random events with average rates α and σ respectively (Cox and Smith, 1961). If queue lengths are to remain finite, it is clearly essential that $\alpha < \sigma$, and the ratio α/σ is called the 'traffic intensity', and denoted by ρ. Let us look at the probability that there might be n in the queue at some time, which we will write P_n. This could arise in three ways:

(a) There were previously $(n + 1)$ in queue, and one has been served.
(b) There were previously $(n - 1)$,, ,, ,, ,, ,, arrived.
(c) There were previously n ,, ,, ,, nobody has arrived or been served.

Writing down the probabilities of these mutually exclusive alternatives, which must together add up to P_n (see Chapter 7 for a discussion of this rule for combining probabilities);

$$P_n = P_{n+1}\,\sigma + P_{n-1}\,\alpha + P_n(1-\alpha-\sigma) \qquad (6.132)$$

Rearranging, we have a second-order linear difference equation,

$$\sigma P_{n+1} - (\alpha+\sigma)P_n + \alpha P_{n-1} = 0. \qquad (6.133)$$

This relationship breaks down at $n = 0$, since possibility (b) above cannot occur, and in possibility (c), service is impossible. Thus at $n = 0$:

$$P_0 = P_1 \cdot \sigma + P_0(1-\alpha) \qquad (6.134)$$

or

$$P_1 = \frac{\alpha}{\sigma}P_0 = \rho_0 \qquad (6.135)$$

This condition removes one arbitrary constant; and the other will be obtained by recalling that there must be *some* number in the queue, so that

$$\sum_{n=0}^{\infty} P_n = 1 \qquad (6.136)$$

Applying the method of section 6.5.2 to equation 6.133

$$[\sigma p^2 - (\alpha+\sigma)\,p + \alpha]\sum_0^{\infty}\frac{1}{p^{n+1}}P_n = \sigma(P_0 p + P_1) - (\alpha+\sigma)\,P_0 \qquad (6.137)$$

so that

$$\sum^{\infty}\frac{1}{p^{n+1}}P_n = \frac{pP_0 + P_1 - \left(1+\frac{\alpha}{\sigma}\right)P_0}{p^2 - \left(1+\frac{\alpha}{\sigma}\right)p + \frac{\alpha}{\sigma}} \qquad (6.138)$$

Replacing α/σ by ρ, the traffic intensity, and P_1 by ρP_0 (from equation 6.135);

$$\sum^{\infty}\frac{1}{p^{n+1}}P_n = \frac{pP_0 - P_0}{(p-1)(p-\rho)} = \frac{P_0(p-1)}{(p-1)(p-\rho)} = \frac{P_0}{p-\rho} \qquad (6.139)$$

Reference to Table 6.5 shows that

$$P_n = P_0\rho^n \qquad (6.140)$$

Summing the P_ns to apply the criterion of equation 6.136

$$\sum_{n=0}^{\infty} P_n = P_0 \sum^{\infty} \rho^n = \frac{P_0}{1-\rho} \text{ (as } \rho < 1) \qquad (6.141)$$

so that
$$P_0 = (1-\rho), \qquad (6.142)$$

and
$$P_n = (1-\rho)\rho^n \qquad (6.143)$$

Table 6.6 shows that long queues become increasingly probable as ρ approaches 1; that is as arrival rate approaches service rate

TABLE 6.6

Queue length probabilities

	P_0	P_1	P_2	P_3	P_4	P_5	mean queue lengths
ρ = 0·1	0·90	0·09	0·0009				0·11
0·5	0·50	0·25	0·125	0·062	0·031	0·016	1·00
0·7	0·30	0·21	0·147	0·103	0·072	0·050	2·33
0·8	0·110	0·09	0·081	0·073	0·066	0·059	9·00

6.7. Partial Differential equations

6.7.1. Some contexts and examples

Models of a developing form, whether a city or a hillside, a traffic flow or a glacier, are concerned with a quantity which varies in at least one spatial dimension, *and* over time. There are therefore two or more independent variables, and a differential equation connecting them is almost certain to be a partial differential equation, containing partial derivatives relating to both space and time variables. Very few partial differential equations can be completely solved, and instead we tend to look for special classes of solution which are appropriate to particular problems. This is not as restrictive as it sounds because a first-order partial differential equation contains, not one arbitrary constant, but one arbitrary *function.*

It is because of the need to look for solutions in the context of a problem, that this survey of partial equations begins by examining some contexts in which they commonly arise for geographers. At present, most of the examples are from physical geography and are variants on the 'continuity equation', and this provides a framework on which to hang particular contexts.

6.7.2. The continuity equation in one-dimension

The continuity equation is a statement of conservation of matter, as it is transferred through space over time. As such it is a statement of the formal relationship between space and time distributions, which is central to many geographical problems.

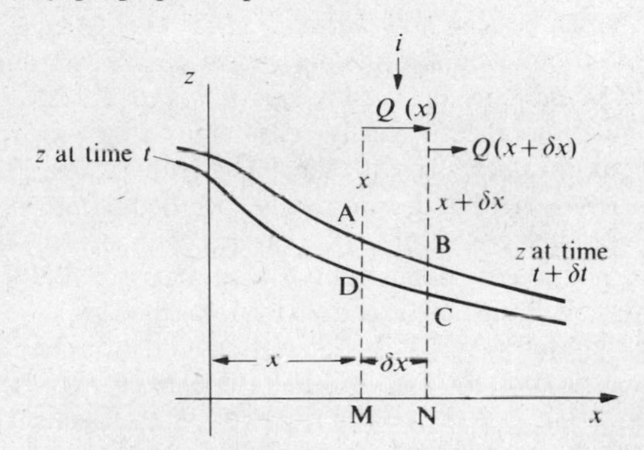

FIG. 6.6. The one-dimensional continuity equation

In Figure (6.6) q(x) is the quantity of matter being transported in the positive x-direction and i is the rate at which material is being added from outside the system (which may vary with x and t). Then for the small unit ABCD,

$$\text{input} - \text{output} = \text{increase of matter within ABCD.} \qquad (6.144)$$

In symbols, over the time period δt;

$$[Q(x) - Q(x + \delta x)]\ \delta t + i\delta x\delta t = \delta z\delta x. \qquad (6.145)$$

On the left-hand side, the first term represents the net excess of material transported in $Q(x)\delta t$, over material transported out, $Q(x + \delta x)\delta t$. The second term represents the accumulation of material from outside over the length of the unit (MN $= \delta x$). The term on the right-hand side is the increase of matter within the unit, associated with the growth of z. Dividing through by δx and δt;

$$\frac{Q(x) - Q(x + \delta x)}{\delta x} + i = \frac{\delta z}{\delta t} \qquad (6.146)$$

If δx and δt are now allowed to become small and tend to zero, it can be seen that the equation tends towards the exact form:

$$-\frac{\partial Q}{\partial x} + i = \frac{\partial z}{\partial t} \qquad (6.147)$$

or

$$\frac{\partial Q}{\partial x} + \frac{\partial z}{\partial t} = i \tag{6.148}$$

This equation is the standard form of the continuity equation, to which we will refer. It can be seen however, that it has two dependent variables, z and Q, and so is not even in principle a soluble partial differential equation until Q and i are specified in a particular context.

The accumulation rate i is usually considered to vary with position x and time t but not normally with the dependent variable z. The transport rate Q on the other hand, is normally specified in terms of x, z and its derivates, although commonly not in terms of time – that is to say that transport rate usually depends on the geometry of the system.

Alternatively the transport rate may itself be the focus of interest, and z expressed in terms of Q. These additional relationships specify the systems we are actually studying, and the exact form of the partial differential equation, but the continuity equation imposes a common framework, which produces some similarities in the types of solution. Before examining ways of solving this equation, it will help to look more closely at some of the physical contexts in which the continuity equation arises.

6.7.3. Hillslope profile development

The development of a slope profile must closely follow the continuity equation, with conservation of mass as a necessary accounting procedure. In this case, in the absence of uplift, there is no accumulation of material from outside, so that $i = 0$, giving a simplification of the equation. In Figure 6.7, the transport rate S, is the rate of debris movement, which

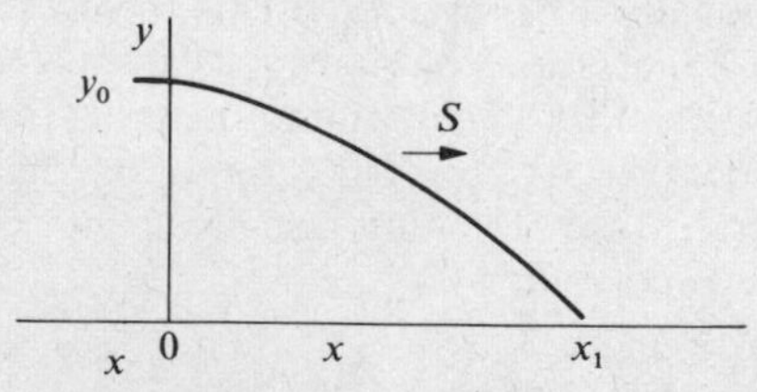

FIG. 6.7. Slope profile development

is thought to depend on slope gradient, $-(\partial y/\partial x)$, and on distance of overland flow, (x), so that a fairly general form for S is:

$$S = f(x)\left(\frac{-\partial y}{\partial x}\right) \tag{6.149}$$

The continuity equation then becomes

$$\frac{\partial S}{\partial x} + \frac{\partial y}{\partial t} = \frac{\partial}{\partial x}\left[\mathrm{f}(x)\left(-\frac{\partial y}{\partial x}\right)\right] + \frac{\partial y}{\partial t} = 0 \qquad (6.150)$$

A variety of 'initial' and 'boundary' conditions might be suggested to obtain a solution to this equation, but a natural and simple set appears to be:
(a) Divide fixed at $x = 0$; that is:

$$S = 0 \text{ at } x = 0 \qquad (6.151)$$

(b) Base-level fixed at $x = x_1$, and base-level elevation specified over time; that is:

$$\text{at } x = x_1, y = y_1(t) \qquad (6.152)$$

The simplest case is of course a fixed base level elevation,

$$y = 0 \text{ at } x = x_1 \qquad (6.153)$$

(c) An 'initial form' is specified at time $t = 0$, that is the profile from which development begins, is

$$y = y_0(x) \text{ at } t = 0 \qquad (6.154)$$

Inspection of the equations and consideration of the geomorphic literature suggest at least two types of solution to look for – a Davisian slope-decline; and a Penck/King parallel retreat (see, for example, Carson and Kirkby, 1972). These might be represented simply in the forms:

$$\text{Slope decline: } y = y(x)\,\mathrm{e}^{-kt} \qquad (6.155)$$

$$\text{Parallel retreat: } y = y(x - ct) \qquad (6.156)$$

A third type of simple solution can be seen for a base-level downcutting at a constant rate T; with elevation z measured above the (downcutting) basal point. In this case

$$z = y - Tt \qquad (6.157)$$

Substituting in equation 6.150

$$\frac{\partial S}{\partial x} + \frac{\partial z}{\partial t} - T = 0 \qquad (6.158)$$

and we can look for a *steady-state* solution.

$$\frac{\partial S}{\partial x} = T, \text{ or } S = Tx \qquad (6.159)$$

This case has already been considered as an example of an ordinary differential equation in section 6.6.1 above.

6.7.4. Saturated water flow

In a hydrological context, the continuity equation is well-known as the *storage equation.* For flow down a hillside the continuity equation 6.148 can be interpreted for flow across a unit-width of the surface;

$$\frac{\partial Q}{\partial x} + \frac{\partial z}{\partial t} = i \qquad (6.160)$$

where Q = water discharge/unit-width
z = flow depth
i = net effective precipitation rate (rainfall − evaporation − infiltration).

In this case, the precipitation rate can usually be taken as spatially uniform, but varying over time. The discharge is related to flow depth by a 'rating function', which is usually of the form

$$Q = \alpha z^n \text{ for } 1 \leqslant n \leqslant 3. \qquad (6.161)$$

Boundary and initial conditions are written most simply as:

$$Q = 0 \text{ at } x = 0 \qquad (6.162)$$

$$z \to 0 \text{ as } t \to \infty \text{ when } i = 0 \qquad (6.163)$$

$$z = z_0(x) \text{ at } t = 0 \qquad (6.164)$$

The first and third of these conditions are similar to those for the hillslope case. The second condition has a similar role, in determining the overall level of matter (in this case, water) at the end of the development sequence.

The appropriate solutions in this case appear to be:

Steady state: for average rainfall inputs over a period, the steady-state solution gives average (and hence essentially low-flow) water levels.

$$\frac{\partial Q}{\partial x} = i \text{ or } Q = ix \qquad (6.165)$$

Floodwave: in a rainstorm a wave of water moves down slope as a hydrograph peak, at velocity c, so that we can look for solutions

$$Q = Q(x - ct) \qquad (6.166)$$

The analogue of the decline solution for slope profiles represents the

levelling out of an uneven mass of water. This occurs so rapidly that dynamic effects must also be taken into account, and is of less relevance than in the hillslope case. Equation 6.150 can be modified for streamflow by changing x to drainage area A; z to channel-cross-sectional area a; and including in the i term both channel precipitation, and flows from the stream banks, q.
Thus

$$\frac{\partial Q}{\partial A} + \frac{\partial z}{\partial a} = (i + q) \tag{6.167}$$

6.7.5. Unsaturated infiltration of soil water

During downward vertical infiltration, in the positive direction, there is no accumulation, and moisture content m varies according to the continuity equation

$$\frac{\partial Q}{\partial z} + \frac{\partial m}{\partial t} = 0 \tag{6.168}$$

Flow is controlled by Darcy's law:

$$Q = -K\frac{\partial \psi}{\partial z} \tag{6.169}$$

where K is the (moisture-dependent) hydraulic conductivity, and ψ is total hydraulic potential,

$$\psi = \phi - z \tag{6.170}$$

where z is the gravitational potential and ϕ is the hydraulic potential (negative in tension), so that

$$\begin{aligned} Q &= K - K\frac{\partial \phi}{\partial z} \\ &= K - \left(K\frac{\partial \phi}{\partial m}\right)\frac{\partial m}{\partial z} \end{aligned} \tag{6.171}$$

The quantity $K\,(\partial\phi/\partial m)$ is moisture dependent, and is commonly written as D the hydraulic diffusivity, so that

$$Q = K - D\frac{\partial m}{\partial z} \tag{6.172}$$

Substituting in equation (6.168), we have:

$$\frac{\partial K}{\partial z} - \frac{\partial}{\partial z}\left(D\frac{\partial m}{\partial z}\right) + \frac{\partial m}{\partial t} = 0 \tag{6.173}$$

Boundary and initial conditions usually specify:
(a) Input of water at the soil surface:

$$m = m_1(t) \text{ at } x = 0 \tag{6.174}$$

At its simplest this condition is used for applying water at the surface for an indefinite period starting at time $t = 0$; ie

$$\left.\begin{aligned} m &= 0, z > 0, t < 0 \\ m &= m_{sat}, z = 0, t \geqslant 0 \end{aligned}\right\} \tag{6.175}$$

(b) Behaviour of soil at depth tends to a steady value

$$m \to m_1 \text{ as } x \to \infty \text{ for all } t. \tag{6.176}$$

(c) Initial moisture distribution:

$$m = m_0(t) \text{ at } t = 0 \tag{6.177}$$

The most commonly examined solution is for maximum infiltration starting at time $t = 0$. Although this solution is complex in detail because of the moisture-dependence of K, D, it broadly shows the advance of a sharp wetting front at a declining rate, and with some loss of sharpness of definition. This behaviour can be described by the terms of a solution like:

$$m = m_s \operatorname{erf}\left(\frac{z}{2(Dt)^{\frac{1}{2}}}\right) \tag{6.178}$$

where erf(x) is the error function, which is related to the integral of the normal curve; and is defined by.

$$\operatorname{erf} x = \frac{4}{\pi}\,\tfrac{1}{2}\int_0^x e^{\zeta^2}\,d\zeta \tag{6.179}$$

This type of solution is called, unsurprisingly, an *error-function* solution.

6.7.6. Glacier flow

Glacier mass is conserved if allowance is made for net accumulation, (Nye, 1959), a (negative for ablation).

$$\frac{\partial Q}{\partial x} + \frac{\partial h}{\partial t} = a, \tag{6.180}$$

where h is glacier thickness.
The net accumulation rate can most simply be considered as constant; and the flow rate has been shown to approximate to:

$$Q = -\alpha \cdot h \cdot \frac{\partial h}{\partial x} \qquad (6.181)$$

Substituting into equation (6.180),

$$-\alpha \frac{\partial}{\partial x}\left(h \frac{\partial h}{\partial x}\right) + \frac{\partial h}{\partial t} = a \qquad (6.182)$$

Boundary and initial conditions are most simply:

$$Q = 0 \text{ at } x = 0; \qquad (6.183)$$

$$h \to 0 \text{ as } x \to \infty \text{ or } h = 0 \text{ at } x = x_1 \qquad (6.184)$$

$$h = h_0(x) \text{ at } t = 0 \qquad (6.185)$$

The most useful solutions are either the steady-state profile of the glacier (Nye 1959)

$$-\alpha h \frac{\partial h}{\partial x} = \int a \ dx \qquad (6.186)$$

or the response of the glacier to a sudden or cyclic climatic change (Nye 1960) which can be treated as a travelling wave on the surface of the above equilibrium form.

6.7.7. *Heat conduction in the ground*

As air temperature varies, ground temperature at depth z responds, and heat energy and hence temperature (for a uniform soil) is conserved. Thus the continuity equation is:

$$\frac{\partial Q}{\partial z} + \frac{\partial \theta}{\partial t} = 0 \qquad (6.187)$$

where θ is ground temperature.
In this version of the equation, there is no inflow of heat within the soil, and it is assumed that no freeze-thaw is occurring (cf section 6.6.2. above). The heat flow Q is described by:

$$Q = -K \frac{\partial \theta}{\partial z} \qquad (6.188)$$

where K is a thermal conductivity.

Thus

$$\frac{\partial \theta}{\partial t} = K \frac{\partial \theta}{\partial x^2} \tag{6.189}$$

which is the classic 'diffusion equation'.
Boundary and initial conditions are commonly stated as:

$$\text{(a) } \theta = \theta_1(t) \text{ at } x = 0 \tag{6.190}$$

$$\text{(b) } \theta \rightarrow \bar{\theta} \text{ as } x \rightarrow \infty \tag{6.191}$$

$$\text{(c) } \theta = \theta_0(x) \text{ at } t = 0 \tag{6.192}$$

Since air temperature varies cyclically through the day and through the year, and ground temperature becomes more uniform with depth, the most valuable solution is of the form:

$$\theta = \bar{\theta} + \theta_2 \mathrm{e}^{-\lambda x} \sin(\omega t + x) \tag{6.193}$$

for constant $\bar{\theta}, \theta_2, \lambda, \omega$.

6.7.8. Diffusion processes

The diffusion equation, with its applications over a wide range of physical and human geography applications, generally has implications of continuity, but also arises in the context of a random walk. Random walks are discussed more fully in Chapter 8, but a brief derivation of the diffusion equation is included here for completeness. (Comparisons may also be made with section 6.6.3).

Particles are free to move one step of length δ forwards or back in the x-direction, and take r steps in unit time. The probability of stepping in the positive direction is p, and in the negative direction is q. Then inflow rate − out flow rate = rate of increase in number at x. Thus, the inflow rate is the rate of movement from

$$(x - \delta) \text{ and } (x + \delta) \text{ to } x: \tag{6.194}$$

$$\text{Inflow rate} = r[\mathrm{p}c(x - \delta) + \mathrm{q}c(x + \delta)] \tag{6.195}$$

where $c(x)$ is the concentration of particles at x; and the outflow rate is the rate of movement to $(x - \delta)$ and $(x + \delta)$ from x:

$$\text{Outflow rate} = r[\mathrm{q}c(x) + \mathrm{p}c(x)] \tag{6.196}$$

substituting into equation 6.194;

$$r\{p[c(x-\delta)-c(x)]+q[c(x+\delta)-c(x)]\} = \frac{\partial c}{\partial t}\frac{\partial c}{\partial t} \quad (6.197)$$

Expanding by Taylor's theorem (section 5.10.1);

$$c(x+\delta) = c(x)+\delta\frac{\partial c}{\partial x}+\tfrac{1}{2}\delta^2\frac{\partial^2 c}{\partial x^2}+\ldots \quad (6.198)$$

$$c(x-\delta) = c(x)-\delta\frac{\partial c}{\partial x}+\tfrac{1}{2}\delta^2\frac{\partial^2 c}{\partial x^2}-\ldots \quad (6.199)$$

and substituting into equation 6.197;

$$r\left[p\left\{-\delta\frac{\partial c}{\partial x}+\tfrac{1}{2}\delta^2\frac{\partial^2 c}{\partial x^2}\right\}+q\left\{\delta\frac{\partial c}{\partial x}+\tfrac{1}{2}\delta^2\frac{\partial^2 c}{\partial x^2}\right\}+\ldots\right] = \frac{\partial c}{\partial t} \quad (6.200)$$

or

$$-(p-q)\delta\ r\frac{\partial c}{\partial x}+\tfrac{1}{2}(p+q)\ \delta^2\ r\ \frac{\partial^2 c}{\partial x^2}+\ldots = \frac{\partial c}{\partial t} \quad (6.201)$$

In the limiting case with very many, very small steps, we let $\delta \to 0$, and $r \to \infty$ in such a way that

$$(p-q)\,\delta r \to v;\ (p+q)\,\delta^2 r \to D; \quad (6.202)$$

So that we obtain the general diffusion equation

$$-v\frac{\partial c}{\partial x}+\tfrac{1}{2}D\frac{\partial^2 c}{\partial x^2} = \frac{\partial c}{\partial t} \quad (6.203)$$

in which v is called the 'drift velocity' and D is called the 'diffusivity'. If, v, D are not constants, then the equation should be rewritten as:

$$-\frac{\partial}{\partial x}(vc)+\tfrac{1}{2}\frac{\partial}{\partial x}\left(D\frac{\partial c}{\partial x}\right) = \frac{\partial c}{\partial t} \quad (6.204)$$

6.7.9. Types of solution of the continuity equation

From the previous examples, a number of types of solution can be seen as appropriate to equations of the continuity type. These are explored in the following sections. Each group of methods produces a set of solutions which are appropriate for particular sets of boundary and initial conditions. None produces a complete solution, and all are essentially devices reducing the partial differential equation to a subsidiary ordinary differential equation.

6.7.10. Steady state solutions

If a process reaches a steady state with inflow equal to outflow, then we can see what it will be by setting $\partial z/\partial t = 0$ in the continuity equation.
It then becomes:

$$\frac{\partial Q}{\partial x} = i \qquad (6.148)$$

This is an ordinary differential equation, with solution

$$Q = \int i \mathrm{d}x \qquad (6.205)$$

provided that i is independent of time.
In the glacier case, for example with a constant, from equation 6.182

$$\alpha h \frac{\mathrm{d}h}{\mathrm{d}x} = - ax + \mathrm{B} \qquad (6.206)$$

and integrating again

$$\tfrac{1}{2}\alpha h^2 = -\tfrac{1}{2}ax^2 + \mathrm{B}x + \mathrm{C} \qquad (6.207)$$

With the boundary condition that Q and hence $\mathrm{d}h/\mathrm{d}x$ is zero at $x = 0$ and $h = 0$ at $x = x_1$ the glacier snout (figure 6.8);

$$h = \sqrt{\frac{\mathrm{a}}{\alpha}}\sqrt{x_1^2 - x^2} \qquad (6.208)$$

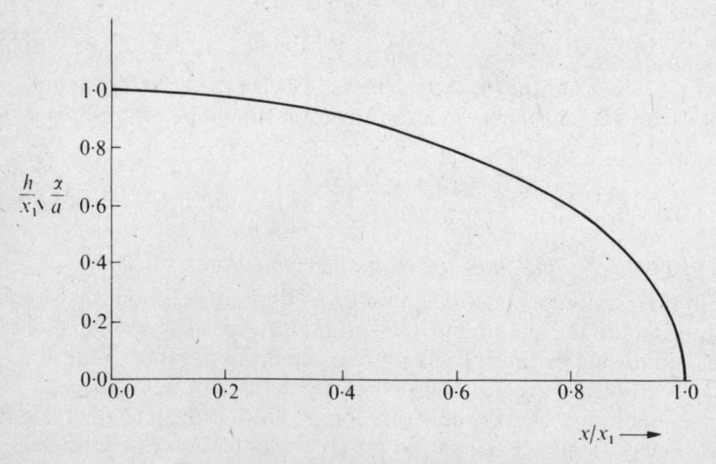

FIG. 6.8. Profile of an equilibrium glacier

In the hillslope case, a type of steady state solution can be attained where the basal point, and hence the entire profile, is downcutting at a uniform rate, T; and the accumulation rate i is zero or constant. We can then put

$$\mathrm{u} = z - Tt. \tag{6.209}$$

and

$$\frac{\partial Q}{\partial x} + \frac{\partial \mathrm{u}}{\partial t} = (T + i) \tag{6.210}$$

Our new quantity u is measured relative to the reducing basal point, and

$$Q = (T + i)\,x \tag{6.211}$$

6.7.11. Separation of variables

Solutions which decay over time, and a few others, may be obtained by seeking solutions to the continuity equation in the form

$$z = \mathrm{X}(x)\,\mathrm{T}(t) \tag{6.212}$$

where X, T are functions of x, t alone.
If the continuity equation leads to a linear equation in z, $\partial z/\partial x$, $\partial^2 z/\partial x^2$ we have:

$$\mathrm{b}_0 \frac{\partial^n z}{\partial x^n} + \mathrm{b}_1 \frac{\partial^{n-1} z}{\partial x^{n-1}} + \ldots + \mathrm{b}_{n-1} \frac{\partial z}{\partial x} + \mathrm{b}_n z + \frac{\partial z}{\partial t} = i \tag{6.213}$$

where the b s are functions of x alone.
Substituting equation 6.212 as a trial solution for the case of zero accumulation rate ($i = 0$):

$$\mathrm{b}_0 \frac{\mathrm{d}^n \mathrm{X}}{\mathrm{d}x^n} + \mathrm{b}_1 \frac{\mathrm{d}^{n-1} \mathrm{X}}{\mathrm{d}x^{n-1}} + \ldots + \mathrm{b}_{n-1} \frac{\mathrm{dX}}{\mathrm{d}x} + \mathrm{b}_n \mathrm{X} \quad \mathrm{T} + \mathrm{X} \frac{\mathrm{dT}}{\mathrm{d}t} = 0 \tag{6.214}$$

or

$$\frac{\mathrm{b}_0 \frac{\mathrm{d}^n \mathrm{X}}{\mathrm{d}x^n} + \mathrm{b}_1 \frac{\mathrm{d}^{n-1} \mathrm{X}}{\mathrm{d}x^{n-1}} + \ldots + \mathrm{b}_{n-1} \frac{\mathrm{dX}}{\mathrm{d}x} + \mathrm{b}_n \mathrm{X}}{\mathrm{X}} = -\frac{\mathrm{dT}}{\mathrm{d}t} \Big/ \mathrm{T} \tag{6.215}$$

Now the left-hand side of this equation is a function of x alone, and the right-hand side a function of t alone. Thus both can only be a constant.

For a time-decaying solution as is usually required, this constant, k will be positive. Thus

$$-\frac{\mathrm{dT}}{\mathrm{d}t}\Big/\mathrm{T} = \mathrm{k} \tag{6.216}$$

and

$$\mathrm{T} = \mathrm{Ae}^{-\mathrm{k}t} \tag{6.217}$$

while

$$\mathrm{b}_0\frac{\dfrac{\mathrm{d}^n x}{\mathrm{d}x^n} + \ldots. + \mathrm{b}_n\mathrm{X}}{\mathrm{X}} = \mathrm{k} \tag{6.218}$$

and

$$\mathrm{b}_0\frac{\mathrm{d}^n\mathrm{X}}{\mathrm{d}x^n} + \ldots + (\mathrm{b}_n - \mathrm{k})\,\mathrm{X} = 0 \tag{6.219}$$

an ordinary linear differential equation, which can be solved by the methods of sections 6.32 to 6.34. A more general solution to equation 6.213 will be built up as a series:

$$z = \sum_{\mathrm{k}} \mathrm{A_k X}(\mathrm{k}, x)\,\mathrm{e}^{-\mathrm{k}t} \tag{6.220}$$

where the constants k are chosen to satisfy the boundary conditions at large x (or at $x = x_1$) and the constants A are chosen to satisfy the initial condition $z = z_0(x)$ at $t = 0$. Where the accumulation rate i is non-zero, then a particular solution $z_0(x, t)$ to the equation must first be found. Substitution of $\mathrm{u} = z - z_0(x, t)$ will then reduce the equation to the form of equation 6.213.

As an example of this type of solution, consider the hillslope development equation 6.141 for the case:

$$S = -\alpha(\mathrm{u}^2 + x^2)\frac{\partial y}{\partial x} \tag{6.221}$$

Substitution into equation 6.150 gives:

$$\frac{\partial}{\partial x}\left\{-\alpha\,(\mathrm{u}^2 + x^2)\frac{\partial y}{\partial x}\right\} + \frac{\partial y}{\partial t} = 0 \tag{6.222}$$

so that

$$(\mathrm{u}^2 + x^2)\frac{\partial^2 y}{\partial x^2} + 2x\frac{\partial y}{\partial x} = \frac{1}{\alpha}\frac{\partial y}{\partial t} \tag{6.223}$$

Substituting $y = X(x).\ T(t)$ as above leads to:

$$\frac{(u^2 + x^2)\dfrac{d^2X}{dx^2} + 2x\dfrac{dX}{dx}}{X} = -\frac{1}{\alpha}\frac{dT}{dt}\Big/T \qquad (6.224)$$

or

$$T = t_0 e^{-A\alpha t} \qquad (6.225)$$

for positive A, and

$$(u^2 + x^2)\frac{d^2X}{dx^2} + 2x\frac{dX}{dx} + AX = 0 \qquad (6.226)$$

This is equation 6.54 above, solutions to which are illustrated in Fig. 6.2. Relevant values of A are those which satisfy the boundary conditions at the base-level point.

Thus, for example, for a base-level of constant elevation at $x_1/u = 0{\cdot}72$, reference to Figure 6.2 shows that successive values of A for which $z = 0$ at $x = x_1$ are approximately 6 and 50. Since the exponential decay terms are thus $e^{-6\alpha t}$ and $e^{-50\alpha t}$, it can be seen that the first term very rapidly comes to dominate the overall solution to the equation.

6.7.12. Laplace Transform solutions

The general linear equation 6.213 can also be approached using Laplace Transforms. As before (section 6.3.2) we multiply through by e^{-pt} (using t instead of x as before) and integrate over the range $t = 0$ to ∞. This leaves the terms in x on the left-hand side almost alone, as

$$\int_0^\infty b_{n-r}\frac{\partial r_z}{\partial x^r}e^{-pt}\,dt = b_{n-r}\frac{\partial r}{\partial x^r}\int_0^\infty ze^{-pt}\,dt \quad \text{and so on} \qquad (6.227)$$

Also

$$\int_0^\infty \frac{\partial z}{\partial t}e^{-pt}\,dt = -z_0(x) + p\int_0^\infty ze^{-pt}\,dt \qquad (6.228)$$

Writing $\bar{z}$ for $\int_0^\infty ze^{-pt}\,dt$; we now have a subsidiary ordinary differential equation to solve:

$$b_0\frac{d^n\bar{z}}{dx^n} + b_1\frac{d^{n-1}\bar{z}}{dx^{n-1}} + \ldots + b_{n-1}\frac{d\bar{z}}{dx} + b_n\bar{z} + p\bar{z} = Z_0(x) + \int_0^\infty i\cdot e^{-pt}\,dt$$

(6.229)

This equation differs from equation 6.219 in the sign of the final term ($p\bar{z}$ in 6.229; $-kX$ in 6.219, and in the inclusion of an accumulation term in the subsidiary equation ($\int_0^\infty i \cdot e^{-pt}\, dt$). If this subsidiary equation can be formally solved for $\bar{z}$, then the Laplace transform can be inverted to give z.*

This method and the method of separating variables, though producing similar subsidiary equations, are alternatives which are appropriate for somewhat different boundary and initial conditions. The Laplace Transform solution, although more general, has the disadvantage that it cannot readily be inverted for all functions.

6.7.13. Kinematic wave solutions

In hydrology especially, we wish to concentrate on a travelling wave solution, and we seek a solution to the continuity equation which follows the wave form (Wooding, 1965). In general, we have the continuity equation 6.148.

$$\frac{\partial Q}{\partial x} + \frac{\partial z}{\partial t} = i$$

together with a transport or rating function

$$Q = Q\left(x, z, \frac{\partial z}{\partial x}, \text{etc}\right) \qquad (6.230)$$

In the very simplest linear case, where

$$Q = cz, i = 0 \qquad (6.231)$$

we have one of the rare partial differential equations for which a complete solution exists:

$$c\frac{\partial z}{\partial x} + \frac{\partial z}{\partial t} = 0 \qquad (6.232)$$

and the solution is $\qquad z = f(x - ct) \qquad (6.233)$

where f is an arbitrary *function.* The idea of an arbitrary function is difficult to grasp, because it seems to allow total freedom. A moment's

*Formally, the Laplace transform can be inverted by contour integration of the complex variable

$$z = \frac{1}{2\pi i}\int_{\gamma-1\infty}^{\gamma+1\infty} e^{\lambda t} z(\lambda)\, d\lambda$$

See for example, Carslaw and Jaeger, 1941, Chapter 4.

thought, however, shows that the function is specified by the distribution at $t = 0$;

$$z = \mathrm{f}(x) \tag{6.234}$$

At time t later, every value of z at $x = x_0$ is to be found at a new position, x' such that $x = x' - ct$. Equation 6.224 thus represents a

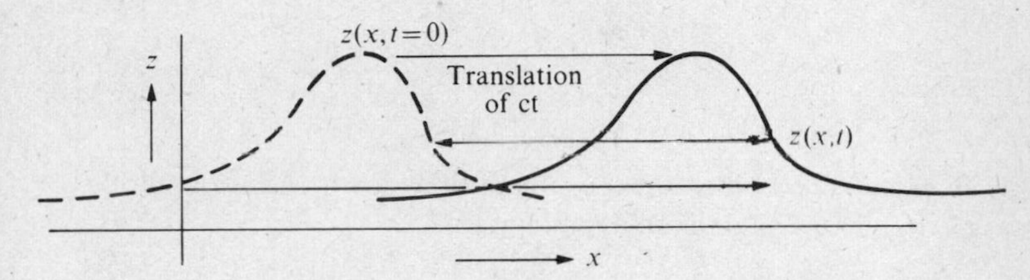

FIG. 6.9. A wave solution to a partial differential equation

uniform translation at velocity c in the direction of increasing x (Fig 6.9). By analogy with this solution, a kinematic wave solution has been developed (Lighthill and Whitham, 1955) for the more general case. Comparing equations 6.148 and 6.232, we write intuitively,

$$\frac{\partial Q}{\partial x} + \frac{\partial z}{\partial t} = i \tag{6.148}$$

so that

$$\frac{\partial Q}{\partial z}\frac{\partial z}{\partial x} + \frac{\partial z}{\partial t} = i \tag{6.235}$$

or

$$\mathrm{c}\frac{\partial z}{\partial x} + \frac{\partial z}{\partial t} = i \tag{6.236}$$

where

$$\frac{\mathrm{d}x}{\mathrm{d}t} = c = \frac{\partial Q}{\partial z} \tag{6.237}$$

in which $\partial Q/\partial z$ indicates partial differentiation of $Q(z, x)$, keeping x constant. Equation 6.236 describes the 'characteristics', that is the position of points on the wave over time. Each characteristic is associated with a value of the arbitrary constant. On the characteristic *only*, and not generally, we can then obtain the form of the wave from the four ordinary differential equations:

$$\frac{dQ}{dt} = ic \tag{6.238}$$

$$\frac{dQ}{dx} = i \tag{6.239}$$

$$\frac{dz}{dt} = i \tag{6.240}$$

$$\frac{dz}{dx} = i/c \tag{6.241}$$

As an example, consider linear overland flow down a convex hillside (Fig 6.10), with spatially uniform rainfall, i.

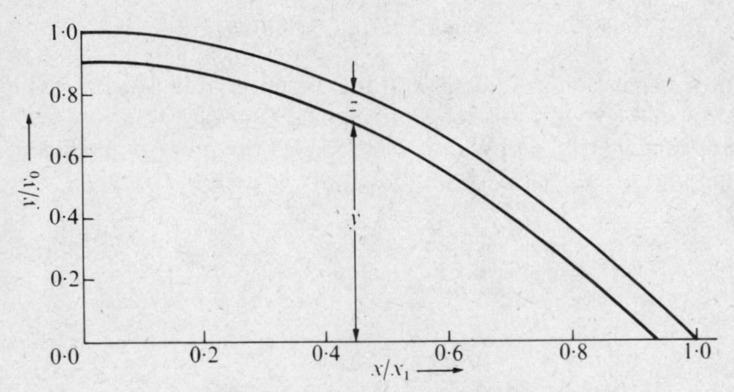

FIG. 6.10. Flow on a convex hillside

Foı a linear flow, Q can be taken as

$$-Kz\,\frac{dy}{dx},$$

so that

$$Q = \frac{2Ky_0}{x_1^2}\,zx. \tag{6.242}$$

The characteristics are given by:

$$\frac{dx}{dt} = c = \frac{\partial Q}{\partial z} = \frac{2Ky_0}{x_1^2}\,x \tag{6.243}$$

or

$$x = x_0 \, e^{\frac{2Ky_0}{x_1^2} t} \tag{6.244}$$

Each associated with a value of the arbitrary constant x_0.
Discharge is given by (equation 6.239):

$$\frac{dQ}{dx} = i, \tag{6.239}$$

From equation (6.244)

$$dx = \frac{2Kx_0y_0}{x_1^2} \, e^{\frac{2Ky_0}{x_1^2} t'} \, dt' \tag{6.245}$$

so that

$$Q = \int_{-\infty}^{x} i dx = \frac{2Kx_0y_0}{x_1^2} \int_{-\infty}^{t} i \cdot e^{\frac{2Ky_0}{x_1^2} t'} \, dt' \tag{6.246}$$

Substituting in equation 6.235 again to eliminate x_0, and so obtain a general expression for Q, and not one restricted to the characteristics:

$$Q = \frac{2Ky_0}{x_1^2} \cdot x e^{\frac{-2Ky_0}{x_1^2} t} \int_{-\infty}^{t} i \cdot e^{\frac{2Ky_0}{x_1^2} t'} \, dt' \tag{6.247}$$

If we change the time-direction, and refer it *back* from the present, so that t, a(t) refer to time *ago* and rainfall intensity at that time, the discharge can be re-written as

$$Q = \frac{2Ky_0}{x_1^2} \cdot x \int_{0}^{\infty} i e^{\frac{-2Ky_0}{x_1^2} t} \, dt \tag{6.248}$$

If the rainfalls are obtained for a sequence of unit periods (say days) of length t_0, so that average rainfall intensity i_n fell between $[nt_0]$ and $[(n + 1) t_0]$ ago;

$$Q = \frac{2Ky_0}{x_1^2} x \sum_{n} \int_{nt_0}^{(n+1)t_0} i_n e^{\frac{-2Ky_0}{x_1^2} t} \, dt' \tag{6.249}$$

i_n is now a constant in the integration, which becomes:

$$\frac{x_1^2}{2Ky_0} i_n \left[e^{\frac{-2Ky_0 t}{x_1^2} (n+1) t_0} \right] = \frac{x_1^2}{2Ky_0} i_n e^{\frac{-2Ky_0 t_0}{x_1^2}} n \left[1 - e^{\frac{-2Ky_0 t_0}{x_1^2}} \right] \tag{6.250}$$

Substituting into equation 6.249 for Q and putting

$$p = e^{\frac{-2Ky_0 t_0}{x_1^2}}$$

we get

$$Q = (1 - p)\, x\, [i_0 + i_1 p + i_2 p^2 + \ldots + i_n p^n + \ldots] \tag{6.251}$$

Such an expression may be used as a basis for an 'antecedent precipitation index'.

6.7.14. Some examples of solutions for $\partial^2 z/\partial x^2 = K\partial z/\partial t$

To illustrate the range of types of solution to a partial differential equation for different sets of boundary and initial conditions, this chapter is concluded with examples of some appropriate solutions in the context of heat conduction and slope development under soil creep, both of which give rise to the basic diffusion equation;

$$\frac{\partial^2 z}{\partial x^2} = K \frac{\partial z}{\partial t} \tag{6.252}$$

In both contexts x represents distance along a one-dimensional system and t time. In heat conduction z is temperature, and in slope profiles, z is elevation.

Case 1: Boundary conditions: $z(0, t) + Z(t)$

$z(x, t)$ finite everywhere

Initial condition $z(x, 0) = 0$

The solution is

$$z(x, t) = \frac{x}{2(K\pi)^{\frac{1}{2}}} \int_0^t \frac{z(t - t')}{t'^{3/2}} e^{-x^2/4Kt'} dt' \tag{6.253}$$

Figure 6.11 sketches the effect of dropping z suddenly to $-z_0$ and then keeping it there, as might occur during incision of a gorge. This case

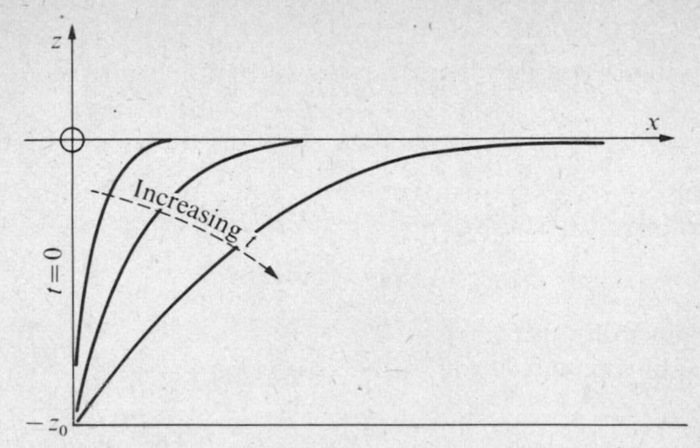

FIG. 6.11. Heat conduction, or slope development under soil creep $z = z_0 \text{erfc}\left[\frac{x}{(2kt)}^{-\frac{1}{2}}\right]$

shows the transient response to changes in $z(0, t)$.

Case 2: For a steady cyclic change in $z(0, t)$ to which the medium is responding cyclically, we have:

Boundary conditions: $z(0, t) = z_0 \sin \omega t$
$z(x, t)$ has zero mean
and $\to 0$ as $x \to \infty$

The solution is

$$z(x, t) = z_0 \sin\left(\omega t + x\sqrt{\frac{\omega}{2K}}\right) e^{-x\sqrt{\omega/2K}} \qquad (6.254)$$

Figure 6.12 illustrates this steadily cyclic solution, which is applicable to the penetration of daily and annual temperature waves into the ground.

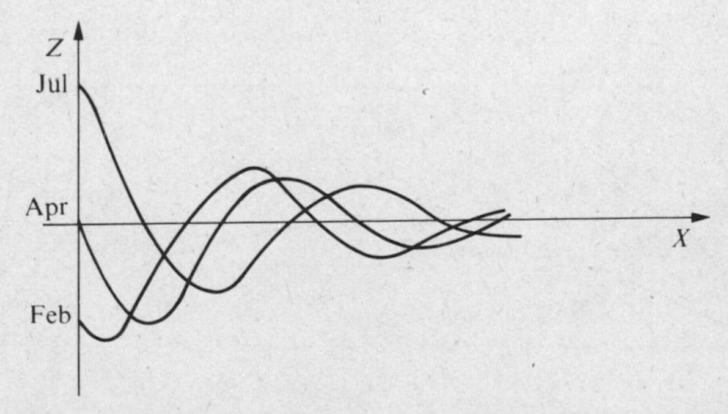

FIG. 6.12. Steady cyclic solution to heat conduction equation

Case 3:
The total quantity of debris/heat in the system m is given by:

$$m = \int_0^\infty z\,dx \tag{6.255}$$

If we wish to specify boundary conditions in terms of rate of addition or removal of substance;

$$m(0, t) = M(t)$$

with initial condition $z(x, 0) = 0$,
then the appropriate solution differs from case 1, and is:

$$z(x, t) = \frac{1}{\sqrt{(\pi k)}} \int_0^t \frac{M(t - t')}{t'^{\frac{1}{2}}} e^{-x^2/4Kt'}\,dt' \tag{6.256}$$

Case 4:
For a soil of finite thickness, or a slope of finite length, $2x_1$, case 1 is modified as follows.

Boundary condition $z(0, t) = z(2x_1, t) = Z(t)$

Initial condition $z(x, 0) = 0$

Solution as a series:

$$z = \sum_{n=0}^{\infty} \left[(-1)^n \frac{K(2n + 1)\pi}{x_1^2} e^{\frac{-K(2n+1)^2\pi^2 t}{4x_1^2}} \sin \frac{(2n + 1)\pi x}{2x_1} \int_0^t Z(t') e^{\frac{K(2n+1)^2\pi^2 t'}{4x_1^2}}\,dt' \right] \tag{6.257}$$

This is illustrated for a sudden drop to $-z_0$ in Figure 6.13.

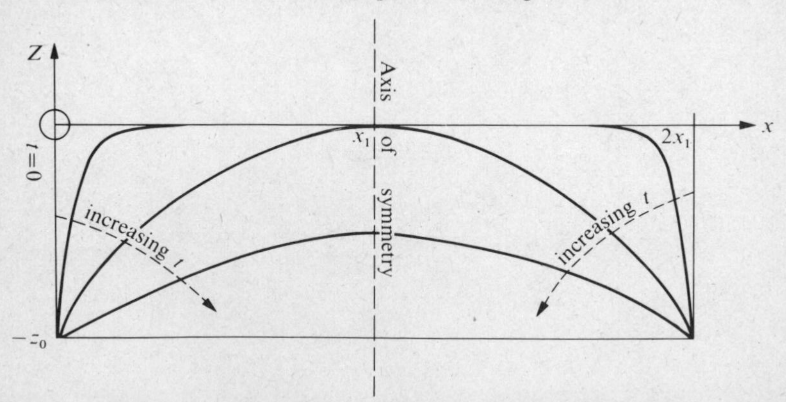

FIG. 6.13. Heat conduction following a drop in temperature

These cases show some of the range of solutions which are possible, and relevant to particular problems. This equation 6.252 has been very fully studied in the heat conduction context, and Carslaw and Jaeger (1959) is recommended for many further examples.

Exercises

Section 6.2

(1) Show by differentiation that the expressions for y on the left are solutions to the given differential equation on the right.

(a) $y = ax^2 + 1$ to $x\dfrac{dy}{dx} = 2(y-1)$

(b) $y = A \sin x$ to $\dfrac{dy}{dx} = \sqrt{1-y^2}$

(c) $y = \sin(Ax)$ to $x\dfrac{dy}{dx} = \sqrt{1-y^2}\ \sin^{-1} y$

(d) $y = A + Bx^2$ to $\dfrac{dy}{dx} = x\dfrac{d^2y}{dx^2}$

(2) Find the general solution to the following first-order linear differential equations

(a) $\dfrac{dy}{dx} + xy = x$

(b) $x\dfrac{dy}{dx} + y = x$

(c) Find the solutions to the above equations which satisfy
for (a): $y = 2$ at $x = 0$
for (b): $y = 1$ at $x = 1$

Section 6.3

(3) Solve the following equations using Laplace Transforms.

(a) $\dfrac{d^3y}{dx^3} = 1$

(b) $\dfrac{d^2y}{dx^2} + a^2y = \cos ax$

(4) Obtain a series solution for the equation:

$$\frac{d^2y}{dx^2} + y = 0$$

given that $y = 1$ at $x = 0$

and $\frac{dy}{dx} = 0$ at $x = 0$

(5) Obtain an approximate solution by iteration to the equation:

$$(x)\frac{d^2y}{dx^2} + \frac{dy}{dx} + y = 0$$

given that $y = 1$ at $x = 0$

and $\frac{dy}{dx} = -1$ at $x = 0$

Hint; Rearrange the equation as $\frac{d}{dx}\left[x\frac{dy}{dx}\right] = -y,$

or $\frac{dy}{dx} = -\int\frac{y\,dx}{x}$ and start with a trial value of $y = 1$

Section 6.4

(6) For the function, specified numerically as follows:

x	0	0·1	0·2	0·3	0·4	0·5	0·6	0·7	0·8
y	·000	·110	·243	·399	·581	·790	1·029	1·297	1·596

(a) Tabulate the finite differences up to third differences

(b) Calculate $\frac{dy}{dx}$ at $x = 0{\cdot}4$

(c) Calculate $\frac{dy}{dx}$ at $x = 0{\cdot}55$

(d) Calculate $\frac{d^2y}{dx^2}$ at $x = 0{\cdot}4$

(e) Calculate $\int_{0\cdot2}^{0\cdot3} y\,dx$

(f) Calculate $\int_{0.3}^{0.7} y\,dx$

(7) Approximate with a second-order difference equation to the differential equation:

$$x\frac{dy}{dx} + y = 0$$

Section 6.5

(8) Solve the difference equation

$$y_{n+2} + a^2 y_n = 0$$

Section 6.6

(9) Find the slope profile corresponding to a steady-state equilibrium with a constant rate of down-cutting, if the sediment transport, S is given by:

$$S \propto -x^2\left(\frac{dy}{dx}\right)^2$$

Section 6.7

(10) Show that a solution to the slope continuity equation,

$$\frac{\partial S}{\partial x} + \frac{\partial y}{\partial t} = 0$$

$$\text{for } S = -K\frac{\partial y}{\partial x}$$

is:

$$y = \sum_{n=0}^{\infty} A_n e^{\frac{-(2n+1)^2\pi^2}{4x_1^2}} \cos\left\{\frac{(2n+1)\pi x}{2x_1}\right\}$$

Sketch the first two terms of this series for $x = 0$ to x_1

(11) Using the method of section 6.7.13, obtain an antecedent precipitation index for a hillside of uniform gradient.

7. The mathematics of probability and statistics

7.1. Introduction

The need of geographers and planners for statistical methods is perhaps more widely accepted than their need for other mathematical techniques. In this book, our concern has been not to write another book or part of a book on statistical methods for geographers, but to look at some of the mathematics at the foundation of statistical methods. This mathematics is essentially the study of probability theory. It is hoped that readers will pursue the basic probability theory of this chapter by reference to books on both statistical techniques and mathematical statistics. Our aim will be to follow a direction which leads not towards statistics but instead offers alternatives to the models elsewhere in this book. The preceding chapters are essentially aimed at the construction of models of geographic reality in which the effects of chance variation are ignored or averaged out. Any given input will produce an exactly predictable output – that is the models are deterministic. The alternative is to allow random variations in the model, and look at the probabilities of different outcomes. This method, of stochastic modelling, tends to concentrate our attention on the variability among possible outcomes, and not just on the average outcome. Where, as is usual, the real world does not exactly 'fit' a deterministic model, we have no way of knowing whether the fit is good enough, or whether we need a new model. Stochastic models lessen this disadvantage, though at the cost of additional complexity. In this chapter the basic concepts of probability are discussed, and the rules by which probabilities may be combined. These rules are applied to the examination of frequency distributions.

7.2. Basic definitions

7.2.1. Randomness and probability

The ideas of probability theory are basically intuitive. We can describe the relationships between probabilities, but we cannot ultimately define probability more closely than to say that an event having a probability of 0·5 will occur, on average, 50 times in every hundred trials. Within this statement, three concepts can be separated out which contain our ideas of probability and randomness:

(ii) Random events may, in principle, be repeated.
(ii) No prediction can be made for a single event, but in a long series of n trials, in which the event occurs m times, m/n tends to some sort of limit, which is the *probability* of the event.
(iii) This limit is the same for any sub-series in which inclusion of the rth trial is decided by a rule involving only data on the $(r-1)$ previous trials.

If these conditions are satisfied, then the event can be said to be *random.*

The probability of a random event only has a meaning in terms of the total set of events considered, called the 'event space'. In simple cases we can list all possible events, but in more complex situations we often have to fall back on the notion that we could, in principle, do so if we had enough information. For example our event space might be defined as the sub-set of outcomes of spinning a coin which consists of the events *heads* and *tails.* This definition excludes other outcomes, such as getting jammed in a crack in the floor, which in this case are rare, but need not be. If the event space consists of the two events 'green eyes' and 'blue eyes' for a group of people, then for the brown-eyed majority, the event space is simply meaningless. The probability of all events in the event space is defined as equal to one. The probabilities of single events, A, B, ... within the space are written as p(A), p(B), ... , and intuitive ideas of combining probabilities can be seen very readily using set theory notation and Venn diagrams. In Fig. 7.1 the rectangle represents the event space, and A, B are events within it. The union of A *or* B, is denoted by $A \cup B$ and corresponds to the shaded area in figure incorporating the whole of both A and B. The junction, C *and* D is denoted by $C \cap D$ and corresponds to the lower shaded area in which events C and D have *both* occurred.

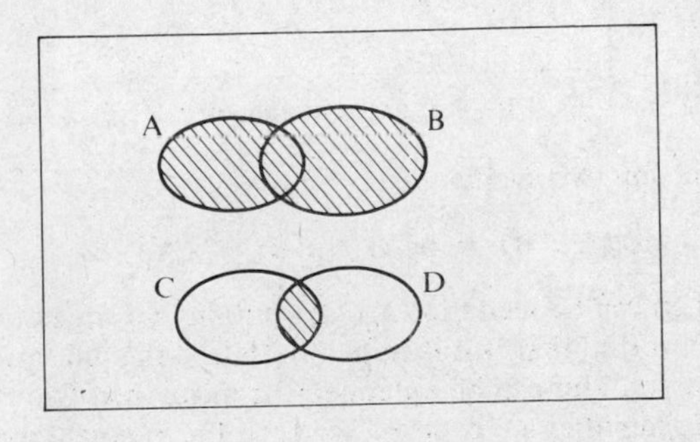

FIG. 7.1. Venn diagram for events and probabilities

7.2.2. Adding probabilities: exclusive and exhaustive events

We can write down a series of propositions which are self-evident from the diagram. They *define* more closely what we mean by probability.

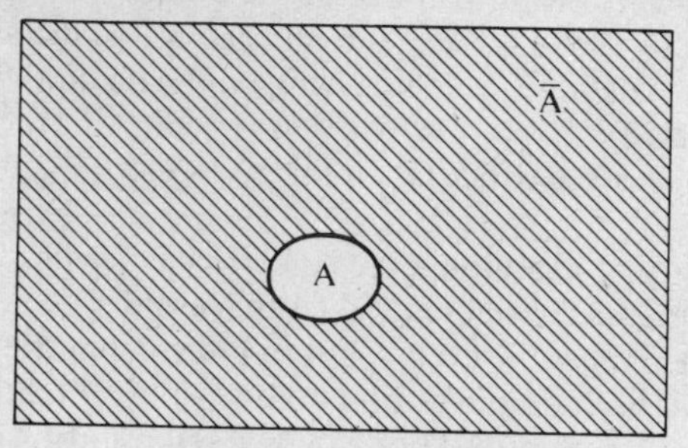

FIG. 7.2. The event space

(i) $\bar{A}$ is used to denote the event 'not A' within the event space, shown by the shaded area in figure 7.2. Then

$$p(\bar{A}) = 1 - p(A). \tag{7.1}$$

(ii) If when A has occurred, B must also have occurred, we can say that A implies B, and denote it by $A \subset B$ as in Fig. 7.3. In this case:

$$p(A) \leqslant p(B) \tag{7.2}$$

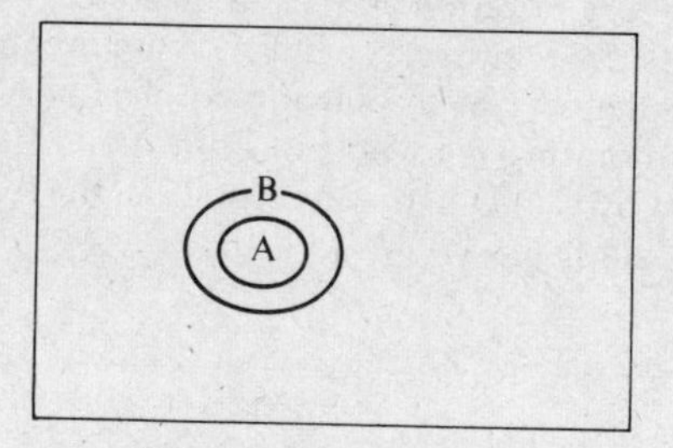

FIG. 7.3. Set containment

(iii) if A, B are any two events:

$$p(A \cup B) = p(A) + p(B) - p(A \cap B) \tag{7.3}$$

In Fig. 7.1, it can be seen that $A \cup B$ contains the area of overlap, $A \cap B$, twice over, so that it is intuitive to define the joint probability as in equation 7.3. This can be extended for more than two events, and, because all probabilities are positive, leads to the inequalities:

$$p(A \cup B) \leqslant p(A) + p(B) \tag{7.4}$$

$$p(\bigcup_i A_i) \leqslant \Sigma\, p(A_i) \qquad (7.5)$$

where the ∪-notation is extended to cover the union of many events, $\bigcup_i A_i$. Where A and B do not overlap,

$$p(A \cap B) = 0$$

and

$$p(A \cup B) = p(A) + p(B) \qquad (7.6)$$

In this case we say that A, B are mutually exclusive. That is, if A has occurred, B cannot have occurred. If the set of A_is together completely cover the event space with or without repetition, then the A_is are said to be *exhaustive.* If the A_is are both exhaustive and mutually exclusive, then

$$\sum_i p(A_i) = 1 \qquad (7.7)$$

7.2.3 Multiplying probabilities: conditional probability and independence

In many situations, we are concerned with more than one dimension of variation. For example we may be concerned with (i) type of employment and (ii) means of travel to work, for a group of people. Within our total population we can distinguish a number of sub-populations like students, or taxi-riders. The probability of riding to work on a bicycle has meaning not only for the population as a whole, but for sub-populations by employment (such as students). Conversely we can give meaning to the probability of being a student for the sub-population of cyclists. These probabilities for sub-populations are called *conditional*

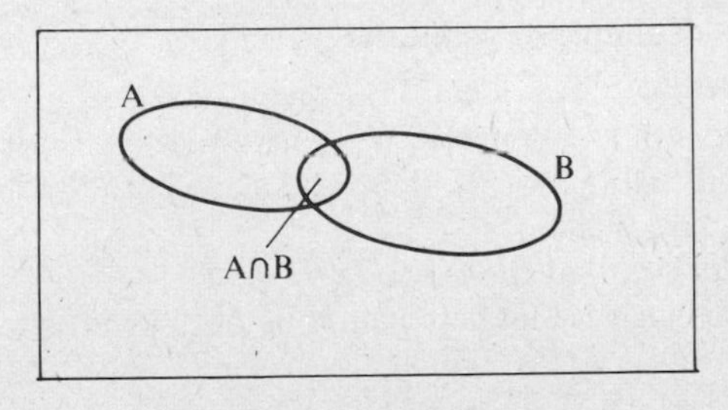

FIG. 7.4. Set intersection

probabilities. If A is the event 'being a student', and B is the event 'being a cyclist', then the probability of A, conditional on B, is the probability of being a student, given that you cycle to work, and is denoted by p (A|B). The converse probability of cycling to work, given that you are a student is p (B|A).

From figure 7.4, we write intuitively

$$p(A|B) = \frac{p(A \cap B)}{p(B)} \quad (7.8)$$

$$p(B|A) = \frac{p(A \cap B)}{p(A)} \quad (7.9)$$

or
$$p(A \cap B) = p(A)\ p(B|A) = p(B)\ p(A|B) \quad (7.10)$$

If the proportion of students who ride taxis to work is the same as for the population as a whole, then we can write:

$$p(B|A) = p(B) \quad (7.12)$$

and define A and B as *independent* events.

If and only if A and B are independent, equation 7.10 reduces to:

$$p(A \cap B) = p(A)\ p(B) \quad (7.12)$$

Notice that equation 7.10 shows that p (A|B) must also equal p(A). Independence is the criterion which must be satisfied to allow probabilities to be multiplied together simply in this way. When independence cannot be assumed, the right-hand side of equations (7.10) can be rewritten as Bayes' theorem:

$$p(A|B) = \frac{p(B|A)\ p(A)}{p(B)} \quad (7.13)$$

7.3. Examples of simple probabilities

7.3.1. Coin tossing

In a single throw of a coin, the event space consists of just two events, 'heads' (A) and 'tails' (B), to which we can assign the probabilities (not necessarily equal), p and q.

We can say, *a priori,* that these two events are

(i) Mutually exclusive: that is it cannot be both heads and tails on one throw. Thus:

$$p(A \cap B) = 0 \quad (7.14)$$

(ii) Exhaustive: that is there are no other possibilities

$$p(\overline{A \cup B}) = 0 \text{ or } p(A \cup B) = 1 \quad (7.15)$$

Combining these proportions, we can use equation 7.3;

$$p(A \cup B) = p(A) + p(B) - p(A \cap B) \tag{7.3}$$

to show that:

$$1 = p + q \tag{7.16}$$

This example is fairly trivial, but shows how the rules for summing probabilities are strictly used. The events, 'heads' (A) and 'tails' (B) are clearly *not* independent, because

$$p(A|B) = \frac{p(A \cap B)}{p(B)} = 0 \tag{7.17}$$

so that

$$p(A|B) \neq p(A) \tag{7.18}$$

In a series of n trials of throwing a coin, the event space is more complex. For three throws ($n = 3$), for example, the following are all events which can be assigned a probability: (i) 2 heads, 1 tail, (ii) more heads than tails, (iii) alternating faces of the coin, (iv) all three the same, (v) heads on the first throw, (vi) tails on the second throw. These events are neither exhaustive nor mutually exclusive. In order to work out probabilities, we need a set of events which are. Using the obvious notation, we have one possible set as:

HHH HHT HTH HTT THH THT TTH TTT

To assign probabilities to one of these events, say HHT, we have to ask the probability of heads on the second throw given that we have heads on the first throw. In other words, are the outcomes of successive throws *independent* events? It is only by making this *a priori* assumption, as we do, that we can proceed.
In this case:

$$p(HHH) = p(H_1)\,p(H_2)\,p(H_3) = p \cdot p \cdot p = p^3 \tag{7.19}$$

Similarly

$$p(HHT) = p(HTH) = p(THH) = p^2q \tag{7.20}$$

$$p(HTT) = p(THT) = p(TTH) = pq^2 \tag{7.21}$$

$$p(TTT) = q^3 \tag{7.22}$$

Returning to the orginal events (i) to (vi) above, they can be decomposed into the list of exclusive and exhaustive events as follows:

(i) p (2 heads, 1 tail) $=$ p(HHT + p(HTH) + p(THH) $= 3p^2q$
(ii) p (More heads than tails) $=$ p(HHT) + p(HTH) + p(THH) + p(HHH) $= 3p^2q + p^3$
(iii) p (Alternating faces) $=$ p(HTH) + p(THT) $= p^2q + q^2p$
(iv) p (All three the same) $=$ p(HHH) + p(TTT) $= p^3 + q^3$
(v) p(Heads on first throw) $=$ p(HHH) + p(HHT) + p(HTH) + p(HTT)

$$= p^3 + 2p^2q + pq^2$$
$$= p(p^2 + 2pq + q^2)$$
$$= p(p + q)^2 = p$$

(vi) p (Tails on second throw) $=$ p(HTH) + p(HTT) + p(TTH) + p(TTT)

$$= p^2q + 2pq^2 + q^3$$
$$= q(p^2 + 2pq + q^2) = q$$

The last two examples could also have been written down directly, by appealing to the notion of independence.
Thus: p(heads on first throw | outcomes in other throws)

$$= \text{p (heads on first throw)} = p$$

It may be noted that the probability of all events

$$= p^3 + 3p^2q + 3pq^2 + q^3 = (p + q)^3 = 1$$

7.3.2. Examples of real-world data: the relation to statistics

If we knew as much about our data as we do about the behaviour of an unbiassed coin, we would have little need to collect more. Real data sets often lack an explicit list of events making up the space; or may have a continuous range of values, which can only be arbitrarily broken up into a finite number of classes, which can be treated as distinct events. For example, human ages form a set of events which are continuous and have no theoretical upper limit; though in practise we may break them down into one-year units, and curtail the possibilities at around 150 years without appreciable loss of information!

The simplest type of data is concerned with attributes; for example heads or tails, possession of blue eyes or not, weighing more than 100 kg and so on. Provided that we can assume that the probabilities of possessing the attributes are constant for the population we are studying, then the methods used for analyzing coin tossing above are strictly applicable, and we only need to estimate the probability p of possessing the attribute.

The natural generalization of attribute data is into more than two categories which may be different in kind (nominal scale) or form a ranked sequence (ordinal scale); for example species of tree and examination grades/classes respectively. The logical limit of this process of generalization is to continuous data (interval and ratio scales), like temperatures or lengths, where the number of possible events is strictly infinite along a continuum. These data categories are more fully explored in texts on statistics, for example Siegel (1956), Blalock (1960) and many others. In this book we concentrate on attribute data for its simplicity, and consider other types of data as generalizations arising from it.

As important as the scale of the data (nominal or ordinal etc) is its sample structure. Above, we are implicitly dealing with one sample, and it is legitimate to ask questions like:

(i) Given that we have thrown HHH, what is the probability that the coin is biassed towards tails (that is $p < \frac{1}{2}$)?

(ii) Given that we have thrown HHH, what is the 'expected' value of p and its spread about that value?

It can be seen that these questions are within the scope of probability theory, but are the sort of questions which arise in statistics.

Perhaps more common than a single sample of a single attribute are cases where we observe two or more attributes for members of a sample; or take two or more samples. In the former case we are immediately concerned with problems of independence in probability of the two attributes; and in the latter case with whether the probability p could be the same for the two samples.

If we sample two or more attributes, we are concerned with problems of contingency and correlation. In the simplest case, of possession of attributes A and B, we have a contingency table. For example.

	A	$\bar{A}$
B	$A \cap B$ 1	$A \cap B$ 3
$\bar{B}$	$A \cap \bar{B}$ 1	$A \cap \bar{B}$ 15

A is having blonde hair

B is having blue eyes

If these events are independent, then

$$p(A \cap B) = p(A)\, p(B) \tag{7.12}$$

In this case, our intuitive estimates of these probabilities are respectively 1/20, 1/10 and 1/5 and it appears that $p(A \cap B)$ may be greater than expected, and that the events are therefore not independent. It is a legitimate probability and statistical question to ask whether we can *reliably* say that the events A, B are non-independent. By extension, if, instead of attributes A and B, we have measurements (say length and weight) x and y for a number of individuals, independence in probability is identical to the question of whether x and y are correlated.

A sort of independence which has particular significance in a geographic context is the independence of successive members of a time or space sequence. For example, does possession of a landscape attribute at a point depend on its possession at neighbouring points; or does rain tomorrow depend in probability on whether it rains today? In this case we are considering 'rain yesterday' and 'rain today' as the events A and B in the contingency table; and are involved in problems of auto-correlation. More generally we can ask whether the value of x at the nth point x_n is correlated to x_{n+r}: an r – lag-auto-correlation.

In the case of more than one sample, we are concerned with whether the samples can be treated as members of a homogeneous population. That is we ask if the probability of possessing the test attribute is the same for both samples. We can, for instance, ask whether the probability of blue eyes or car ownership differs appreciably between ethnic or social groups. This question too has auto-analogues, in which we ask whether, say, the probability of rainfall on a day differs between epochs or between regions.

The theory of significance testing is outside the scope of this book, but a simple example can show the sort of link which can be made, in turning from probability theory to statistical testing. We will consider the earlier coin throwing questions as an example. They are repeated here:

(i) Given that we have thrown HHH, what is the probability that the coin is biassed towards tails (i.e. $\mathrm{p} < \frac{1}{2}$)?

(ii) Given that we have thrown HHH, what is the 'expected' value of p and its spread about that value?

To solve the first of these problems, we will define events:

$$\mathrm{A} = (\mathrm{p} < \tfrac{1}{2}) \text{ and } \mathrm{B} = (\text{HHH in three throws}).$$

Application of Bayes' theorem (equation 7.13) gives:

$$\mathrm{p}\,(\mathrm{p} < \tfrac{1}{2} | \mathrm{HHH}) = \frac{\mathrm{p(HHH\ p} < \tfrac{1}{2})\,\mathrm{p(p} < \tfrac{1}{2})}{\mathrm{p(HHH)}} \tag{7.23}$$

The quantity on the left is what we require, namely the 'posterior' probability that the coin is biassed towards tails, *given* the outcome of the experiment. On the right hand side, the probability theory presented above gives $\mathrm{p}\,(\mathrm{HHH} | \mathrm{p} < \frac{1}{2})$ and p (HHH), while $\mathrm{p}\,(\mathrm{p} < \frac{1}{2})$ is the 'prior' probability that the coin is biassed towards tails. If we accept a 'null typothesis', then this probability will be ½; but we are free to choose its value according to the strength of our preconceptions.
For a given value of p,

$$\mathrm{p}\,(\mathrm{HHH}) = \mathrm{p}^3 \tag{7.24}$$

Thus over the full range of values of p, we must sum or integrate:

$$p\,(HHH) = \frac{\int_0^1 p^3\,dp}{\int_0^1 dp} = \frac{\frac{1}{4}[p^4]_0^1}{[p]_0^1} = \tfrac{1}{4} \qquad (7.25)$$

the bottom term acting as a scaling factor to ensure that the sum of probabilities for *all* possibilities = 1. Similarly for the range of values of $p < \frac{1}{2}$

$$p\,(HHH|p < \tfrac{1}{2}) = \frac{\int_0^{\frac{1}{2}} p^3 dp}{\int_0^{\frac{1}{2}} dp} = \frac{\frac{1}{4}[p^4]_0^{\frac{1}{2}}}{[p]_0^{\frac{1}{2}}} = \frac{\frac{1}{4}(\frac{1}{2})^4}{\frac{1}{2}} = \frac{1}{32} \qquad (7.26)$$

Thus, given the null hypothesis, from equation 7.23.

$$p\,(p < \tfrac{1}{2}|HHH) = \frac{\frac{1}{32}\cdot\frac{1}{2}}{\frac{1}{4}} = \frac{1}{16} \qquad (7.27)$$

This is what we require, namely the posterior probability that we should believe the coin is biassed towards tails, given both the outcome of the experiment and also our 'prior' probability of $\frac{1}{2}$ for the hypothesis. Turning to the second problem, we can see that we could generalize the above to give:

$$p\,(p < p_0|HHH) = \frac{\frac{1}{2}\,p_0^4\,\frac{1}{2}}{\frac{1}{4}} = p_0^4 \qquad (7.28)$$

This is the cumulative probability of values of $p < p_0$ (Fig. 7.5a). Differentiating with respect to p_0, we obtain the probability density of a particular value of p_0 (see section 7.4.1 below) that is

$$\frac{d}{dp_0}(p_0^4) = 4p_0^3 \qquad (7.29)$$

This gives us the answer we require to the second problem. The 'expected' value of p_0 is most simply associated with the value for which it has 'maximum likelihood' in Figure 7.5b, in this case with $p_0 = 1$, although we could adopt other plausible criteria. Figure 7.5 also shows us the spread about this value, and the probability of deviating from 1·0 by a given amount.
For example, from Fig. 7.5a, it can be seen that

$$p\,(p_0 > 0{\cdot}8) = 1 - (0{\cdot}8)^4 = 0{\cdot}59$$

An alternative approach to this problem, which is normally used in statistical work, but which is in our view much less satisfactory, is to

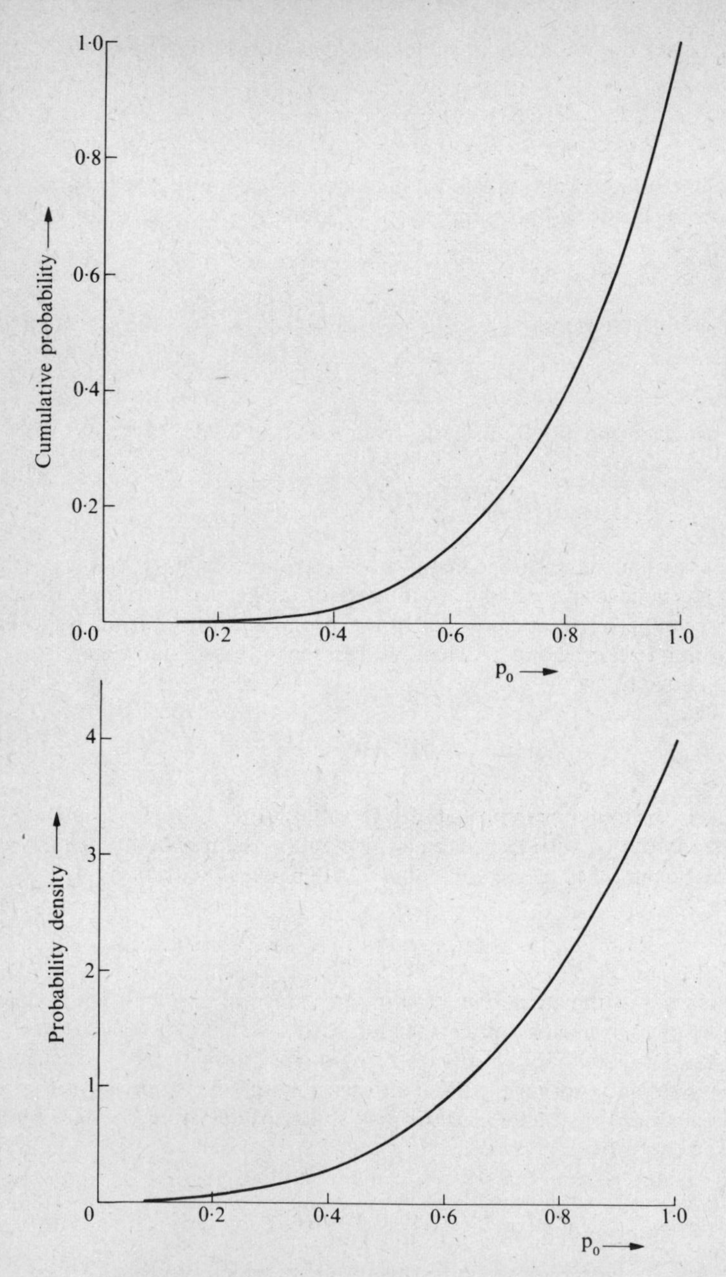

FIG. 7.5. Cumulative probability and probability density

consider not the probability or 'degree of belief' in the *hypothesis,* which we have worked with above; but to use the probability

$$p\,(\mathrm{HHH} \mid p < \tfrac{1}{2})$$

as a direct measure of significance. This approach makes no explicit use of past experience, and contradicts our intuition in not answering the question, 'how much does this experimental outcome *add* to our knowledge?'.

7.4. Compound probabilities; The binomial distribution

7.4.1. Probability distribution

The concept of a probability distribution was introduced in the last example. It is simply the probability associated with each of a series of discrete values, or with values on a continuum. For example, in sampling attributes we can assign a probability to 0, 1, 2, 3 etc occurrences in a sequence of trials.

Thus for three throws, we have seen that the probabilities of 0, 1, 2, 3 heads are respectively 1/8, 3/8, 3/8, 1/8. This probability distribution can be shown in the form of a histogram, (Figure 7.6).

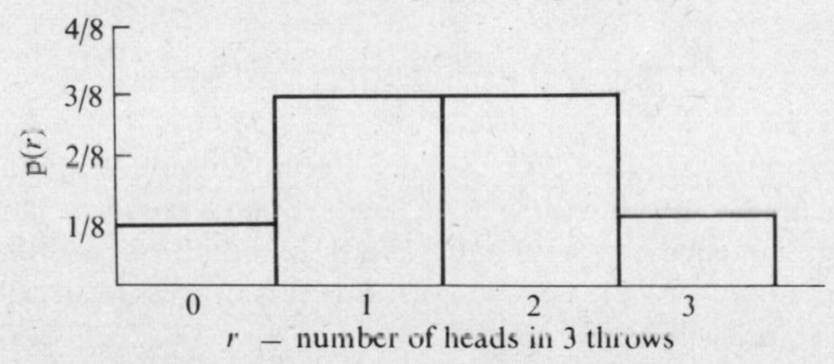

FIG. 7.6. A probability distribution

Where the value, r or x, can take any of a *continuous* range of values, then the probability associated with any exact value is very small, and we use instead the 'probability density', that is the probability per unit change in x. If we say that the probability density of x is $p(x)\delta x$. Alternatively we can talk of *cumulative* probabilities, that is the probability that $x > x_0$ which is obtained by summing over all $x > x_0$.
In this case

$$P(x_0) = p\,(x > x_0) = \int_{x0}^{\infty} p(x)\mathrm{d}x. \tag{7.30}$$

Differentiating both sides, we can see that:

$$\frac{\mathrm{d}P(x)}{\mathrm{d}x} = p(x) \tag{7.31}$$

a result we have used above in equation 7.29.

7.4.2. The binomial distribution

The simplest distribution to study is the general case of which Fig. 7.6 is an example: the probability of r occurrences in n trials of an attribute (such as heads) for which the probability of occurrence is p. As we have seen above (equation 7.19 to 7.22) this problem resolves itself into the probability of each particular sequence of n containing r occurrences of the attribute, and of calculating the number of such sequences. For example (equation (7.20)) the probability of two heads in three throws is compounded from the three events

$$\text{THH, HTH, HHT,}$$

each with probability p^2q.
Let us denote the number of sequences of length n which contain r heads by $\binom{n}{r}$*. Since heads and tails are interchangeable, it is clear that

$$\binom{n}{r} = \binom{n}{n-r} \tag{7.32}$$

In general we will show that

$$\binom{n}{r} = \frac{n!}{r!\,(n-r)!} \tag{7.33}$$

where $n! = n(n-1)(n-2)\ldots\ldots 3.2.1$ and 0! is conventionally 1. Consider sequences of n with $(r+1)$ heads. Such a sequence can be made from sequence of n with any r heads, by adding an additional head in place of any one of the $(n-r)$ tails. But this process leads to repetition as can be seen from Figure 7.7.

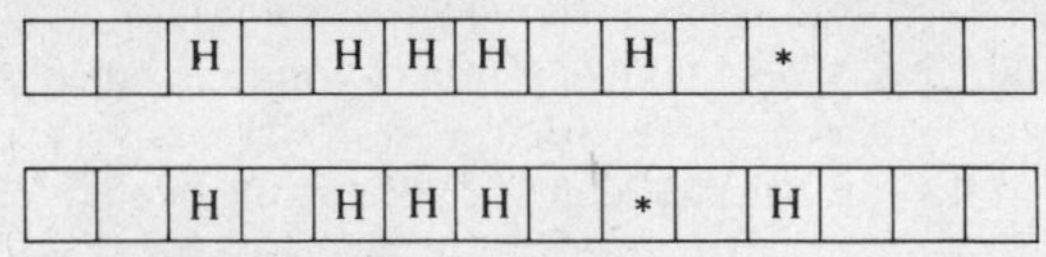

H = r original 'heads'
* = additional 'head'

FIG. 7.7. Repetition of events

Each sequence of $(r+1)$ heads leads to exactly $(r+1)$ repetitions, corresponding to the $(r+1)$ possible positions of the additional 'head'. Thus

$$\binom{n}{r+1} = \frac{n-r}{r+1}\binom{n}{r};\quad n>r\geqslant 0. \tag{7.34}$$

* The notation ${}_nC_r$ and ${}^nC_r$ are also used for this quantity.

This is a first order difference equation for r. It can be seen that for $r = 0$, there is only one possibility (all 'tails'), so that the arbitrary constant is given by

$$\binom{n}{0} = 1 \tag{7.35}$$

It can be seen by substitution that equation 7.33 satisfies both of these equations, and is therefore the unique solution. We have

$$\binom{n}{r} = \frac{n!}{r!\,(n-r)!} \tag{7.33}$$

for r 'heads' and $(n-r)$ 'tails', and the probability for any particular sequence is $p^r q^{n-r}$. Thus the total probability of r occurrences in n trials is

$$p_{r,n} = \binom{n}{r} p^r q^{n-r} \tag{7.36}$$

which defines the probability distribution.

It may be recognized that the expression is the coefficient of p^r in the binomial expansion of $(p+q)^n$. It is for this reason that the distribution is called the 'binomial distribution'. This property also means that the joint probability of 0, 1, 2 . . . $(n-1)$ or n occurrences

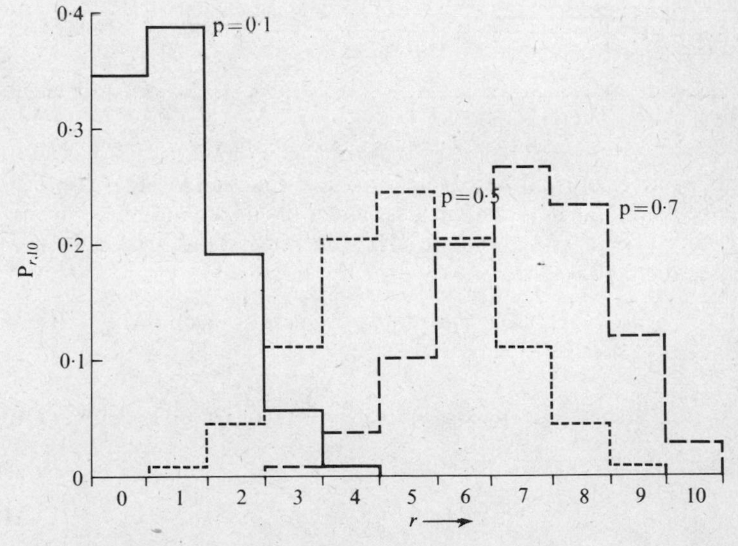

FIG. 7.8. Examples of the binomial distribution for $n = 10$

$$= \sum_{r=0}^{n} p_{r,n} = \sum_{n=0}^{n} \binom{n}{r} p^r q^{n-r} = (p+q)^n = 1^n = 1 \qquad (7.37)$$

Figure 7.8 shows some examples of the distribution for $n = 10$.

7.4.3. Stirling's formula

Calculation of the values for $p_{r,n}$ can be seen to become lengthy where large numbers are concerned, and it is often desirable to approximate to $n!$ This is done by means of Stirling's formula.

$$n! \approx \sqrt{(2\pi)}\, n^{n+\frac{1}{2}} e^{-n} \qquad (7.38)$$

where the sign $\approx$ indicates an approximation which improves as n becomes larger.

We can go some way to showing that this is true by considering

$$\log(n!) = \log 1 + \log 2 + \ldots + \log n$$

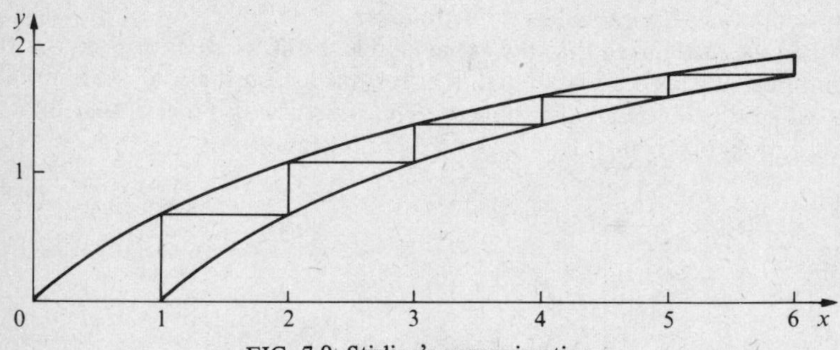

FIG. 7.9. Stirling's approximation

The required sum is the area under the step function in Fig. 7.9 which can be seen to lie between the areas under the upper and lower curves $y = \log(1+x)$ and $y = \log x$. Thus the integral mean value theorem (*cf* section 5.10.4) gives:

$$\int_1^n \log x \, dx < \log(n!) < \int_0^n \log(1+x)\, dx \qquad (7.39)$$

or

$$n \log n - n < \log(n!) < (n+1)\log(n+1) - n \qquad (7.40)$$

This inequality suggests the form chosen

$$\log(n!) \approx (n + \tfrac{1}{2}) \log n - n \qquad (7.41)$$

and in fact the difference between the left- and right-hand sides tends to a limit which is $\log_e \sqrt{(2\pi.)}$

Thus

$$\log(n!) \sim \log_e \sqrt{(2\pi)} + (n + \tfrac{1}{2}) \log n - n$$

or

$$n! \sim \sqrt{(2\pi)}\; n^{n+\frac{1}{2}}\, e^{-n} \tag{7.38}$$

which is the required form.

TABLE 7.1

n! using Stirling's formula

n	$n!$	Stirling's formula eqn (7.38) for $n!$	Equation 7.42 for $n!$
1	1.	·92	0·37
2	2.	1·92	·54
3	6.	5.84	1·34
5	120	118·02	21·1
10	3,628,800	3,598,695	454,000

If $\log_e (2\pi) + \frac{1}{2} \log n$ can be neglected relative to $n \log n - n$, the following simpler formula can be used to give orders of magnitude:

$$\log n! = n \log n - n \tag{7.42}$$

and this will be used in another context in Chapter 9.

7.5. Moments of Distributions

7.5.1. Definitions

As may be seen from Figure 7.8, probability distributions differ considerably in shape and position. Their central position may be defined by a 'maximum likelihood' or modal value for which the probability (or probability density) is greatest. Alternatively, their centre may be defined by a centre of gravity, about which a cut-out of the distribution would balance. At this point, x,

$$\int_{-\infty}^{\infty} p(x) \cdot (x - \bar{x})\, dx = 0 \tag{7.43}$$

Since $\bar{x}$ is a constant, we can write:

$$\int_{-\infty}^{\infty} x\, p(x)\, dx - \bar{x} \int_{-\infty}^{\infty} p(x)\, dx = 0 \tag{7.44}$$

or since

$$\int_{-\infty}^{\infty} p(x)\, dx = 1; \bar{x} = \int_{-\infty}^{\infty} x\, p(x)\, dx \tag{7.45}$$

If discrete intervals were used, we could use a summation sign to give

$$\bar{x} = \sum_{\text{all } x} x\, \mathrm{p}(x) \tag{7.46}$$

This is called the *mean*, or the *first moment* of the distribution about $x = 0$. It is also possible to define higher moments, which are usually taken about the mean. For example the second moment about the mean is:

$$\sigma^2 = \int_{-\infty}^{\infty} (x - \bar{x})^2\, \mathrm{p}(x)\, \mathrm{d}x \text{ or } \Sigma\, (x - \bar{x})^2\, \mathrm{p}(x) \tag{7.47}$$

Expanding the integral:

$$\begin{aligned} \sigma^2 &= \int_{-\infty}^{\infty} x^2\, \mathrm{p}(x)\, \mathrm{d}x - 2\bar{x} \int_{-\infty}^{\infty} x\, \mathrm{p}(x)\, \mathrm{d}x + \bar{x}^2 \int_{-\infty}^{\infty} \mathrm{p}(x)\, \mathrm{d}x \\ &= \int_{-\infty} x^2\, \mathrm{p}(x)\, \mathrm{d}x - \bar{x}^2 \text{ or } \Sigma\, (x^2)\, \mathrm{p}(x) - \bar{x}^2 \end{aligned} \tag{7.48}$$

since $\int_{-\infty}^{\infty} x\, \mathrm{p}(x)\, \mathrm{d}x = \bar{x}$. This is a common expression for calculating the variance. This second moment describes the spread of the distribution about the mean. Since neither $(x - \bar{x})^2$ nor $\mathrm{p}(x)$ can ever be negative, their product, and so the integral, must be *positive.* It is found that the second moment is *least* when it is taken about the mean, and is then called the variance σ^2. Its square root σ is the standard deviation, the most usual measure of the 'spread' of a distribution. The third and fourth moments also help to describe the shape of a distribution. The assymmetry (skewness) of a distribution is often particularly diagnostic. The usual measures used are:

$$\text{Skewness} = \frac{\int_{-\infty}^{\infty} (x - \bar{x})^3\, \mathrm{p}(x)\, \mathrm{d}x}{\sigma^2} \text{ or } \frac{\Sigma\, [(x - \bar{x})^3\, \mathrm{p}(x)]}{\sigma^2} \tag{7.49}$$

$$\text{Kurtosis} = \frac{\int_{-\infty}^{\infty} (x - \bar{x})^4\, \mathrm{p}(x)\, \mathrm{d}x}{\sigma^4} \text{ or } \frac{\Sigma\, (x - \bar{x})^4\, \mathrm{p}(x)}{\sigma^4} \tag{7.50}$$

These measures have been used by sedimentologists in particular to describe the form of a grain size distribution. (Folk and Ward 1957).

7.5.2. Moments of the binomial distribution

For the binomial distribution, the mean and variance can be calculated as a sum (since r takes only integer values).

$$\begin{aligned} \bar{x} &= \sum_{r=0}^{n} r\, \mathrm{p}_{r,n} \\ &= \sum_{r=0}^{n} r \frac{n!}{r!\,(n-r)!} \mathrm{p}^r \mathrm{q}^{n-r} \end{aligned} \tag{7.51}$$

The factorial term can be re-arranged as:

$$r\frac{n\,(n-1)!}{r\,(r-1)!\,(n-r)!} = n\frac{(n-1)!}{(r-1)!\,(n-r)!}$$

$$= n\binom{n-1}{r-1} \tag{7.52}$$

Thus

$$\bar{x} = n\sum_{r=1}^{n}\binom{n-1}{r-1}\mathrm{p}^r\mathrm{q}^{n-r}$$

$$= n\mathrm{p}\sum_{r=1}^{n}\binom{n-1}{r-1}\mathrm{p}^{r-1}\,\mathrm{q}^{n-r} \tag{7.53}$$

The terms inside the Σ are now the terms in the binomial expansion of $(\mathrm{p}+\mathrm{q})^{n-1}$, so that:

$$\bar{x} = n\mathrm{p}\,(\mathrm{p}+\mathrm{q})^{n-1} = n\mathrm{p} \tag{7.54}$$

This result for the mean number of occurrences seems reasonable since they occur a proportion p of the time in n trials.
Likewise for the variance

$$\sigma^2 = \sum_{r=0}^{n} r^2\,\mathrm{p}_{r,n} - \bar{x}^2 \tag{7.55}$$

which leads by a similar, though lengthier, argument to

$$\sigma^2 = n\mathrm{p}\;\mathrm{q} \tag{7.56}$$

These are the standard results for the binomial distribution.

$$\left.\begin{aligned}\bar{x} &= n\mathrm{p}\\ \sigma^2 &= n\mathrm{pq}\end{aligned}\right\}\;\ldots(7.57)$$

7.6. Limiting cases of the binomial distribution

7.6.1. Introduction

For large n, the binomial distribution becomes difficult to calculate, even using Stirling's formula. It is therefore common to approximate it by a continuous distribution (r or x taking any value). This can readily be done in two very important cases, leading to the Poisson and Normal distributions.

7.6.2. The Poisson distribution

In many cases the number of sample points becomes infinite, because a sample could be made at *any* point on a continuous line, but the total number of occurrences remains finite. For example, people may join a queue at any moment, and in a simple model, their probability of joining is constant over time, but the number of possible moments is very large. In a given time, representing n such moments, we allow

n to become infinite, while
$np \to z$, the expected number joining the queue.

Then the probability of r joining the queue in the given time is given by the binomial distribution

$$p_r = \frac{n!}{r!(n-r)!} p^r q^{n-r} \tag{7.36}$$

For the limiting case, we apply Stirling's formula; and replace p by z/n, q by $1 - z/n$:

$$\begin{aligned} p_r &\approx \frac{n^{n+\frac{1}{2}} e^{-n}}{r!(n-r)^{n-r+\frac{1}{2}} e^{-(n-r)}} \\ &\approx \frac{e^{-r}}{r!} \frac{n^r}{(1-r/n)^{n-r+\frac{1}{2}}} (z/n)^r (1-z/n)^{n-r} \\ &\approx \frac{e^{-r}}{r!} z^r \frac{(1-z/n)^{n-r}}{(1-r/n)^{n-r+\frac{1}{2}}} \end{aligned} \tag{7.58}$$

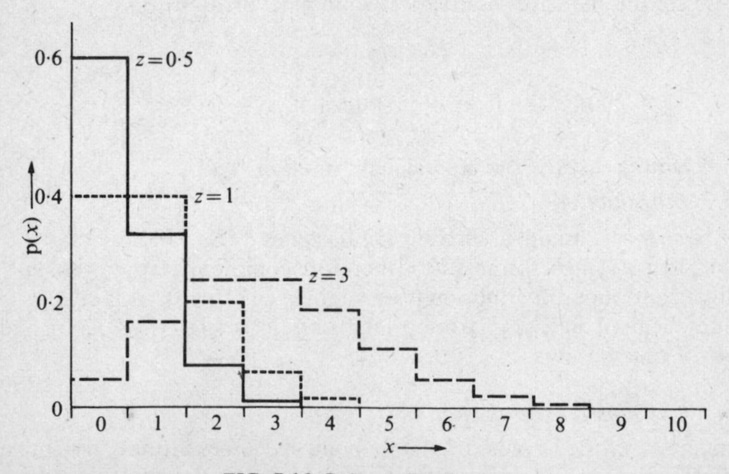

FIG. 7.10. Poisson distribution

Now $(1 - z/n)^n \rightarrow e^{-z}$ as $n \rightarrow \infty$, so that

$$p_r \rightarrow \frac{e^{-r}}{r!} z^r \frac{e^{-z}}{e^{-r}} = e^{-z} \frac{z^r}{r!} \tag{7.59}$$

which is the Poisson distribution

$$\begin{aligned} \bar{x} &= np = z \\ \sigma^2 &= npq = z(1 - z/n) \rightarrow z \text{ as } n \rightarrow \infty \end{aligned} \tag{7.60}$$

so that the moments are given by:

$$\bar{x} = \sigma^2 = z \tag{7.61}$$

This distribution is markedly right-skew for small z, as is shown in Figure 7.10.

7.6.3. The normal distribution

Where both n and np become large, we can again look at an extreme case case of the binomial distribution if we calculate the distribution about its mean ($r = np$). Again applying Stirling's formula, rather as for the Poisson distribution, but applying it to $r!$ as well; and observing that $x = (r - np)$ has magnitude $\sqrt{n}$, we finally arrive at the result

$$p_{r,n} \rightarrow \frac{1}{\sqrt{(2\pi npq)}} e^{\frac{-(r-np)^2}{-2npq}} \tag{7.62}$$

$$= \frac{1}{\sigma\sqrt{(2\pi)}} e^{\frac{-(x-\bar{x})^2}{-2\sigma^2}} \tag{7.63}$$

This is the standard form for a normal distribution. It can be seen that $(x - \bar{x})$ has magnitude $\sigma \approx n^{\frac{1}{2}}$, as assumed above.

7.6.4. The central limit theorem

The real power of the normal distribution, however, lies not in its relation to the binomial distribution, but on account of the central limit theorem, which is here stated without proof:

If the x_i are a sequence of independent values, with identical probability distributions of *any* form (provided that $\bar{x}$, σ^2 are finite), then the distribution of:

$$\frac{\sum_{i=1}^{n}\left(\frac{x_i}{n}-\bar{x}\right)}{\sigma/\sqrt{n}} \qquad (7.64)$$

tends towards the *normal* distribution as $n \to \infty$. In practice the approximation is very close for $n > 30$. This theorem is the basis of sampling theory. It means that if we take a number of samples from the same population, and calculate the mean of each, then these *means* are normally distributed about the population mean, whatever the original distribution within the population. It is for this reason that the normal distribution occurs very frequently, both in the distribution of many aggregate properties, and in the theory of sampling (Stuart, 1962) and statistical testing for which the reader is referred to standard texts.

Exercises

Section 7.1.

(1) Write an essay on the concepts underlying the notions of 'probability' and 'randomness'.

Section 7.2.

(2) Check the truth or falsity of the following relationships, using Venn diagrams:

(a) $p[A|(B \cup A)] = \dfrac{p(A)}{p(B \cup A)}$

(b) $p[A \cup (B \cap \bar{A})] = p(A)$

(c) If $p(A \cap B) = p(A)$, then $A \subset B$

(d) $p[A \cap (B \cup C)] = p[(A \cap B) \cup (A \cap C)]$

(e) $p[\bar{A} \cup B) \cap (\bar{B} \cup A)] = p[A \cap B]$

Section 7.3

(3) In a series of 4 throws of an unbiassed coin ($p = q = \frac{1}{2}$)
What are the probabilities of
(a) At least one of both heads and tails
(b) More heads than tails
(c) Two heads, not counting the first throw
(d) Two heads in all, conditional on at least *one* of both heads and tails

(4) After the banker has thrown 5 sixes in a row, you feel 90% sure that the dice is loaded. How sure should you feel after he throws yet another six? Did you start the game with an open mind?
Hint: Calculate the probability

$$p(6|q < 5/6)$$

Section 7.4.

(5) Calculate the value of $\binom{n}{r}$ for $n = 6$ and $r = 0, 1, 2, 3, 4, 5$ and 6. Check that they add up to $(1 + 1)^6 = 64$.

(6) Calculate $\binom{20}{10}$ using Stirling's formula in its normal form:

$$\log_e (n!) = (n + \tfrac{1}{2}) \log_e n - n + \log_e \sqrt{(2\pi)}$$

Compare the values obtained using less and more accurate forms.
(a) $\log_e (n!) \simeq n \log_e n - n$

(b) $\log_e (n!) \simeq (n + \tfrac{1}{2}) \log_e n - n + \dfrac{1}{12n} + \log_e \sqrt{(2\pi)}$

Section 7.5 and 7.6

(7) Derive the moments of the Poisson distribution

$$p_r = e^{-z} \frac{z^r}{r!}$$

The results are given in equation 7.61.

(8) By summing the digits in each row of a table of random numbers, obtain an empirical distribution for the row sums. Compare it with the distribution of individual random numbers:

$$p(0) = p(1) = \ldots = p(9) = 0{\cdot}1$$

(9) By plotting the cumulative distributions, compare the Poisson distribution for $z = 20$ with a normal distribution with the same mean and variance.

8. Stochastic Processes

8.1. Random walks

8.1.1. Introduction

In Chapter 7, we have been primarily concerned with successive trials or replications as independent events. Here we consider them as a space or time sequence, mainly in one dimension. The idea of a random walk is classically associated with the 'drunkard's walk' in which a man takes a forward or backward step at random. His position after a series of such steps is the *sum* of the outcomes. Similarly the 'gambler's ruin' problem of classical theory is concerned with the exhaustion of stake money by one of two players who play a long sequence of games in which each has a fixed probability of winning. The capital of one player, or the drunkard's position (on the y-axis) is thus subject to a one-dimensional 'random walk', which is commonly plotted against the number of steps or games on the x-axis.

8.1.2. The binomial process

The simplest form of random walk, and the one most fully studied, is a process in which each step may be either forwards, with fixed probability p, or backwards with probability q(= 1 − p). The total number, r of forward steps in a total of n steps clearly has a probability given by the binomial distribution:

$$\mathrm{p}_{r,n} = \binom{n}{r}\mathrm{p}^r\mathrm{q}^{n-r} \tag{7.36}$$

In the random walk representation (Figure 8.1) the position, y, after r positive steps in a total of n [and thus $(n-r)$ negative steps] is

$$y = r-(n-r) = 2r-n \tag{8.1}$$

The binomial coefficient $\binom{n}{r}$ represents the number of distinct routes to this point, as can be checked by trial and error.

Because the route paths (though not their probabilities) are symmetrical with respect to up and down, then for every possible route in Figure 8.1,

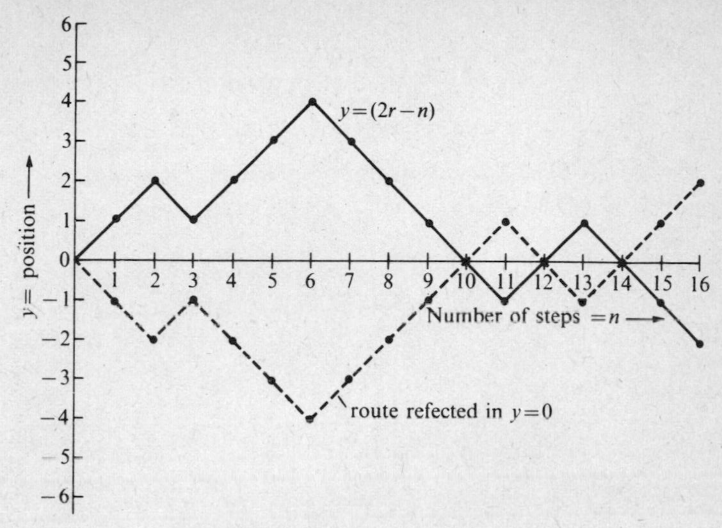

FIG. 8.1. Representation of a random walk

there is another route obtained by reflecting in the $y = 0$ axis or any other horizontal line, giving for example the broken-line route in Figure 8.1. This is called the 'reflection principle'.

8.1.3. The ballot theorem

The reflection principle may often be used to show more complex results. For example, we will prove the following:

> 'In a ballot, candidates A and B score respectively r and $(n-r)$ votes, where $r > (n-r)$. The probability that throughout the counting there are always more votes for A than B equals $(2r-n)/n$.

A path which is admissible according to the above criterion is one which does not touch or cross the $y = 0$ axis in Figure 8.2. It must thus start through $y = 1, n = 1$.

Consider the set of inadmissible routes through (1, 1), for example the broken line in Figure 8.2. By the reflection principle, we can reflect the left-hand portion to pass through $y = -1, n = 1$ for *all* these inadmissible routes, and *only* these routes. Thus the number of admissible routes

= Total no of routes from $(1, 1)$ to $(2r-n, n)$

− Total no of routes from $(-1, 1)$ to $(2r-n, n)$

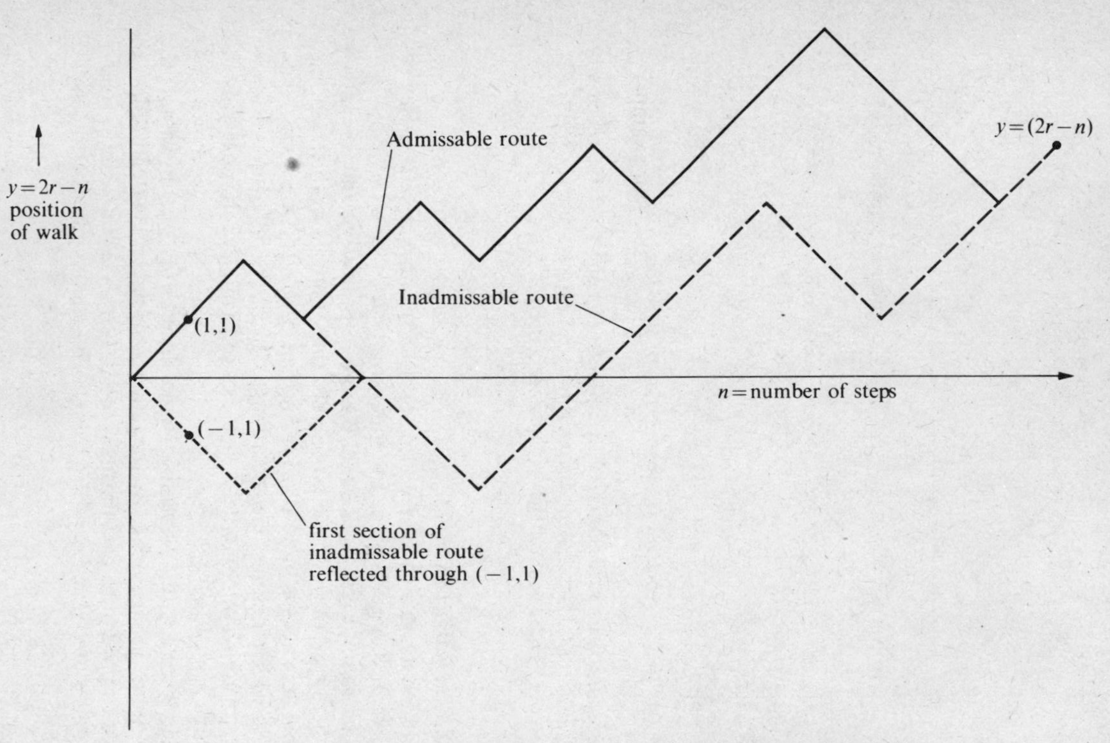

FIG. 8.2. Admissible and inadmissible paths in a random walk

$$= \text{Total no of routes from } (0, 0) \text{ to } (2r - n - 1, n - 1)$$

$$- \text{Total no of routes from } (0, 0) \text{ to } (2r - n + 1, n - 1)$$

$$= \binom{n-1}{r-1} - \binom{n-1}{r}$$

$$= \frac{(n-1)!}{(n-r)!(r-1)!} - \frac{(n-1)!}{(n-r-1)!r!}$$

$$= \frac{n!}{r!\,(n-r)!}\left(\frac{r}{n} - \frac{n-r}{n}\right)$$

$$= \frac{2r-n}{n}\binom{n}{r} \qquad (8.2)$$

Thus the probability of such an all-positive route is this number divided by the total number of routes, $\binom{n}{r}$, giving the desired result:

$$\text{p} = \frac{2r-n}{n} \qquad (8.3)$$

8.1.4. Returns to zero

A return to $y = 0$ can only occur at n even $(= 2r)$ and its probability is simply

$$\text{u}_{2r} = \binom{2r}{r}(\text{pq})^r \leqslant \binom{2r}{r} 2^{-2r} \qquad (8.4)$$

since $\text{pq} \leqslant \frac{1}{4}$ for all $\text{p} + \text{q} = 1$ (with equality at $\text{p} = \text{q} = \frac{1}{2}$). Stirling's formula (equation 7.38) shows that

$$\text{u}_{2r} \sim \frac{1}{\sqrt{(2\pi)}}\frac{(2r)^{2r+\frac{1}{2}}}{r^{2r+1}} \cdot 2^{-r}\,(4\text{pq})^r \sim \frac{(4\text{pq})^r}{\sqrt{(\pi r)}} \qquad (8.5)$$

If we consider *first* returns to the origin, then the number of first returns at $n = 2r$ is double the number of all-positive routes from zero to

$$r' = r; n' = 2r - 1$$

giving in all:

$$2 \cdot \frac{2(r) - (2r-1)}{2r-1} \cdot \binom{2r-1}{r}$$

$$= \frac{2}{2r-1}\binom{2r-1}{r} \qquad (8.6)$$

routes.

The probability of a *first* return after $n = 2r$ is thus the ratio of this number of routes to the total number of routes returning to zero at $n = 2r$, and multiplied by u_{2r}. Thus:

$$f_{2r} = \frac{\dfrac{2}{2r-1}\dbinom{2r-1}{r}}{\dbinom{2r}{r}} u_{2r} \tag{8.7}$$

$$= \frac{\dfrac{2}{2r-1} \dfrac{(2r-1)!}{r!\,(r-1)!}}{\dfrac{(2r)!}{r!\,r!}} u_{2r}$$

so that

$$f_{2r} = \frac{2r}{2r\,(2r-1)} u_{2r} = \frac{1}{2r-1} u_{2r} \tag{8.8}$$

Using this approach we can prove the arc-sine law, which is stated without proof:

The probability that up to and including the $2r$ th step, the last visit to zero was at the $2k$ th steps.

$$\alpha_{2k,\,2r} = u_{2k}\, u_{2r-2k} \sim \frac{1}{n\pi\sqrt{x(1-x)}}$$

for n large where

$$x = k/r \tag{8.9}$$

This probability is greatest near $x = 0$ or 1, and least near $x = \frac{1}{2}$.

This result shows an example of a probability distribution which is not simply 'peaked in the middle' (Figure 8.3)

It also contradicts intuition in showing that the random walk is likely to stay almost all the time on *one* side of the central position. (The probability that $2k$ will be spent on the positive side out of $2r$ steps is equal to $\alpha_{2k,\,2r}$ as given in equation 8.9 above). The cumulative frequency is given by

$$p(x > k/r) = \frac{2}{\pi} \arc\sin \sqrt{x}, \tag{8.10}$$

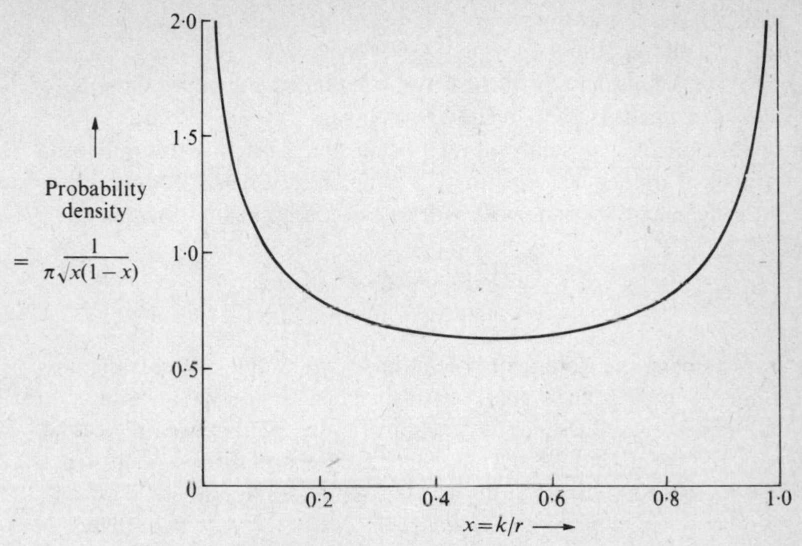

FIG. 8.3. Probability distribution of steps in a random walk

explaining why this is called the 'arc-sine law'.* The extension of these first-return laws is to the probability that in a run of $(2r + 1)$ steps there will be k changes of sign (that is changes of lead). This can be shown to be:

$$\begin{aligned} \beta_{k,2r+1} &= 2\mathrm{p}_{r+k,\,2r+1} \\ &= 2\binom{2r+1}{r+k}\mathrm{p}^{r+k}\,\mathrm{q}^{r-k+1} \end{aligned} \tag{8.11}$$

For an unbiassed random walk in which p = q, this quantity decreases as k increases because

$$\frac{\beta_{k+1,2r+1}}{\beta_{k,2r+1}} = \frac{r-k+1}{r+k+1} \tag{8.12}$$

8.1.5. *Examples of random walks*

Random walks have clear implications for sampling a sequence of individuals for some attribute. After any number of individuals (trials) the random walk path is the history of the sampling sequence. We can ask the question: 'Given the run of trials, what is the probability that the proportion having the attribute differs from 50% (or any other prior hypothesis)?' What we have shown very clearly is that a long period in

*arc-sin θ is an alternative notation to $\sin^{-1}\theta$

which 'haves' exceed 'have nots' is very poor evidence for departure from a null hypothesis (p = q = 50%).

Statistical tests to distinguish two samples on the basis of a series of ranked values are closely related to the random walk. In this case, the two samples are consolidated into a single rank-list, and this rank order is taken as the sequence of trials. If samples A (up) and B (down) have the same mean, then the walk will be indistinguishable from one in which:

$$p \propto \text{number of values in A}$$
$$q \propto \text{number of values in B}$$

For example, the Kolmogorov–Smirnov two-sample test uses the maximum *deviation* of the walk from its mean path. The Mann–Whitney test, on the other hand, uses as its test statistic the *area* between the actual path and the mean path. Experience has shown that these quantities are more reliable indicators of departure from a null hypothesis than is the number of times which the *actual* path crosses its *mean* path (which is what first-return laws are about).

In terms of modelling, the random walk provides a compromise between independence and dependence which is frequently relevant to the geographical and planning context. Sequences over time and space, even if not directly accumulative in the strict random-walk sense, are practically constrained from too great a 'step' at any point; and the mean path or 'most likely' value at the *next* measurement point is commonly strongly dependent on the value at the *current* point. We are clearly dealing here with auto-correlation (section 7.3.2.) and Markovian properties which are generalized below (section 8.2).

8.1.6. Stream networks as random walks

As an example of a simple random walk, we can take the topological structure of a dendritic stream network. Beginning from the downstream end of the trunk stream, we can scan the links of such a network in one of several ways, denoting points at which links branch by a '1' and points at which links terminate (stream heads) by a '0'. Each 1 increases the number of free ends by 1 and each 0 reduces the number by 1. Starting with a single trunk stream, denoted by an initial 1, we continue until the number of 0 s equals the number of 1 s, at which point there are no free links left, and the network is complete. Perhaps the most functional way of scanning the network is by 'generations', as in Figure 8.4.

Scanning each 'generation' (gen below) clockwise, the network in Figure 8.4 is denoted by

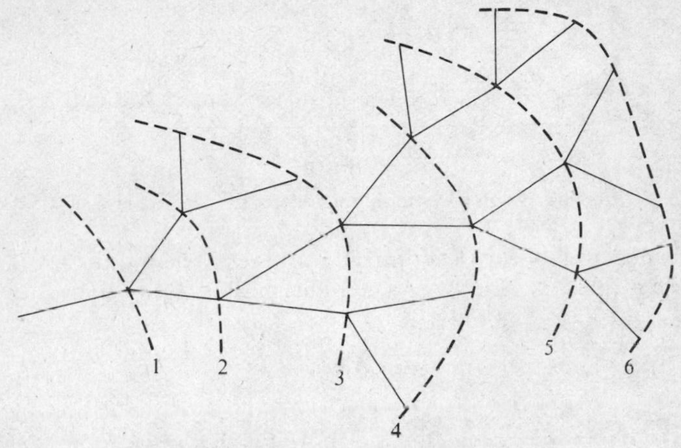

FIG. 8.4. Stream network as a random walk

1,	1,	11,	0011,	1100	0111,	000000
Initial trunk stream Gen. 0	Branch at Gen. 1.	Gen. 2.	Gen. 3.	Gen. 4.	Gen. 5.	Gen. 6

Each generation after the 1st has an even number of values in it (free links). The number of stream *heads,* the 'magnitude' of the network, is equal to the number of 0 s, in this example 11. Ignoring the commas between generations, this sequence is represented by a random walk in which 1 s are a step 'up' and 0 s a step 'down'. The y-value is then the number of links across the width of the network. The network of Figure 8.4 can thus be converted into the random walk of Figure 8.5. The number at the start of each new generation is the number of branches (or descendents) in that generation.

Because the number of 1 s and 0 s in the complete network are equal only when the network is complete, the behaviour of the network is described by the 'first return' of the random walk. If each left- and right-hand branch is treated as distinct, then each random walk path defines a 'topologically distinct channel network' (TDCN) using the terminology of Shreve (1969).

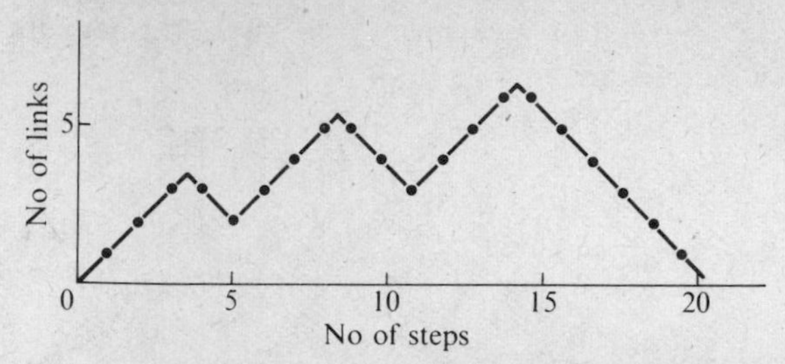

FIG. 8.5. Graph of 'random walk steps' for stream network

The random walk theory leads directly to several relevant results. First that the number of TDCNs for a given magnitude N is given by

$$\frac{1}{2N-1}\binom{2N-1}{N} \tag{8.13}$$

which is the version of equation 8.6 above for all-positive routes only. Secondly we can ask the probabilities of networks of different magnitude, given probabilities p of branching and q of terminating. This prediction may be used to give a partial test for a model of stream-net development by headward growth.

This probability is (for all positive routes which must start with a 1);

$$\text{p}(N) = \frac{1}{2N-1}\binom{2N-1}{N-1}\text{p}^{N-1}\text{q}^{N} \tag{8.14}$$

Since after the initial 1, there are $(n-1)$ 1 s and n 0s. By comparing coefficients of $(\text{pq})^n$ in the binomial expansion of

$(1-4\text{pq})^{\frac{1}{2}}$, the sum of the $\text{p}(N)$ is:

$$\begin{aligned}\Sigma\text{p}(N) &= \frac{1}{2\text{p}}[1-(1-4\text{pq})^{\frac{1}{2}}]\\ &= 1 \text{ for } 0 < \text{p} \leqslant \tfrac{1}{2}\\ &= \frac{1-\text{p}}{\text{p}} \text{ for } \tfrac{1}{2} \leqslant \text{p} \leqslant 1\end{aligned} \tag{8.15}$$

Thus, as intuition suggests, all networks must come to an end if $\text{p} < \frac{1}{2}$, that is if bifurcation is less frequent than termination; but if $\text{p} > \frac{1}{2}$ a proportion of networks never end at all. Figure 8.6 shows this theoretical distribution of network magnitudes for a range of values of p. The mean network magnitude;

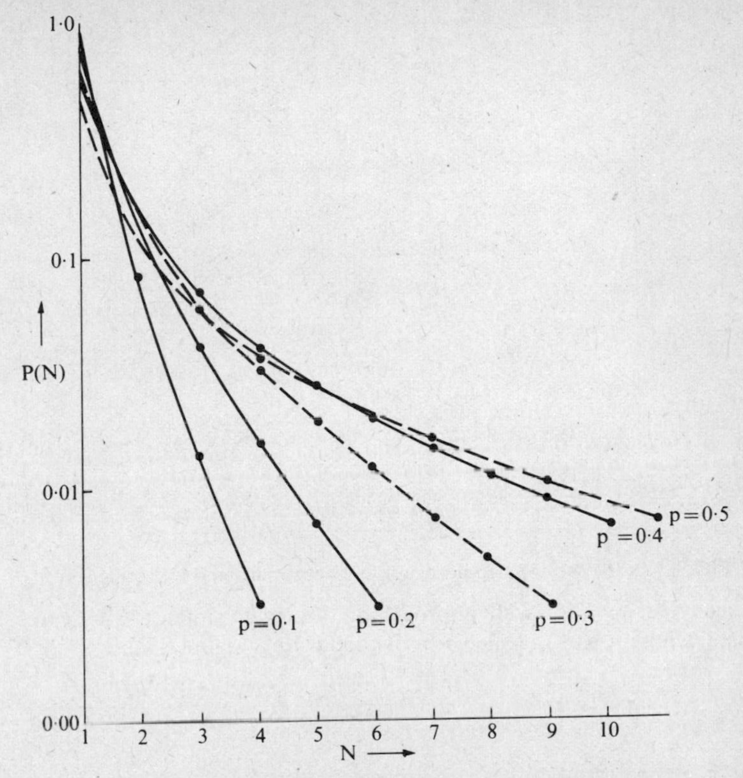

FIG. 8.6. Theoretical distribution of network magnitudes

$$\bar{N} = \frac{q}{1-2p} \tag{8.16}$$

which becomes infinite as $p \to \frac{1}{2}$.

8.1.7. Gambler's ruin: absorbing and reflecting barriers

In the simple random walk, steps have been assigned equal probabilities at all positions of the walk (y-values). To generalize random walks in a useful way, we can relax this uniformity. A variation which is commonly appropriate is where a walk, having once reached zero (or some other y-value), remains at zero – that is, the walk is *absorbed* in the $y = 0$ level (Figure 8.7). For example a gambler who has exhausted his original stake money (and is allowed no credit) cannot continue to play and remains with zero capital from then on. A game between two players similarly has two of these 'absorbing states', corresponding to bankruptcy for each of the players. On the other hand, a level which cannot

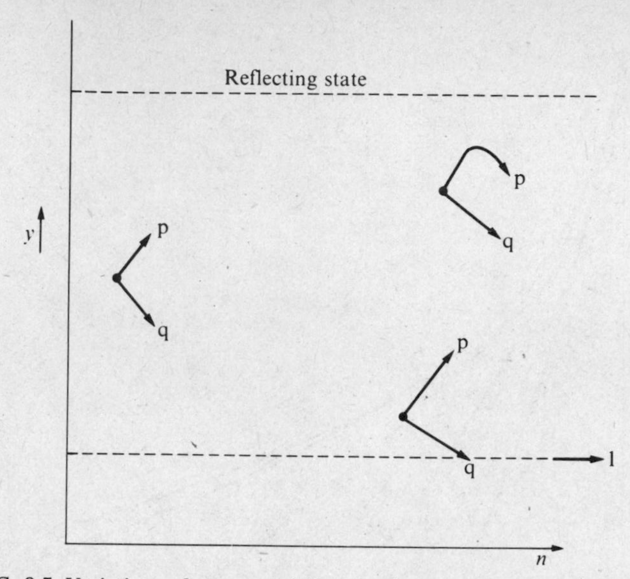

FIG. 8.7. Variations of step probabilities at absorbing and reflecting states

be occupied, so that a walk approaching this level is reflected back to remain where it was, is called a 'reflecting state' (Figure 8.7).

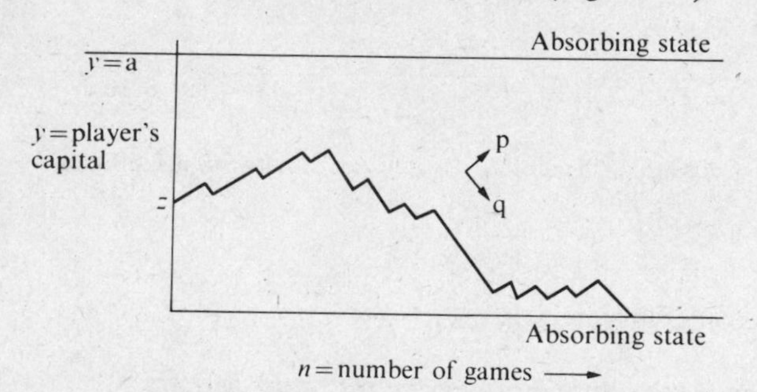

FIG. 8.8. Gambler's ruin

The classic 'gambler's ruin' problem is concerned with two players, with combined capital a (Figure 8.8), the random walk follows the progress of one player, whose initial capital is z $(z < a)$.

In each individual game, the probability of winning is p, and of losing q. Let p_z, q_z be the respective probabilities of winning and losing overall. After one game the player's capital is either $(z - 1)$ or $(z + 1)$, so that:

$$q_z = p\ q_{z+1} + q\ q_{z-1} \tag{8.17}$$

This is a difference equation for q_z, satisfying the boundary conditions $q_0 = 1$, $q_a = 0$. Using the methods of section 6.5 above, the solution is:

$$\left.\begin{aligned} &\text{for } p \neq q \qquad q_z = \frac{(q/p)^a - (q/p)^z}{(q/p)^a - 1} \\ &\text{for } p = q \qquad q_z = 1 - z/a \end{aligned}\right\} \qquad (8.18)$$

8.1.8. The effect of base-level – a geomorphological example

The development of a slope profile is constrained by particular processes which define the rate or probability of erosion at a point; but also by a general dependence on gravity as a driving force, requiring that gradients must lead steadily down from divides, and that it is impossible to erode appreciably below sea level (or base-level). A simple random walk which illustrates the influence of these factors in isolation is one in which base-level is an absorbing state at $y = 0$ (Figure 8.9). At

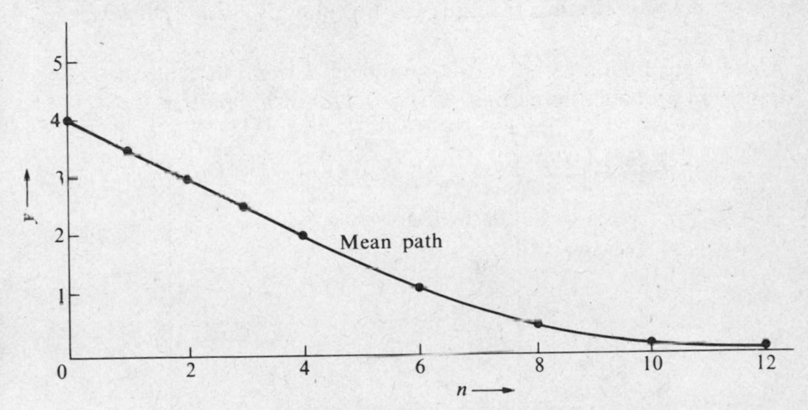

FIG. 8.9. Slope profile, showing effect of base level

higher elevations, the probabilities of remaining at *same* elevation is p, and of reducing by one step in elevation q. Using the binomial process directly, after n steps from a start at elevation z, the probability of being at elevation y (> 0);

$$p(y, n) = \binom{n}{z-y} p^{n-z+y}\, q^{z-y} \qquad (8.19)$$

For $n \leqslant z$ the mean path of this walk exactly follows the route:

$$y = z - nq \qquad (8.20)$$

At larger n, the sun is truncated by the walks which are adsorbed in zero, and the mean path approaching closer and closer to the zero line, as shown in Figure 8.9.
This curve, with a suitable scale for the unit step, shows one effect of base-level: the effect of preventing erosion below it. It ignores subtler effects, for example in reducing erosion on lower gradients, and so perhaps reflects only the *short*-term adjustment to a new base-level.

8.2. Markov chains

8.2.1. Definitions

A natural generalization of random walks, which includes the concepts of absorbing and reflecting states, is to a walk in which movement can be between any two states. In a simple Markov chain, as it is called, the probability of moving from the i th to the j th state is called a transition probability, and denoted by p_{ij}. Normally the transition probabilities are considered to remain constant from step to step, so that the probabilities of moving from the i th state depend only on the *present* position (in the i th state) and not directly on any *previous* positions.
Where there are n possible states, numbered 1 to n, the table of transition probabilities defines an $(n \times n)$ *transition matrix;*

$$\mathbf{P} = (p_{ij}) = \begin{bmatrix} p_{11} & p_{12} & p_{13} \cdots\cdots & p_{1n} \\ p_{21} & p_{22} & & \\ p_{31} & & & \\ \cdots\cdots & \cdots\cdots & \cdots\cdots & \cdots\cdots \\ \cdots\cdots & \cdots\cdots & \cdots\cdots & \cdots\cdots \\ p_{n1} \cdots\cdots & \cdots\cdots & \cdots\cdots & p_{nn} \end{bmatrix} \quad (8.21)$$

For example, the transition matrix in the last (base-level) example, for the five states, $y = 0, 1, 2, 3, 4$ (Figure 8.9), is:

$$\mathbf{P} = \begin{bmatrix} 1 & 0 & 0 & 0 & 0 \\ \frac{1}{2} & \frac{1}{2} & 0 & 0 & 0 \\ 0 & \frac{1}{2} & \frac{1}{2} & 0 & 0 \\ 0 & 0 & \frac{1}{2} & \frac{1}{2} & 0 \\ 0 & 0 & 0 & \frac{1}{2} & \frac{1}{2} \end{bmatrix} \quad (8.22)$$

We can also use matrix notation to define a column vector containing the probabilities of being in the n states at time t:

$$\boldsymbol{\pi}_t = \begin{bmatrix} \pi_{1,t} \\ \pi_{2,t} \\ \cdot \\ \cdot \\ \cdot \\ \cdot \\ \pi_{n,t} \end{bmatrix} \tag{8.23}$$

where $\pi_{i,t}$ is the probability of being in state i at time t.

8.2.2. Change in probabilities over time

The probability of being in state i at the next time interval, $t + 1$ can be seen to be made up of the possibilities:

$$\sum_k \text{p (being in state } k \text{ at time } t) \times \text{p (moving from } k \text{ th to } i \text{ th state)} \tag{8.24}$$

or

$$\pi_{i,t+1} = \sum_k \pi_{k.t}\, p_{ki} \tag{8.25}$$

Repeating for all values of i, we have:

$$\boldsymbol{\pi}_{t+1} = \mathbf{P}' \cdot \boldsymbol{\pi}_t \tag{8.26}$$

where $\mathbf{P}'$ is the transpose of the transition matrix (rows and columns exchanged). It can be seen, by the rules of matrix multiplication (section 4.2), that the product on the right hand side is a

$$(n \times n) . (n \times 1) = (n \times 1) \text{ matrix}$$

for which the i th element in the column is:

$$\sum_k (\mathbf{P}')_{ik} \cdot \pi_{k,t}$$
$$= \sum_k p_{ki} \cdot \pi_{k,t} \tag{8.27}$$

confirming the deduction from equation 8.25.

Applying equation 8.26 repeatedly:

$$\boldsymbol{\pi}_m = \mathbf{P}' \boldsymbol{\pi}_{n-1} = (\mathbf{P}')^2 \boldsymbol{\pi}_{m-2} = \ldots = (\mathbf{P}')^m \boldsymbol{\pi}_0 \tag{8.28}$$

where π_0 is the distribution of probabilities at $t = 0$.
For the base-level example starting at $y = 4$ (5th state) at time 0.

$$\pi_1 = P'\pi_0 = \begin{bmatrix} 1 & \frac{1}{2} & 0 & 0 & 0 \\ 0 & \frac{1}{2} & \frac{1}{2} & 0 & 0 \\ 0 & 0 & \frac{1}{2} & \frac{1}{2} & 0 \\ 0 & 0 & 0 & \frac{1}{2} & \frac{1}{2} \\ 0 & 0 & 0 & 0 & \frac{1}{2} \end{bmatrix} \cdot \begin{bmatrix} 0 \\ 0 \\ 0 \\ 0 \\ 1 \end{bmatrix} = \begin{bmatrix} 0 \\ 0 \\ 0 \\ \frac{1}{2} \\ \frac{1}{2} \end{bmatrix} \begin{matrix} \text{Level } 0 \\ 1 \\ 2 \\ 3 \\ 4 \end{matrix}$$

$$\pi_2 = P'\pi_1 = \begin{bmatrix} 0 \\ 0 \\ \frac{1}{4} \\ \frac{1}{2} \\ \frac{1}{4} \end{bmatrix} \quad \pi_3 = P'\pi_2 = \begin{bmatrix} 0 \\ \frac{1}{8} \\ \frac{3}{8} \\ \frac{3}{8} \\ \frac{1}{8} \end{bmatrix} \quad \text{etc} \tag{8.29}$$

8.2.3. Distribution after long times

The states of a Markov chain may be of various types, of which absorbing states are an extreme example. An important group of states is a 'closed set', within which movement can continue indefinitely, but from which there is no escape. An absorbing state is really a closed set with only one number. Any state which does not belong to a closed set (and is not an absorbing state) must be a 'transient state' *from* which it is possible to get into a closed set, but with no prospect of returning. A reflecting state is just one example of a transient state.

If a Markov chain is continued through a large number of steps, then all the transient states will become empty, and the walk will circulate within one of the closed sets (or be fixed in an absorbing state).

Each absorbing state and each closed set gives rise to one eigenvalue of + 1 for the transition matrix. The simplest irreducible matrix consists of just one closed set, within which every state communicates with every other state (indirectly). Within such a matrix, the probabilities of being in the various states persist indefinitely, either as fixed values (aperiodic), or cycling through a repeated sequence. A single example of the latter (periodic) behaviour is seen for the simple drunkard's walk (seetion 8.1.2. *et seq*). For a zero starting position, (say), even positions can only be reached at even numbers of trials; and odd positions at odd numbers of trials, a cyclic movement of period 2.

For the aperiodic case there is a set of probabilities $\boldsymbol{\pi}_\infty$ for which

$$\mathbf{P}'\boldsymbol{\pi}_\infty = \boldsymbol{\pi}_\infty \tag{8.30}$$

which corresponds to the + 1 eigenvalue and to the equilibrium distribution of probabilities after a long time period (independent of starting position). Convergence towards this distribution is rapid when the next largest eigenvalue is of magnitude much less than 1.

For example, the following transition matrix shows the probability for next year of growing corn (state 1), maguey (state 2), or fallowing (state 3) following each this year's crop, for a semi-arid area in Mexico (Kirkby, 1973, p. 80).

$$\mathbf{P} = \begin{bmatrix} \cdot 25 & \cdot 21 & \cdot 54 \\ \cdot 22 & \cdot 67 & \cdot 11 \\ \cdot 22 & \cdot 02 & 1\cdot 76 \end{bmatrix} \tag{8.31}$$

This means that the probability of a rotation from corn to maguey is 0·21 etc. If this rotation is allowed to settle down to equilibrium proportions, they will be given by $\mathbf{P}'\boldsymbol{\pi}_\infty = \boldsymbol{\pi}_\infty$:

$$\begin{bmatrix} \cdot 25 & \cdot 22 & \cdot 22 \\ \cdot 21 & \cdot 67 & \cdot 02 \\ \cdot 54 & \cdot 11 & \cdot 76 \end{bmatrix} \cdot \begin{bmatrix} \pi_1 \\ \pi_2 \\ \pi_3 \end{bmatrix} = \begin{bmatrix} \pi_1 \\ \pi_2 \\ \pi_3 \end{bmatrix} \tag{8.32}$$

but these equations are redundant, since both sides add to $\pi_1 + \pi_2 + \pi_3 = \pi_1 + \pi_2 + \pi_3$. We therefore need an additional equation, which is that each field must be in one of the three states;

$$\pi_1 + \pi_2 + \pi_3 = 1\cdot 0 \tag{8.33}$$

Solving equations 8.32 and 8.33, we have the equilibrium proportions:

$$\boldsymbol{\pi}_\infty = \begin{bmatrix} \cdot 227 \\ \cdot 180 \\ \cdot 593 \end{bmatrix} \tag{8.34}$$

The rapid convergence to this value can be seen both by the low 2nd eigenvalue of 0·65 (λ = 1·00, 0·65, 0·03), which converges as

$$\begin{aligned} &0\cdot 65^0, 0\cdot 65^1, 0\cdot 65^2, 0\cdot 65^3, \ldots. \\ &= 1, 0\cdot 65, 0\cdot 42, 0\cdot 27, \ldots \end{aligned} \tag{8.35}$$

or by the sequence of probabilities, starting with an all-corn pattern (say);

$$\pi_0 = \begin{bmatrix} 1 \\ 0 \\ 0 \end{bmatrix} \quad \pi_1 = \begin{bmatrix} \cdot 25 \\ \cdot 21 \\ \cdot 54 \end{bmatrix} \quad \pi_2 = \begin{bmatrix} \cdot 229 \\ \cdot 204 \\ \cdot 568 \end{bmatrix} \quad \pi_3 = \begin{bmatrix} \cdot 227 \\ \cdot 196 \\ \cdot 517 \end{bmatrix} \qquad (8.36)$$

8.2.4. Other types of chain

The Markov chain is a very simple type of variation over time in which the development at any time is controlled entirely by the present state of the system, and not by the history which led up to the present state. It is thus the probabilistic analogue of the simple deterministic models of classical physics, and a natural extension of deterministic modelling to processes with variable outcomes.

The simple Markov process may be generalized to allow variation of the transition probabilities over time, although the mathematical simplicity is lost. It may also be generalized to allow the transition probabilities to depend not only on the present position, but on a sequence of past positions, for example in the linear relation:

$$_m = \mathbf{P}'\boldsymbol{\pi}_{m-1} + \mathbf{Q}'\boldsymbol{\pi}_{m-2} + \mathbf{R}'\boldsymbol{\pi}_{m-3} + \ldots \qquad (8.37)$$

in which **P**, **Q**, **R**, are appropriate transition matrices. Such a relation with r terms is sometimes described as an rth order Markov process.

8.3. Poisson processes

8.3.1. Derivation

The Poisson distribution has been obtained in section 7.6.2. as a limiting case of the binomial distribution where a finite number of occurences occur on an infinite number of possible occasions. Sampling in time or space is intimately concerned with this distribution, since a finite number of events or items are found in a space or time continuum of one or more dimensions, containing an infinite number points.

Where the probability of an occurrence is the same at every point, then the distribution of occurrences should follow a Poisson distribution. For clarity, consider occurrences along a line, with a constant probability density p of occurrence per unit distance (Figure 8.10).

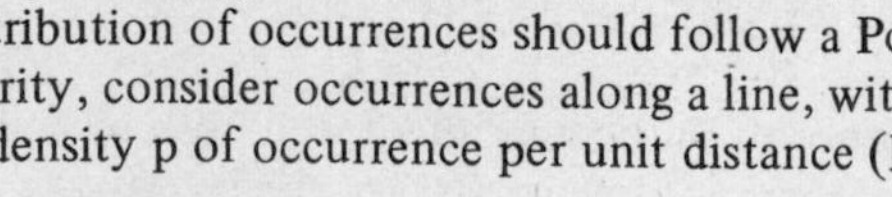

$0 \rightarrow x$ $\quad x \quad (x + \delta x)$

FIG. 8.10. Probability of events along a line

If $p_n(x)$ is the probability of n occurrences in distance x, then:

$$p_n(x + \delta x) - (1 - p \cdot \delta x)\, p_n(x) + p\delta x\, p_{n-1}(x) \tag{8.38}$$

In the limit as $\delta x \rightarrow 0$;

$$\frac{d[p_n(x)]}{dx} = p[p_{n-1}(x) - p_n(x)] \tag{8.39}$$

For $n = 0$:

$$\frac{dp_0(x)}{dx} = p \cdot p_0(x) \tag{8.40}$$

so that the initial condition is

$$p_0(x) = Ae^{-px} \tag{8.41}$$

It can readily be shown by inspection that

$$p_n(x) = A\frac{(px)^n}{n!}\, e^{-px} \tag{8.42}$$

satisfies equations 8.39 and 8.41.
By summing over n, which must have *some* value;

$$1 = \sum_{n=0}^{\infty} p_n(x) - Ae^{-px} \sum_{n=0}^{\infty} \frac{(px)^n}{n!} \tag{8.43}$$

Since the sum on the right-hand side is the series for e^{+px}, the constant $A = 1$, and

$$p_n(x) = \frac{(px)^n}{n!}\, e^{-px} \tag{8.44}$$

It is apparent by comparison with section 7.6.2 that this distribution is the Poisson distribution, and so has mean and variance of:

$$z = px \tag{8.45}$$

Although the distribution has been derived for events along a line, exactly the same distribution applies to sampling in more than one dimension, where p becomes the probability per unit area and so on.

8.3.2. Spacing of occurrences

The Poisson distribution may be looked at in two ways. First as a sampling distribution which describes the frequency distribution of

number of occurrences within a fixed space x. Second as a spatial distribution describing the probability of varying spaces containing a fixed number of events n. Of particular interest is the expected distance apart of events, that is the probability that a space x contains *no* events, which is the same as the cumulative probability that events are spaced more than x apart.
Thus

$$\text{p (spacing} \geqslant x) = \text{p}_0(x) = \text{e}^{-\text{p}x} \tag{8.45}$$

This distribution has a probability density obtained by differentiating with respect to x (along a line);

$$\text{p (spacing} = x) = -\frac{\text{d}}{\text{d}x}[\text{p}(x)] = \text{pe}^{-\text{p}x} \tag{8.46}$$

which is also called the 'exponential distribution'.
In a similar way the cumulative probability for the space occupied by $(n + 1)$ events is

$$\begin{aligned} \text{p}(n + 1 \text{ spacing} \geqslant x) &= \sum_{m=0}^{n} \text{p}_m(x) \\ &= \sum_{m=0}^{n} \frac{(\text{p}x)^m}{m!} \text{e}^{-\text{p}x} \end{aligned} \tag{8.47}$$

with probability density:

$$\begin{aligned} \text{p}(n + 1 \text{ spacing} = x) &= -\sum_{m=0}^{n} \frac{\text{d}}{\text{d}x}[\text{p}_m(x)] \\ &= -\text{p}\sum_{m=0}^{n} \left[\frac{(\text{p}x)^{m-1}}{(n-1)!}\text{e}^{-\text{p}x} - \frac{(\text{p}x)^m}{m!}\text{e}^{-\text{p}x}\right] \\ &= \text{p}\frac{(\text{p}x)^n}{n!}\text{e}^{-\text{p}x} \end{aligned} \tag{8.48}$$

the other terms cancelling out. This distribution is called the Gamma distribution of order $(n + 1)$.

8.3.3. *Nearest neighbour statistics*

The spacing of occurrences in a two-dimensional space is used directly in calculating nearest neighbour statistics. The probability that a circle of radius r around an occurrence contains no other occurrences is the probability that the nearest neighbour distance N_1, is greater than r.

Thus the probability that $N_1 \geqslant r$ is the cumulative probability:

$$p(N_1 \geqslant r) = e^{-p\pi r^2} \qquad (8.49)$$

The probability density is obtained by differentiating with respect to r.

$$p(N_1 = r) = -2p\,\pi\, r\, e^{-p\pi r^2} \qquad (8.50)$$

The mean nearest neighbour distance on a random (Poisson) distribution is then

$$\begin{aligned} \bar{N}_1 &= \int_{r=0}^{\infty} r\ p(N_1 = r)\ dr \\ &= -\int_{r=0}^{\infty} 2p\,\pi\, r^2\, e^{-p\pi r^2}\ dr \qquad (8.51) \end{aligned}$$

Integrating by parts,

$$\begin{aligned} \bar{N}_1 &= [re^{-p\pi r^2}]_0^{\infty} - \int_0^{\infty} e^{-p\pi r^2} dr \\ &= \tfrac{1}{2} p^{\frac{1}{2}} \qquad (8.53) \end{aligned}$$

Replacing p, the density (number/unit area), by d = 1/p, gives the more usual form:

$$\bar{N}_1 = \tfrac{1}{2}\, d^{\frac{1}{2}} \qquad (8.54)$$

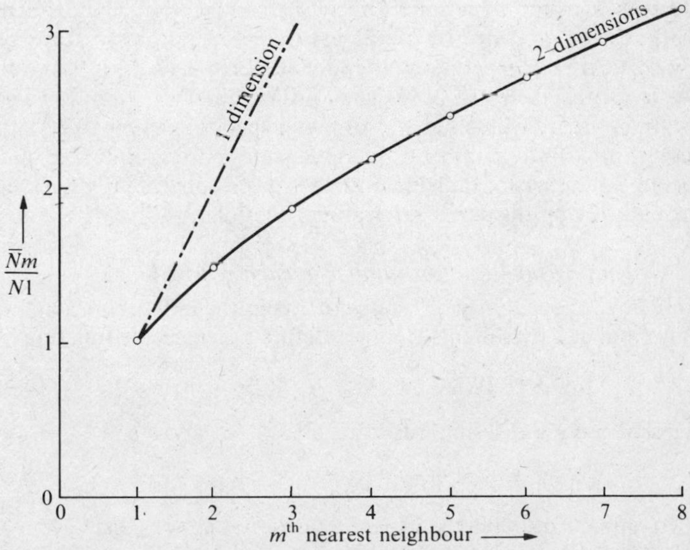

FIG. 8.11. mth–nearest–neighbour distances

The nearest neighbour statistic is the ratio of the actual measured mean nearest-neighbour distance to this random norm. In a similar way the $(m+1)^{\text{th}}$ nearest neighbour in a random distribution is at a distance whose mean is:

$$\bar{N}_{m+1} = \tfrac{1}{2}d^{\frac{1}{2}}\left[1+\tfrac{1}{2}+\frac{1.3}{2.4}+\frac{1.3.5}{2.4.6}+\ldots+\frac{(2m)!}{2^{2m}(m!)^2}\right] \qquad (8.55)$$

This relationship is shown in Figure 8.11. It increases roughly as $(n)^{\frac{1}{2}}$; compared to the exact linear increase ($\bar{N}_m = n\bar{N}_1$) for nearest neighbours in *one*-dimension.

8.3.4. Other examples of Poisson processes

The Poisson process describes the interval between events which are occurring with constant probability in space or time. They therefore occur in the simplest models of events over time, such as arrivals in a queue, or over space, such as branching along a stream, but are by no means universally applicable. One critical reason for this lies in the auto-correlation of natural sequences which leads to clustering of events in time and space.

8.4. Branching processes

8.4.1. Introduction

Processes such as reproduction or nuclear chain reaction, and dendritic networks can be modelled by looking at their generation structure in a family tree. At each generation, an individual may have 0, 1, 2 or more descendents, and these can be assigned probabilities $p_0, p_1, p_2 \ldots$ The total number of individuals at any time is a special example of a Markov chain with an infinite number of possible states corresponding to all the possible numbers of individuals, but is more conveniently handled mathematically by the use of generating functions.

8.4.2. Generating function for number in each generation

Let $p_{r,n}$ be the probability of r descendents in the n th generation, starting from one individual. Then we define a 'generating function'.

$$P_n(x) = p_{0,n} + p_{1,n}x + \ldots + p_{r,n}x^r + \ldots \qquad (8.56)$$

and clearly, using the definitions of $p_0, p_1, p_2, \ldots$ above,

$$P_1(x) = p_0 + p_1x + p_2x^2 + \ldots + p_rx^r + \ldots \qquad (8.57)$$

Then to obtain r descendents in generation $(n+1)$, we must have k ($\leqslant r$) in generation n, and go from k to r in generation $(n+1)$.

The latter probability can be seen to be the coefficient of x^r in the expansion of

$$[p_1(x)]^k \tag{8.58}$$

so that $p_{r,n+1}$ is the coefficient of x^r in the expansion of

$$p_{k,n}\ [p_1(x)] \tag{8.59}$$

which is the coefficient of x^r in

$$P_{n-1}\ [P_1(x)] \tag{8.60}$$

so that

$$P_n(x) = P_{n-1}\ [P_1(x)] \tag{8.61}$$

Applying repeatedly from one end and then the other, we obtain:

$$\begin{aligned} P_n(x) = P_{n-1}\ [P_1(x)] &= P_{n-2}\left\{ P_1[P_1(x)] \right\} \\ &= \dots \\ &= P_1\left(P_1\left\{ \dots [P_1(x)] \right\} \right) \\ &= P_1[P_{n-1}(x)] \end{aligned} \tag{8.62}$$

In this form, the generating functions $p_n(x)$ can be built up in sequence.

For example, consider a stream network which can either bifurcate (with probability 0·5) or terminate (0·5).
Then

$$\left.\begin{aligned} p_0 &= 0{\cdot}5 \\ p_1 &= 0 \\ p_2 &= 0{\cdot}5 \\ p_3 \text{ and so on} &= 0 \end{aligned}\right\} \tag{8.63}$$

We can write

$$P_1(x) = 0{\cdot}5 + 0{\cdot}5\,x^2 \tag{8.64}$$

$$\begin{aligned} P_2(x) &= P_1[P_1(x)] \\ &= 0{\cdot}5 + 0{\cdot}5\,(0{\cdot}5 + 0{\cdot}5\,x^2)^2 \\ &= 0{\cdot}625 + 0{\cdot}25\,x^2 + 0{\cdot}125\,x^4 \end{aligned} \tag{8.65}$$

so that

$$\left.\begin{aligned} p_{0,2} &= 0{\cdot}625 \\ p_{2,2} &= 0{\cdot}25 \\ p_{4,2} &= 0{\cdot}125 \end{aligned}\right\} \qquad (8.66)$$

Again,

$$\begin{aligned} P_3(x) &= P_1[P_2(x)] \\ &= 0{\cdot}5 + 0{\cdot}5\,(0{\cdot}625 + 0{\cdot}25\,x^2 + 0{\cdot}125\,x^4)^2 \\ &= 0{\cdot}695 + 0{\cdot}156\,x^2 + 0{\cdot}109\,x^4 + 0{\cdot}031\,x^6 + 0{\cdot}008\,x^8 \\ &= p_{0,4} + p_{2,4}\,x^2 + p_{4,4}\,x^4 + p_{6,4}x^6 + p_{8,4}x^8 \end{aligned} \qquad (8.67)$$

8.4.3. Extinction probabilities

Two problems are of particular interest. One is the probability of extinction (zero descendents) at a given number of generations, and the second is the behaviour of a network conditional on its extending through a given number of generations. These are problems analogous to those in section 8.1.6, where we were concerned with the properties of a network conditional on a given *magnitude* (number of first-order links). The similarities and differences between the two approaches, one based on random walks and the other on branching models, show how different modelling methods may complement one another.

If x is set equal to zero in the generating function then it can be used to generate $p_{0,n}$, the probability of stopping at or before the generation n as follows:

$$p_{0,n+1} = P_1(p_{0,n}) \qquad (8.68)$$

Thus, in our example above

$$\begin{aligned} p_{0,0} &= 0{\cdot}5 + 0{\cdot}5\,(0)^2 = 0{\cdot}5 \\ p_{0,1} &= 0{\cdot}5 + 0{\cdot}5\,(0{\cdot}5)^2 = 0{\cdot}625 \\ p_{0,2} &= 0{\cdot}5 + 0{\cdot}5\,(0{\cdot}625)^2 = 0{\cdot}695 \end{aligned} \qquad (8.69)$$

and so on.

The probabilities of stopping *at* a particular number of generations is given by the differences between these terms. If π_n is the probability of extinction at n generations:

$$\pi_n = p_{0,n} - p_{0,n-1} \qquad (8.70)$$

The declining sequence of values of π_n is illustrated in Figure 8.12 for probabilities $p_2 = 0{\cdot}5$ (as above) and $p_2 = 0{\cdot}4$ for bifurcation. It should be compared with Figure 8.6 which gives the corresponding probabilities for each network *magnitude.*

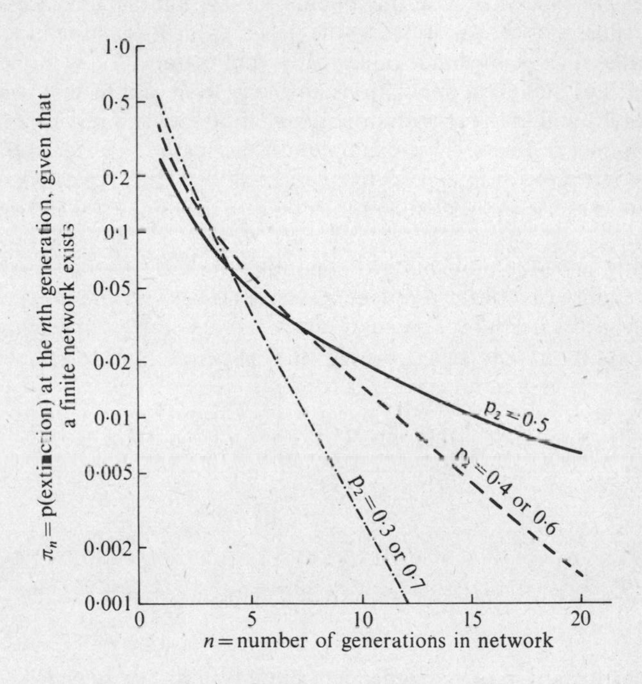

FIG. 8.12. Probabilities of extinction at the nth generation

8.5. Other stochastic processes

8.5.1. Types of stochastic process

This chapter has emphasized random walk processes though these are only one type of stochastic process out of the many which are relevant to geographical models. In an introductory work it has seemed unwise to cover too much ground, and this section can only attempt to summarize some other types of stochastic process. Readers who wish to pursue these topics will need to follow them up in Feller (1950), Cox and Miller (1965), Bailey (1964), Ashton (1966) or other general texts.

8.5.2. Birth and death processes

In simple random walks and Markov chains, the walk passes between many states, but is only in one state at a time. Even when a number of walks are going on simultaneously, it is characteristic that the total of stochastic models is concerned with processes of growth of which the branching processes above is one example, and may be treated generally as birth and death processes. In their simplest form, these processes are concerned with the 'population' within one unit over continuous time interval. Additions to the unit are 'births'; and reductions are 'deaths'. This general model has many applications, for example to the growth of biological populations, in the growth of infected cases in an epidemic, and in the growth or absence of queues at a service counter or telephone exchange. These examples differ greatly however in the factors controlling birth and death rates. In all but the simplest examples, it is simpler to use simulation methods than to attempt an analytical situation.

Let $p(n, t)$ be the probability of a population n at time t. Let α be the probability of a birth (or queue arrival) in unit time and σ the probability of a death (or service of queue member). We can calculate the probability of n at time $(t + \delta t)$ from the possibilities at time t as follows:

$$p(n, t+\delta t) = \delta t p(n+1, t) + 1 - (\alpha + \sigma)\, d t p(n, t) + \alpha \delta t p(n-1, t) \tag{8.71}$$

or

$$\frac{\partial p\,(n, t)}{\partial t} = \sigma p(n+1, t) - (\alpha + \sigma)\, p(n, t) + \alpha\, p(n-1, t) \tag{8.72}$$

This tends towards a time independent equilibrium state in which

$$0 = \sigma\, p\,(n+1) - (\alpha + \sigma)\, p(n) + \alpha\, p(n-1) \tag{8.73}$$

which has been solved as an example of a difference equation in section 6.5 above. For constant $\rho = \alpha/\sigma < 1$. It gives

$$p(n) = (1 - \rho)\, \rho^n \tag{8.74}$$

The case where α and σ are *constant* is an important one, corresponding to a queue in which both arrivals and service times conform to a Poisson distribution, with expectations α, σ respectively.

In simple reproduction processes in which n refers to the population of females, the birth and death rates, α and σ, are not constant, but proportional to the population n. Equation 8.72 is less amenable to analytical solution, though equation 8.75 can be solved to give the time independent solution

$$\left.\begin{aligned} p(0) &= 1 \\ p(n) &= 0 \; n \geqslant 1 \end{aligned}\right\} \qquad (8.75)$$

If the solution is restricted to non-zero populations for which $\rho < 1$, then:

$$p(n) = \frac{-\rho^n}{n \log (1 - \rho)} \qquad (8.76)$$

Other types of dependence of the birth rate and death rate on the population are found where the environment is limiting or in competitive or predator-prey contexts (Maynard Smith, 1973). In the context of epidemics, the growth of infected cases depends jointly on the infected and total populations in an even more complex manner.

These examples of stochastic processes are far from complete, but it is fair to generalize that in most cases, it is preferable to use computer simulation methods rather than mathematical analysis to solve any but the simplest problems. Despite their greater complexity relative to deterministic models, they offer crucial insights especially into the stability of equilibria and the extent to which optima are critically determined.

Exercises

Section 8.1

(1) Using a random number table to generate a random walk with $p = q = \frac{1}{2}$ ('up' for digits 0–4; 'down' for digits 5–9) with 10 steps starting at $y = 0$. By replicating 20 times, obtain a mean path.
(a) for all routes
(b) for all positive routes only

(2) Obtain a dendritic network of *either* stream *or* main road shortest routes to a city from a map. Convert it to a random walk.

(3) Enumerate the 14 'topologically distinct channel networks' of magnitude 5. Classify them into groups on the basis of stream ordering.

Section 8.2.

(4) At home on Sunday mornings I spin a coin to decide whether to go to church or stay another hour in bed. In church on Sunday morning, I spin a coin to decide whether to go to the pub or to lunch.

After each pint in the pub I spin to decide whether to stay for another before lunch. Construct a 4 × 4 transition matrix for the four states. Starting at home, what is the probability that I will need to spin more than four times before lunch?

(5) For the matrix of transition probabilities:

$$P = \begin{bmatrix} 0{\cdot}2 & 0{\cdot}4 & 0{\cdot}4 \\ 0{\cdot}2 & 0{\cdot}8 & 0 \\ 0 & 0{\cdot}8 & 0{\cdot}2 \end{bmatrix}$$

obtain the equilibrium proportions in each of the three states. Comment on the convergence towards these proportions starting from the distribution

$$\begin{bmatrix} 1 \\ 0 \\ 0 \end{bmatrix}$$

(6) Sketch the Gamma distribution of order 2 for $(px) = 2$.

Section 8.3

(7) Show that the mean distance to the n th nearest neighbour along a line with a density d of points is:

$$\bar{N}_m = md$$

Section 8.4

(8) Calculate the generating function $P_3(x)$ for a branching network for which

$$p_0 = 0{\cdot}1$$

$$p_1 = 0$$

$$p_2 = 0{\cdot}9$$

$$p_3, p_4 \ldots = 0$$

Calculate the probabilities of extinction up to the 5th generation.

Section 8.5

(9) Write an essay, outlining how a stochastic model might be constructed for a problem of particular interest to you.

9. Maximization and minimization methods

9.1. Introduction

In Chapter 5, we showed how to use the methods of differential calculus to identify maxima and minima of functions. In this chapter, we consider three more advanced aspects of this topic. We begin, however, by clarifying a number of terms. First, note that if we find the maximum of a function

$$y = f(x) \tag{9.1}$$

say the point (x_0, y_0), then this same point is a *minimum* of

$$y = -f(x) \tag{9.2}$$

Thus, without loss of generality we can, and mostly will, discuss only the search for maxima, for convenience.

Secondly, we will often find it convenient to refer to a maximum (or minimum) value of a function as an 'optimum' value, and to refer to the process as 'optimization'. In the context of geography and planning, it is important to clarify the possible meanings of this word. We will use it here mainly in its mathematical sense – an optimum value of a function is its maximum or minimum, and there are no other connotations. In applied work, an objective function may be defined – say transport cost in a distribution problem – and an extreme value of such a function, such as minimum transport cost, is then an optimum value in both a mathematical and 'planning' sense. There will be many occasions, however, when an optimum value of a function has a meaning in a mathematical sense only, and there is often confusion in the literature when some authors imply other connotations in such cases. Thus, careful use of words in any particular situation is important.

Suppose we have a function of two variables

$$z = f(x, y) \tag{9.3}$$

In Chapter 5, we saw how to find maxima for such functions. A problem which frequently occurs, however, is to find the maximum value of such a function subject to a constraint, for example a budget constraint

$$x + y = B \tag{9.4}$$

This is the problem of *constrained* maximization and it forms the subject of section 9.2.

Another kind of maximization problem can be defined as follows. Suppose

$$y = f(x) \tag{9.5}$$

defines some curve in the (x,y) plane. Define an integral z by

$$z = \int_a^b F(x,y,y')dx \tag{9.6}$$

where

$$y' = \frac{dy}{dx} \tag{9.7}$$

for any given function F. Then, we can find the function f, or the 'path in the (x,y) plane, which maximizes z. This is the problem of the *calculus of variations* and forms the subject matter of section 9.3.

Finally, we identify a number of maximization or minimization problems which cannot be tackled by the conventional methods of algebra and calculus employed in sections 9.2 and 9.3. We tackle such problems by what are known as *algorithmic methods*. These include the solution to the problem of finding the shortest path through a network, and a class of so-called *linear programming* problems which arise partly as important methods in their own right, and partly because the Lagrangian methods for constrained maximization of section 9.2 are inapplicable when the objective function and the constraints are all linear.

We discuss each of these topics in turn, giving examples in each section.

9.2. Constrained maximization

9.2.1. The Lagrangian method

We begin with an example. Consider the function

$$z = x^{\alpha}y^{\beta}, \ \alpha > 0, \ \beta > 0 \tag{9.8}$$

which is to be maximized subject to

$$x + y = B \tag{9.9}$$

α and β are given parameters which satisfy

$$\alpha + \beta = 1 \tag{9.10}$$

This is a simple version of a common economic problem of maximizing utility or production, but for the time being we concentrate on its mathematical aspects only. We also restrict ourselves to the positive octant of a three-dimensional Cartesian co-ordinate system

$$x > 0, \ y > 0, \ z > 0 \tag{9.11}$$

Clearly, z is zero when x or y is zero, and will increase as x and y increase since

$$\frac{\partial z}{\partial x} = \alpha x^{\alpha-1}y^{\beta} > 0 \tag{9.12}$$

and

$$\frac{\partial z}{\partial y} = \beta x^{\alpha}y^{\beta-1} > 0 \tag{9.13}$$

The surface given by equation 9.8 will thus be a 'skewed' cone, the 'skewness' being determined by the relative values of α and β. Its shape is indicated in Fig. 9.1.

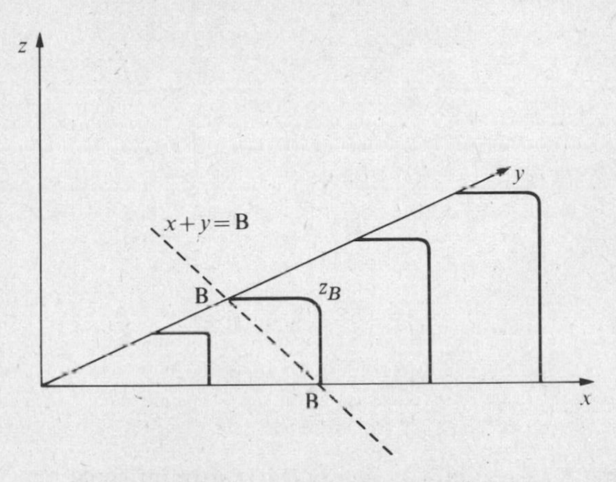

FIG. 9.1. Sketch of $z = x^{\alpha}y^{\beta}$

If we are seeking unconstrained maxima and minima, we can see intuitively that the minima are $z = 0$ at the origin or any point along the x- or y-axes. z takes an arbitrary large value as x or y become arbitrary large (provided neither is zero) and so there is no finite maximum. Suppose now we impose the condition given by equation 9.9. This equation represents a plane in Fig. 9.1 though the line $x + y = \mathrm{B}$ (which is shown) and perpendicular to the (x,y) plane. The surface $z = x^{\alpha}y^{\beta}$ cuts a section on this plane, which is the curve shown joining $y = \mathrm{B}$ and $x = \mathrm{B}$. On this curve, we can see that z has a finite maximum which we have denoted by z_{B} on the figure. Our task, now, is to find z_{B}, and the associated values of x and y. The method used for this was invented by Lagrange and uses what is called a Lagrangian multiplier. First, the constraint must be written in the form 'something equals zero'. That is,

$$x + y - \mathrm{B} = 0 \tag{9.14}$$

Let λ be the Lagrangian multiplier, and we form the expression

$$L = z + \lambda(x + y - \mathrm{B}) \qquad (9.15)$$

$$= x^{\alpha}y^{\beta} + \lambda(x + y - \mathrm{B}) \qquad (9.16)$$

which is known as the Lagrangian. Thus, the Lagrangian is the function to be maximized plus the multiplier times the left-hand side of the constraint equation. L is considered to be a function of x, y and λ. We will now be able to find the maximum value of L, using conventional methods, as a function of x, y and λ, and show that this is the *constrained* maximum of z. The maximum of L will be the solution of the simultaneous equations

$$\frac{\partial L}{\partial x} = 0 \qquad (9.17)$$

$$\frac{\partial L}{\partial y} = 0 \qquad (9.18)$$

$$\frac{\partial L}{\partial \lambda} = 0 \qquad (9.19)$$

That is

$$\alpha x^{\alpha-1}y^{\beta} + \lambda = 0 \qquad (9.20)$$

$$\beta x^{\alpha}y^{\beta-1} + \lambda = 0 \qquad (9.21)$$

$$x + y - \mathrm{B} = 0 \qquad (9.22)$$

Note that the third equation is simply the constraint equation. Further, since $x + y - \mathrm{B}$ vanishes at the maximum, equation 9.16 shows that

$$L = x^{\alpha}y^{\beta} \qquad (9.23)$$

at the maximum *and*

$$x + y - \mathrm{B} = 0 \qquad (9.24)$$

so that the values of x, y and λ which satisfy equations 9.20 to 9.22 give the constrained maximum we require.

Equations 9.20 and 9.21 imply

$$\alpha x^{\alpha-1}y^{\beta} = \beta x^{\alpha}y^{\beta-1} \qquad (9.25)$$

which can be rearranged to give

$$x = \frac{\alpha}{\beta}y \qquad (9.26)$$

Substitute for x in equation 9.22:

$$\left(\frac{\alpha}{\beta}+1\right)y = B \tag{9.27}$$

so

$$y_B = \frac{B}{\frac{\alpha}{\beta}+1} \tag{9.28}$$

Hence,

$$x_B = \frac{B}{1+\frac{\beta}{\alpha}} \tag{9.29}$$

We have added subscripts B to indicate that these are the values of x and y we require. Note that, since equation 9.10 holds,

$$\frac{\alpha}{\beta}+1 = \frac{\alpha+\beta}{\beta} = \frac{1}{\beta} \tag{9.30}$$

and

$$\frac{\beta}{\alpha}+1 = \frac{\beta+\alpha}{\alpha} = \frac{1}{\alpha} \tag{9.31}$$

so that equations 9.28 and 9.29 can be written

$$x_B = \alpha B \tag{9.32}$$

$$y_B = \beta B \tag{9.33}$$

The maximum value of z is

$$z_B = (\alpha B)^{\alpha} (\beta B)^{\beta} \tag{9.34}$$

$$= \alpha^{\alpha}\beta^{\beta}B \tag{9.35}$$

since

$$B^{\alpha+\beta} = B \tag{9.36}$$

because of equation 9.10 again.

Note that it has not been necessary to calculate λ explicitly, though we could do so if required using equations 9.20 and 9.21. Equation 9.20, for example, gives

$$\alpha(\alpha B)^{\alpha-1}(\beta B)^{\beta}+\lambda = 0 \tag{9.37}$$

so

$$\lambda = -\alpha^{\alpha}\beta^{\beta} \tag{9.38}$$

since

$$B^{\alpha+\beta-1} = B^0 = 1 \tag{9.39}$$

again using equation 9.10. Because we do not need to calculate λ explicitly, it is sometimes known as the 'undetermined multiplier'. However, we will find in some of the examples below that a useful interpretation can be given to the multiplier.

The result derived above can easily be presented in a general form for any number of variables and constraints. Suppose

$$z = f(x_1, x_2 \ldots x_n) \tag{9.40}$$

is a function of n variables $x_1, x_2, \ldots x_n$, and it is to be maximized subject to m constraint equations

$$g_j(x_1, x_2, \ldots x_n) = 0, \quad j = 1, 2, \ldots m \tag{9.41}$$

We must, of course, have $m < n$, otherwise any n of the constraint equations can be used directly to find $x_1 \ldots x_n$. We form a Lagrangian L as follows:

$$L = f(x_1, x_2, \ldots x_n) + \sum_j \lambda_j g_j (x_1, x_2 \ldots x_n) \tag{9.42}$$

We now have a multiplier λ_j for each constraint, ranging from $j = 1, 2, \ldots m$. L is a function of $m + n$ variables, $\lambda_1, \lambda_2 \ldots \lambda_n$ and, $x_1, x_2, \ldots x_n$, and its maximum is found by solving the $m + n$ simultaneous equations

$$\frac{\partial L}{\partial x_i} = 0, \; i = 1, 2, \ldots n \tag{9.43}$$

$$\frac{\partial L}{\partial \lambda_j} = 0, \; j = 1, 2, \ldots m \tag{9.44}$$

It can easily be seen that the equations 9.44 are repetitions of the constraint equations. We can write the simultaneous equations to be solved in the form

$$\frac{\partial f}{\partial x_i} + \sum_j \lambda_j \frac{\partial g_j}{\partial x_i} = 0 \tag{9.45}$$

$$g_j(x_1, x_2, \ldots x_n) = 0 \tag{9.46}$$

and we should then have to explore the various possibilities for solving them, given f and the g_j s, as we did for our specific example earlier. We give a number of examples in the following subsections.

We conclude with one note of caution. In Chapter 5, we gave conditions for the identification of maxima and minima in terms of the second derivatives. It is possible to carry out the same task for constrained optimization problems, but to do so would take us beyond the scope of the book. In the example which follows, it is clear from the context (as in our first example above) whether maxima or minima are being achieved, but in cases of doubt, the reader should consult a more advanced text.

9.2.2. Example 1: utility maximizing and profit maximizing

Two of the most important branches of micro-economic analysis are the theory of consumer's behaviour (based on utility maximizing) and the theory of the firm (based on profit maximizing). These theories are important in their own right, but are also of particular significance in geography and planning since they can be seen as forming the basis of location theory.

Suppose some individual consumer purchases quantities $x_1, x_2, \ldots x_n$ of goods $1, 2, \ldots n$ at prices $p_1, p_2, \ldots p_n$, and in so doing achieves an amount of utility given by

$$\mathrm{u} = \mathrm{u}(x_1, x_2, \ldots x_n) \tag{9.47}$$

If his total income is M, he will aim to maximize u subject to his budget constraint

$$\sum_{i=1}^{n} x_i p_i - \mathrm{M} = 0 \tag{9.48}$$

We can therefore introduce a Lagrangian multiplier λ in the usual way, and say that $x_1, x_2, \ldots x_n$ will take values which maximize a Lagrangian L given by

$$L = \mathrm{u}(x_1, x_2, \ldots x_n) + \lambda(\mathrm{M} - \sum_{i=1}^{n} x_i p_i) \tag{9.49}$$

These $x_1, x_2, \ldots x_n$ are therefore the solutions of

$$\frac{\partial L}{\partial x_i} = \frac{\partial \mathrm{u}}{\partial x_i} - \lambda p_i = 0 \tag{9.50}$$

and

$$\frac{\partial L}{\partial \lambda} = \mathrm{M} - \Sigma x_i p_i = 0 \tag{9.51}$$

We cannot proceed further, of course, except in a formal manner, until we specify the form of the utility function u. Formally, we can assume that a solution to equations 9.50 and 9.51 exists and write it in the form

$$x_i = x_i(p_1, p_2, \ldots p_n, \mathrm{M}) \tag{9.52}$$

which merely states that each x_i will be a function of all the prices, and the income $M. x_i$ in equation 9.52 is called a *demand* function. To be more specific, we assume, say, that

$$\mathrm{u}(x_1, x_2, \ldots x_n) = x_1^{\alpha_1} x_2^{\alpha_2} \ldots x_n^{\alpha_n} \tag{9.53}$$

Substitution in equation 9.50 then gives

$$\alpha_i x_1^{\alpha_1} x_2^{\alpha_2} \ldots x_i^{\alpha_i - 1} \ldots x_n^{\alpha_n} - \lambda p_i = 0 \tag{9.54}$$

We can solve this set of equations by a trick: multiply by x_i and re-arrange to give

$$\alpha_i x_1^{\alpha_1} x_2^{\alpha_2} \ldots x_n^{\alpha_n} = \lambda_i x_i p_i \tag{9.55}$$

For any other variable, say x_j, we have similarly

$$\alpha_j x_1^{\alpha_1} x_2^{\alpha_2} \ldots x_n^{\alpha_n} = \lambda x_j p_j \tag{9.56}$$

Divide equations 9.55 and 9.56:

$$\frac{\alpha_i}{\alpha_j} = \frac{x_i p_i}{x_j p_j} \tag{9.57}$$

so that

$$x_j = \frac{\alpha_j}{\alpha_i} \frac{p_i}{p_j} \cdot x_i \tag{9.58}$$

Substitute for all x s except x_i in equation 9.51:

$$\sum_{j \neq i} \frac{\alpha_j p_i x_i}{\alpha_i} + p_i x_i = \text{M} \tag{9.59}$$

so that, rearranging

$$\left(\sum_j \alpha_j\right) \frac{p_i x_i}{\alpha_i} = \text{M} \tag{9.60}$$

giving

$$x_i = \frac{\alpha_i \text{M}}{p_i (\sum_j \alpha_j)} \tag{9.61}$$

For this utility function, therefore, x_i is dependent on p_i and M, but not on any other prices, as would be the case in general. The value of λ could be calculated using any of the equations 9.55, or possibly in a more convenient form by summing over i in that equation to give

$$(\sum_i \alpha_i) x_1^{\alpha_1} x_2^{\alpha_2} \ldots x_n^{\alpha_n} = \lambda \sum_i x_i p_i \tag{9.62}$$

which reduces to

$$\lambda = \frac{(\sum_i \alpha_i) \text{u}_{\max}}{\text{M}} \tag{9.63}$$

where

$$\text{u}_{\max} = x_1^{\alpha_1} x_2^{\alpha_2} \ldots x_n^{\alpha_n} \tag{9.64}$$

is the value of u at the maximum, and we have replaced $\sum_i x_i p_i$ by M. If we differentiate equation 9.49 partially with respect to M, we get

$$\frac{\partial L}{\partial \mathrm{M}} = \lambda \tag{9.65}$$

and since

$$L = \mathrm{u} \tag{9.66}$$

at the maximum, we can write this as

$$\frac{\partial \mathrm{u}}{\partial \mathrm{M}} = \lambda \tag{9.67}$$

Thus, for constant x_i's, but a change $\delta\mathrm{M}$ in income M,

$$\delta \mathrm{u} = \frac{\partial \mathrm{u}}{\partial \mathrm{M}}\, \delta\mathrm{M} = \lambda \delta \mathrm{M} \tag{9.68}$$

This leads to the interpretation of λ as the marginal utility of money: it is the gain in utility for a unit increase in income. A full discussion can be found in Green (1972).

A similar analysis could be given for the theory of the firm. Suppose a firm produces a quantity x of some good, which it can sell at price p, given inputs $y_1, y_2, \ldots y_n$ which it purchases at prices $q_1, q_2, \ldots q_n$. The technology of the firm will be specified by a production function

$$x = x(y_1, y_2, \ldots y_n) \tag{9.69}$$

which says how much can be produced for given inputs. The inputs y_1, $y_2, \ldots y_n$ are then chosen to maximise profits

$$\mathrm{G} = xp - \sum_i y_i q_i \tag{9.70}$$

subject to equation 9.69 as a constraint. This is a similar kind of constrained maximization problem, and we will not take it any further except in the locational analysis context below.

The next step, in fact, is to show how these frameworks can be used in location theory. We follow the method of Alonso (1964), but use a slightly more general notation which helps us to avoid the assumption of a single employment centre only. We begin by considering the consumer again, and now assume that he purchases quantities of three goods, land $x_1(L)$, transport $x_2(L)$ and a composite 'other good', x_3. The first two are assumed to be functions of location, L. Utility is

$$\mathrm{u} = \mathrm{u}[x_1(L),\ x_2(L),\ x_3] \tag{9.71}$$

and this is maximized subject to a budget constraint

$$x_1(L)p_1(L) + x_2(L)p_2 + x_3p_3 = M \qquad (9.72)$$

in which the price of land, p_1(L) is also assumed to vary with location.

The problem now is that the prices cannot be taken as given, but will themselves be functions of the demand of all other consumers at each location. Alonso's achievement was to show how the resulting 'market' process operated. It is important for the purpose to introduce his concept of 'bid rent'. This is an extension of von Thunen's rent, introduced in section 3.8.6. Each consumer has a bid rent at each location for each given level of utility, say u_0. If u_0 is given, equation 9.71 can be written as

$$u(x_1(L), x_2(L), x_3) = u_0 \qquad (9.73)$$

and this determines a so-called 'indifference surface' in $x_1(L), x_2(L)$ and x_3. Alonso (1964, Appendix 4) has shown that, given u_0 there is a unique value of land rent $p_1(L)$ which would satisfy equation 9.72 and which this consumer would be prepared to offer to locate at L to achieve utility level u_0. Formally, we perhaps ought to write this as $p_1(L, u_0)$ to show its dependence on u_0 as well as L.

Similarly (see Wilson, 1974, Chapter 10, for a more detailed account) a set of bid rents for land can be constructed for other uses of land, based on the theory of the firm which was sketched above. Thus, all possible users of land are deemed to construct a set of bid rent functions for each utility level (or profit level in the case of other users) at each location.

The market clearing operation then works as follows. The land owner at each location attempts to maximize the rent he collects. That is, he lets his land to the highest bidder. The consumer (or firm) attempt to locate so as to maximize their bid rent and their utility (or profit). We will show later that, strange though it may sound, this means that the actual rent which is paid is a minimum. In this way, a market equilibrium is established.

In this case, the introduction of mathematical concepts of constrained maximization has enabled us to describe a market process but not to give a fully explicit mathematical analysis of the outcome, which is a more difficult objective to achieve. We will see in a later section, however, (9.4.3) that in one particular case, such an explicit presentation can be given.

9.2.3. Example 2: entropy maximizing models

We illustrate entropy maximizing methods by reference to the doubly constrained spatial interaction model first introduced in section 3.8.7. It will be seen that a quite different perspective on that and similar models can be gained in this way (and for a full account the reader should refer to Wilson, 1970–B, Chapters 1 and 2 and Gould, 1972).

First, we need to identify three different levels of description of a system state (Figure 9.2).

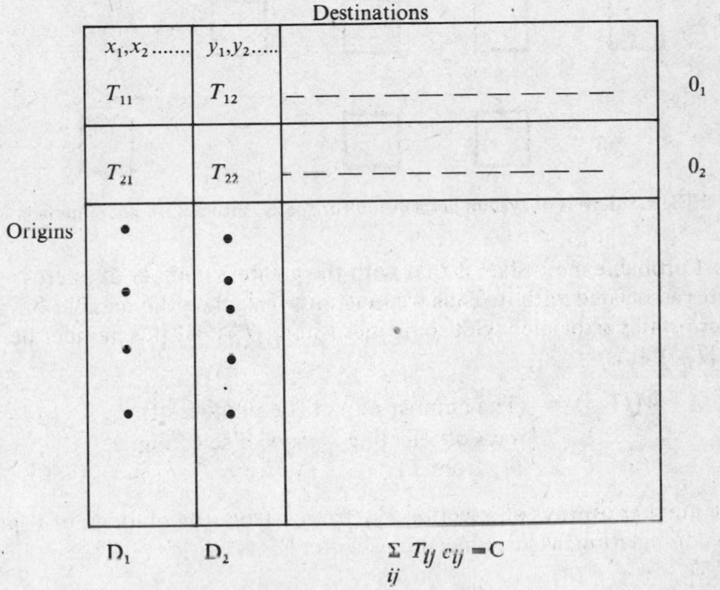

Destinations

	x_1, x_2 .... T_{11}	y_1, y_2 ..... T_{12}	— — — — — — — — — — — — —	O_1
	T_{21}	T_{22}	— — — — — — — — — — —	O_2
Origins	• • • •	• • • • •		
	D_1	D_2	$\sum_{ij} T_{ij} c_{ij} = C$	

FIG. 9.2. An origin – destination table

At the finest level of detail, we could take, in turn, the name of each person making a trip and record their name in the appropriate origin-destination cell, as indicated by $x_1, x_2, \ldots; y_1, y_2, \ldots$ in the first two boxes, which are supposed to represent names. This can be called the micro-level description. At a medium level of detail, we can total up the number of names in each cell, which is of course T_{ij} for the (i, j)th cell, to give the matrix $\{T_{ij}\}$. This can be called the meso-level description. At the coarsest level, we can record only the row and column totals, O_i and D_j, and also, the total expenditure on transport, C. This is the macro-level description. Clearly, there are many possible micro-states associated with each meso-state, and many meso-states with each macro-state. Such relationships are indicated in Fig. 9.3, in which each box represents a state at the appropriate level of description.

Suppose our prime interest lies in the meso-level of description. We make the fundamental assumption that, unless new information is made available to us, each micro-state is equally probable and therefore the

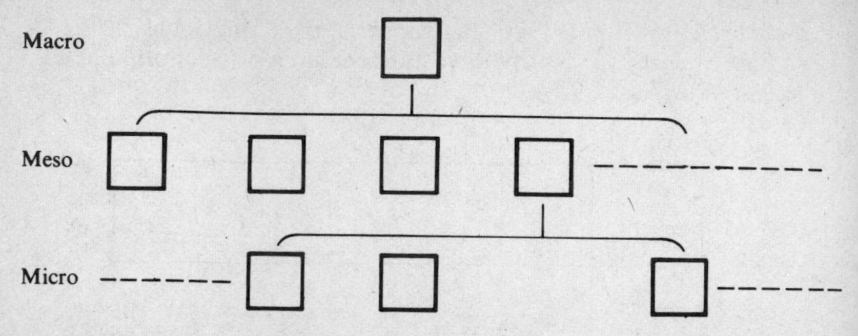

FIG. 9.3. Levels of system description for spatial interaction phenomena

most probable meso-state is that with the greatest number of micro-states associated with it. Thus we must now calculate the number of micro-states associated with some meso-state $\{T_{ij}\}$; let this number be $W(\{T_{ij}\})$. It is

$$W(\{T_{ij}\}) = \begin{pmatrix}\text{The number of}\\ \text{ways of selecting}\\ T_{11} \text{ from } T\end{pmatrix} \times \begin{pmatrix}\text{The number of}\\ \text{ways of selecting}\\ T_{12} \text{ from } T-T_{11}\end{pmatrix} \times \ldots \tag{9.74}$$

The number of ways of selecting T_{11} from T (the total number of trips) is a combination, as introduced in Chapter 7. It is $\begin{pmatrix} T \\ T_{11} \end{pmatrix}$ or $\frac{T!}{(T-T_{11})!\,T_{11}!}$. Thus, equation 9.74 can be written

$$W(\{T_{ij}\}) =$$

$$\frac{T!}{(T-T_{11})!T_{11}!} \times \frac{(T-T_{11})!}{(T-T_{11}-T_{12})!T_{12}!} \times \frac{(T-T_{11}-T_{12})!}{(T-T_{11}-T_{12}-T_{13})!T_{13}!} \times \ldots \tag{9.75}$$

Thus, cancelling the first term of successive denominators with following numerators, we have

$$W(\{T_{ij}\}) = \frac{T!}{T_{11}!T_{12}! \ldots} = \frac{T!}{\prod\limits_{ij} T_{ij}} \tag{9.76}$$

using the product notation. We need to find the $\{T_{ij}\}$ which maximizes this, or any monotonic function of it, and for convenience we maximize

$$S = \log W(\{T_{ij}\}) \tag{9.77}$$

subject to the macro-level constraints

$$\sum_j T_{ij} = O_i \tag{9.78}$$

$$\sum_i T_{ij} = D_j \tag{9.79}$$

$$\sum_{ij}\sum T_{ij}c_{ij} = C \tag{9.80}$$

to find the most probable meso-state compatible with a given macro-state as specified by equations 9.78 to 9.80. The method is called an entropy maximizing method because when a function such as S in equation 9.77 is formed in an equivalent analysis in physics it forms the entropy of the system. More details of this are given in Wilson (1970-B, Chapter 1) and Gould (1972).

We now see that we have a constrained maximization problem in a large number of variables and constraints. If there are N zones, S is a function of N^2 variables, T_{ij}, and is to be maximized subject to $2N + 1$ constraints, equations 9.78 to 9.80. We begin by forming a Lagrangian L given by

$$L = S + \sum_i \lambda_i^{(1)}(O_i - \Sigma T_{ij}) + \sum_j \lambda_j^{(2)}(D_j - \sum_i T_{ij}) + \beta(C - \sum_i \sum_j T_{ij}c_{ij}) \tag{9.81}$$

where the $\lambda_i^{(1)}$ s are the Lagrangian multipliers associated with constraints 9.78, $\lambda_j^{(2)}$ those with 9.79 and β that with (9.80).

As a preliminary, we must get S into a more convenient form so that we can find $\frac{\partial S}{\partial T_{ij}}$ which we will need in calculating $\frac{\partial L}{\partial T_{ij}}$.

$$\log W(\{T_{ij}\}) = \log \frac{T!}{\prod_{ij} T_{ij}!} \tag{9.82}$$

$$= \log T! - \sum_i \sum_j \log T_{ij}! \tag{9.83}$$

We can now use Stirling's approximation, introduced in Chapter 7, that

$$\log N! = N \log N - N \tag{9.84}$$

so

$$\log T_{ij}! = T_{ij} \log T_{ij} - T_{ij} \tag{9.85}$$

Then, using the product rule,

$$\frac{\partial}{\partial T_{ij}}(T_{ij} \log T_{ij} - T_{ij}) = \log T_{ij} \tag{9.86}$$

and so

$$\frac{\partial S}{\partial T_{ij}} = -\log T_{ij} \tag{9.87}$$

Thus, we must now find $\{T_{ij}\}$ to satisfy

$$\frac{\partial L}{\partial T_{ij}} = -\log T_{ij} - \lambda_i^{(1)} - \lambda_j^{(2)} - \beta c_{ij} = 0 \tag{9.88}$$

$$\frac{\partial L}{\partial \lambda_i^{(1)}} = O_i - \sum_j T_{ij} = 0 \tag{9.89}$$

$$\frac{\partial L}{\partial \lambda_j^{(2)}} = D_j - \sum_i T_{ij} = 0 \tag{9.90}$$

and

$$\frac{\partial L}{\partial \beta} = C - \sum_i \sum_j T_{ij} c_{ij} = 0 \tag{9.91}$$

Equations 9.89 to 9.91 are, as usual, repetitions of the constraint equations. From equation 9.88, we can obtain an expression for each T_{ij} in terms of the multipliers as

$$T_{ij} = e^{-\lambda_i^{(1)} - \lambda_j^{(2)} - \beta c_{ij}} \tag{9.92}$$

We can find $\lambda_i^{(1)}$ by substitution in equations 9.89, $\lambda_j^{(2)}$ by substitution in 9.90 and β by substitution in 9.91. For convenience, we retain the β term in its present form. Before solving for $\lambda_i^{(1)}$ and $\lambda_j^{(2)}$, we make a transformation, and use instead variables A_i and B_j defined by

$$A_i O_i = e^{-\lambda_i^{(1)}} \tag{9.93}$$

$$B_j D_j = e^{-\lambda_j^{(2)}} \tag{9.94}$$

Then, the main equation 9.92 can be written as

$$T_{ij} = A_i B_j O_i D_j e^{-\beta c_{ij}} \tag{9.95}$$

and substitution in equations 9.89 and 9.90 shows that

$$A_i = 1/\sum_j B_j D_j e^{-\beta c_{ij}} \tag{9.96}$$

and

$$B_j = 1/\sum_i A_i O_i e^{-\beta c_{ij}} \tag{9.97}$$

Thus, the model is given by equations 9.95 to 9.97 and the reader can easily check that these are the same as equations 3.268, 3.270 and 3.272 of Chapter 3, except that the power function, $c_{ij}^{-\beta}$, is replaced by the negative exponential function, $e^{-\beta c_{ij}}$. This particular difference is not of great significance.

Thus, we have now generated the doubly constrained spatial interaction model by what is best seen as a process of statistical averaging over micro-states to obtain the most probable meso-state. Here, we have been most interested in using the basic mathematics of the process as an illustration and the reader is referred to the references cited earlier for more applications and for a more detailed interpretation.

Physical geographers have used combinatorial theory in the calculation of number of *topologically distinct channel networks* (T.D.C.N. s) in the analysis of river networks, and consideration of most probable sets of micro-states leads to theoretical derivation of Horton's laws (cf. section 3.8.4, 8.1.6 and 8.4). These concepts are described by Haggett and Chorley (1969), for example, and by Shreve (1966) and Werritty (1972). Thus, this may be considered as another example of entropy maximizing methods although without the use of constraints to date. A more direct use of entropy, in the study of landscape evolution, can be seen in the work of Leopold and Langbein (1962).

9.2.4. A note on maximum likelihood methods

Although the methods presented in the previous section can reasonably be characterized as 'entropy maximizing', we should also note that they are equivalent to maximum likelihood methods (cf. section 7.3.2). In particular, if maximum likelihood methods are used for parameter estimation with such models as spatial interaction models, the equations to be solved for these parameters turn out to be just the entropy maximizing constraint equations. For example, to estimate β in equation 9.95, equation 9.91 must be solved. For a detailed discussion of these points, the reader is referred elsewhere (Batty and Mackie, 1972, Wilson, 1974, Chapter 12).

9.3. The 'calculus of variations' method for finding optimum paths

9.3.1. The problem and the method

The basic problem was outlined in equations 9.5 and 9.6 of the introductory section. We have an integral z defined by

$$z = \int_a^b F(x, y, y')dx \tag{9.98}$$

where F is some given function, and y is a function of x:

$$y = f(x) \tag{9.99}$$

y' is the derivative of y with respect to x. The problem is to find the function f – the path in the (x, y) plane – which minimizes z.

In giving the method for this, we make more extensive use of the differential notation than hitherto for this kind of problem. For a simple maximizing problem, say to maximize y in equation 9.99, we would solve

$$\frac{dy}{dx} = 0 \tag{9.100}$$

This condition implies that the variation in y variables at the maximum is zero for arbitrary small variation in x. That is

$$\delta y = \frac{dy}{dx}\delta x = 0 \tag{9.101}$$

for small δx at the maximum. In the calculus of variations problem, suppose $y = f(x)$ is the optimum path, and $y + \delta y$ a *variation* as it, as shown in Figure 9.4. The *variational* principle to be applied now is that δz

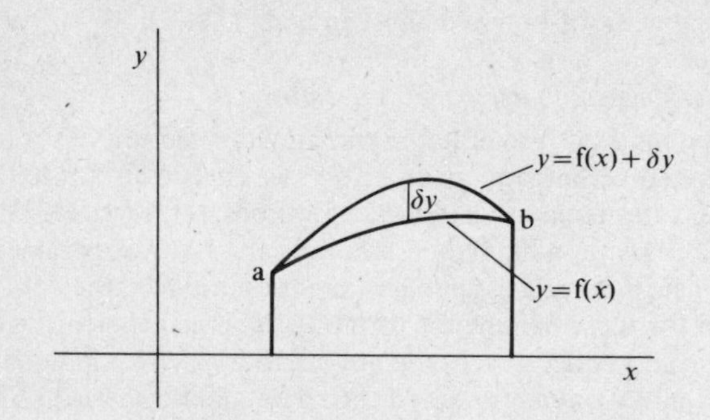

FIG. 9.4. Small variation of the curve $y = f(x)$

should vanish at the maximum for arbitrary variation in y. Equation 9.98 shows that

$$\delta z = \delta \int_a^b F(x,y,y')dx = \int_a^b \delta F(x,y,y')dx \tag{9.102}$$

$$\delta F(x,y,y') = \frac{\partial F}{\partial y}\delta y + \frac{\partial F}{\partial y'}\delta y' \tag{9.103}$$

But

$$\delta y' = \delta\frac{dy}{dx} = \frac{d}{dx}\delta y \tag{9.104}$$

so

$$\delta F(x,y,y') = \frac{\partial F}{\partial y}\delta y + \frac{\partial F}{\partial y'}\frac{d}{dx}\delta y \tag{9.105}$$

Thus, substituting in equation 9.102:

$$\delta z = \int_a^b \left[\frac{\partial F}{\partial y}\delta y + \frac{\partial F}{\partial y'}\frac{d}{dx}(\delta y)\right]dx \tag{9.106}$$

The second term of the integral can be integrated by parts (section 5.8.2), taking $u = \frac{\partial F}{\partial y'}$ $v' = \frac{d}{dx}(\delta y)$, so that $v = \delta y$. Hence

$$\int_a^b \frac{\partial F}{\partial y'}\frac{d}{dx}(\delta y)\,dx = \left[\frac{\partial F}{\partial y'}\delta y\right]_a^b - \frac{d}{dx}\left(\frac{\partial F}{\partial y'}\right)\delta y dx$$

But note from Figure 9.4 that $\delta y = 0$ at $x = a$ and $x = b$, the end points, so that

$$\left[\frac{\partial F}{\partial y'}\delta y\right]_a^b = 0 \tag{9.108}$$

Therefore

$$\delta z = \int_a^b \left[\frac{\partial F}{\partial y} - \frac{d}{dx}\left(\frac{\partial F}{\partial y'}\right)\right]\delta y dx \tag{9.109}$$

We can only have $\delta z = 0$ for arbitrary small δy if the integrand vanishes:

$$\frac{\partial F}{\partial y} - \frac{d}{dx}\frac{\partial F}{\partial y'} = 0 \tag{9.110}$$

Hence, y as a function of x, which maximizes z, is the solution to the second-order differential equation 9.110, which is known as Euler's equation. We give examples of its use in the next section.

9.3.2. Minimum distance and time paths

The most obvious application of the calculus of variations in geography and planning is in computing minimum distance and time paths. We give one trivial example, to illustrate the method, and to provide a basis for the discussion of Angel's and Hyman's (1970) concept of velocity fields. First, however, we must show how distance is measured as an integral. Consider Figure 9.5. we show an increment of distance, δs along a curve, first in Cartesian co-ordinates and then in polar co-ordinates. From Figure 9.5(a).

$$(\delta s)^2 = (\delta x)^2 + (\delta y)^2 \tag{9.111}$$

and from Figure 9.5(b)

$$(\delta s)^2 = (\delta r)^2 + (r\delta\theta)^2 \tag{9.112}$$

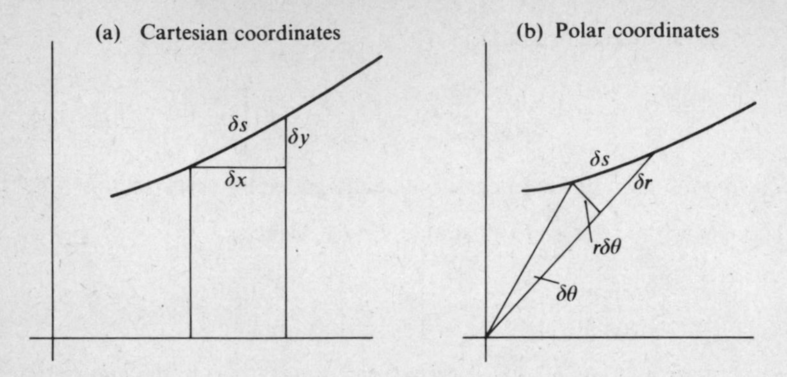

FIG. 9.5. Distance and time paths in Cartesian and polar co-ordinates

Let $s(\mathrm{a}, \mathrm{b})$ be the distance from $x = \mathrm{a}$ to $x = \mathrm{b}$ along $y = \mathrm{f}(x)$, or $s(r_1, r_2)$ from $r = r_1$ to $r = r_2$ along $r = r(\theta)$. Then,

$$s(\mathrm{a}, \mathrm{b}) = \int_{x=a}^{x=\mathrm{b}} \mathrm{d}s \tag{9.113}$$

or

$$s(r_1, r_2) = \int_{r=r_1}^{r=r_2} \mathrm{d}s \tag{9.114}$$

From equation 9.111,

$$\frac{\delta s}{\delta x} = \left[1 + \left(\frac{\delta y}{\delta x}\right)^2\right]^{\frac{1}{2}} \tag{9.115}$$

so

$$\frac{\mathrm{d}s}{\mathrm{d}x} = \left[1 + \left(\frac{\mathrm{d}y}{\mathrm{d}x}\right)^2\right]^{\frac{1}{2}} \tag{9.116}$$

Equation 9.113 can now be written

$$s(\mathrm{a}, \mathrm{b}) = \int_{\mathrm{a}}^{\mathrm{b}} \frac{\mathrm{d}s}{\mathrm{d}x} = \int_{\mathrm{a}}^{\mathrm{b}} \left[1 + \left(\frac{\mathrm{d}y}{\mathrm{d}x}\right)^2\right]^{\frac{1}{2}} \mathrm{d}x \tag{9.117}$$

and equation 9.114 can be written, by a similar piece of manipulation using equation 9.112

$$s(r_1, r_2) = \int_{\mathrm{a}}^{\mathrm{b}} \frac{\mathrm{d}s}{\mathrm{d}r}\,\mathrm{d}r = \int_{r_1}^{r_2} \left[1 + r^2\left(\frac{\mathrm{d}\theta}{\mathrm{d}r}\right)^2\right]^{\frac{1}{2}} \mathrm{d}r \tag{9.118}$$

Now let $\mathrm{v}(x, y)$, or $\mathrm{v}(r, \theta)$ be the velocity along the curve at (x, y) or (r, θ), and define $t(\mathrm{a}, \mathrm{b})$ or $t(r_1, r_2)$ to be travel times. Then

$$\text{time} = \text{distance/speed} \tag{9.119}$$

so that

$$t(a, b) = \int_a^b \frac{1}{v(x, y)} \left[1 + \left(\frac{dy}{dx}\right)^2\right]^{\frac{1}{2}} dx \tag{9.120}$$

or

$$t(r_1, r_2) = \int_{r_1}^{r_2} \frac{1}{v(r, \theta)} \left[1 + r^2 \left(\frac{d\theta}{dr}\right)^2\right]^{\frac{1}{2}} dr \tag{9.121}$$

$v(x, y)$ or $v(r, \theta)$ are velocity fields, since they associate a speed with any point in the plane. We now have a real and interesting calculus of variations' problem: given a velocity field, find the minimum time path through that field. That is, find $y = f(x)$ which minimizes $t(a, b)$ in equation 9.120 (Cartesian co-ordinates) or $r = r(\theta)$ which minimizes $t(r_1, r_2)$ in equation 9.121 (polar co-ordinates).

First, however, let us consider the simpler problem of finding the minimum distance between two points by minimizing $s(a, b)$ in equation 9.117. We begin by proving a general result. If

$$z = \int_a^b F(y, y')dx \tag{9.122}$$

so that F is independent of x, we can integrate Euler's equation directly. Note that

$$\frac{d}{dx}\left(F - y' \frac{\partial F}{\partial y'}\right) = \frac{dy}{dx}\frac{\partial F}{\partial y} + \frac{dy'}{dx}\frac{\partial F}{\partial y'} - \frac{dy'}{dx}\frac{\partial F}{\partial y'}$$

$$- y' \frac{d}{dx}\frac{\partial F}{\partial y'} \tag{9.123}$$

$$= y'\left[\frac{\partial F}{\partial y} - \frac{d}{dx}\frac{\partial F}{\partial y'}\right] \tag{9.124}$$

$$= 0 \tag{9.125}$$

so that we can integrate equation 9.125 directly to give

$$F - y' \frac{\partial F}{\partial y'} = K \tag{9.126}$$

where K is a constant of integration.

Now, in equation 9.117,

$$F = [1 + y'^2]^{\frac{1}{2}} \tag{9.127}$$

So, substituting in equation 9.126

$$[1 + y'^2]^{\frac{1}{2}} - \frac{y'^2}{[1 + y'^2]^{\frac{1}{2}}} = K \tag{9.128}$$

which simply indicates that $y' = \frac{dy}{dx}$ is a constant, so that the shortest distance is, of course, a straight line.

Angel and Hyman (1970) calculate minimum time paths for a number of examples of radially symmetric velocity fields using an equation of the form 9.121. For example, they take

$$v(r, \theta) = \omega r \tag{9.129}$$

where ω is a constant. Since the integral in equation 9.121 is then independent of θ, the differential equation to be solved for minimum time paths $r = r(\theta)$ is of the form 9.126, but with r' replacing y':

$$F - r' \frac{\partial F}{\partial r'} = K \tag{9.130}$$

If we substitute for F, this reduces to

$$\frac{dr}{d\theta} = mr \tag{9.131}$$

where m is a constant given by

$$m = \frac{(1 - k^2\omega^2)^{\frac{1}{2}}}{k\omega} \tag{9.132}$$

The solution is

$$r = r_0 e^{m\theta} \tag{9.133}$$

where r_0 is a constant (assuming that $r = r_0$ when $\theta = 0$). This is the equation of a spiral. Typical paths are indicated in Figure 9.6.

FIG. 9.6. Minimum-time paths in the velocity field $v(r) = \omega r$

9.4. Algorithmic mathematics

9.4.1. Introduction

There is a class of techniques, dissimilar from the kind of mathematics taught in schools, and even in some courses in universities, which is best described as *algorithmic*. They mostly stem from the advent of the computer, and in terms of mathematical notation, they are often rooted in computer programming. To illustrate this in an extremely simple way, consider the statement

$$n = n + 1 \tag{9.134}$$

which, at first sight, seems to represent a contradiction. It *is* a contradiction when interpreted as an algebraic equation; however, it has meaning when interpreted as a statement in a computer program. It means: 'take the number in the computer's store which is labelled n, add one to it, and replace in the same storage location'. There are many mathematical processes which can be represented in this way which cannot be represented easily as algebraic equations.

We illustrate these methods in this section with two rather different kinds of examples: first, the shortest path through a network; and second, a variety of linear programming problems. In the first case, we present an algorithm in detail so that the reader can see what kind of thing is involved. In the second case, we concentrate on a presentation of the main concepts, but we do not give details of the actual algorithms.

9.4.2. The shortest path through a network

Consider the network shown in Figure 9.7. Networks consist of *nodes* (the dots in the figure) and *links* (the lines connecting the nodes). Not all nodes are directly connected by links. A 'distance' (which may be literally distance, or travel time, or cost) is assumed to be associated with each link. The problem is to find the shortest distance, and the route,

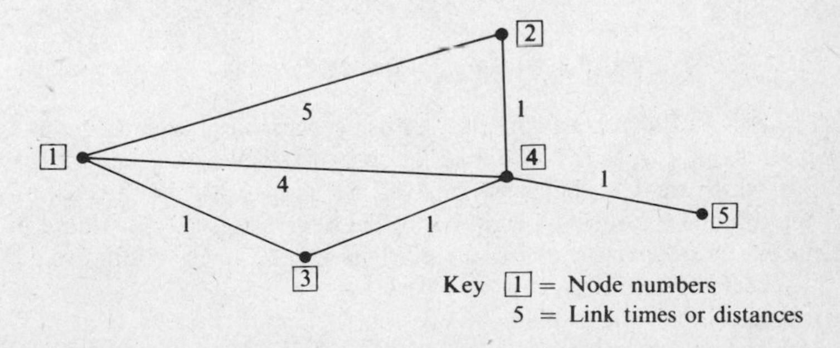

FIG. 9.7. An illustrative network

between each pair of nodes. In the example, we can see by inspection, that although nodes 2 and 4 are directly connected to node 1, the shortest path from node 1 to these nodes is via nodes 3 and 4, and via node 3, respectively. Given a very large network, this sort of calculation becomes extremely laborious and tedious when carried out manually, and so we now present an *algorithm* (a scheme of calculation) for doing the calculation systematically, and which is suitable for high speed computer calculation. We will use the example of Figure 9.7 to illustrate how the algorithm works. Note, incidentally, that in this example, it has implicitly been assumed that the links are two-way links, and that the 'distances' between nodes are the same each way. This is to simplify the presentation. The algorithm to be described here could be applied to networks defined with one-way links, which allowed for the possibility of different distances in each direction. This would be useful, for example, if a highway network was being analysed, and 'distance' was travel time. Note also, in this case, a one-way link could be omitted entirely if the connection represented a one-way street in the opposite direction.

Perhaps the two most famous algorithms for finding shortest paths in a network are those discovered by Dantzig (1957) and by Moore (1959). Dantzig's algorithm is the easiest to describe, and will be exhibited here in detail; Moore's is probably more efficient computationally for some kinds of networks and so will be described briefly also. In each case, the problem is tackled by finding the shortest paths from one node to all other nodes, and repeating for each node in turn.

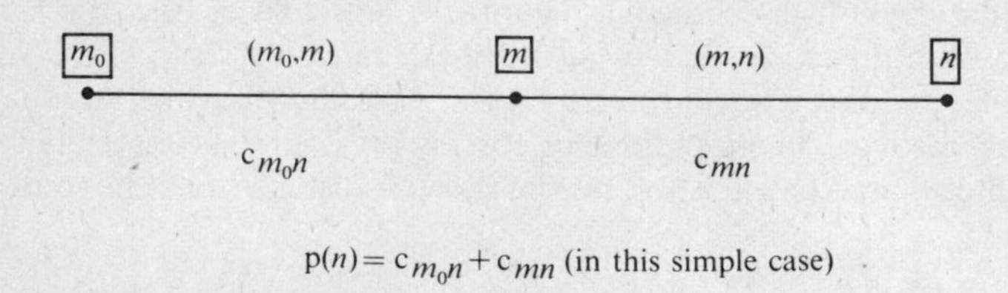

FIG. 9.8. Notation for Dantzig algorithm

To describe Dantzig's algorithm, let us use variables like m and n as node numbers, c_{mn} as the 'distance' or cost of travelling on the link from m to n, which we describe as the link (m, n), and let $p(n)$ be the cost of getting to node n from the origin node. These variables are illustrated in Figure 9.8., where m_0 is taken as the origin node. The algorithm can be stated succinctly as follows:

(i) Assign all nodes n labels of the form
$[m, p(n)]$ where $p(m_0) = 0$, and
$p(n) = \infty$ for $n \neq m_0$

(ii) Set $m = m_0$ initially.
(iii) Search for a link (m, n) such that

$$p(m) + c_{mn} < p(n) \tag{9.135}$$

(iv) If such a link is found, change the label on node n to $[m, p(m) + c_{mn}]$ (that is, set $p(n) = p(m) + c_{mn}$).
Continue steps (iii) and (iv) until no such links are found
(v) Find the node k such that
(a) Node k has never played the role of m in steps (iii) and (iv), and
(b) $p(k)$ is minimal, subject to v(a) above.
(vi) If no such k can be found, terminate. Otherwise, set $m = k$ and repeat from (iii) above.

At termination, the shortest path from m_0, the origin to any node m can be found by reverse tracing through the component of the individual labels, and the length of the shortest path is $p(m)$. That is, in the case of an example, if the origin is zone 1, and we wish to find the shortest path to zone 4, we examine node labels as follows:

Node 4: label [3,2], this means that the length of the minimum path is 2 units, and that node 3 is the next node towards 1 on the shortest path.

Node 3: label [1,1], which indicates that the next link to the path is from node 3 to node 1.

The whole procedure will now be illustrated, using the network from Fig. 9.7, and we calculate all shortest paths from node 1.

Steps (i) and (ii): Node labels initially are

Node	Label
1	$[1, 0]$
2	$[1, \infty]$
3	$[1, \infty]$
4	$[1, \infty]$
5	$[1, \infty]$

Steps (iii) and (iv): Examine links $(1, m)$ and see if

$$p(1) + c_{ml} < p(1) \tag{9.136}$$

This is true for links (1,2), (1,3) and (1,4), and so we amend the labels of nodes 2, 3 and 4 in the list

Node	Label
1	$[1, 0]$
2	$[1, 5]$
3	$[1, 1]$
4	$[1, 4]$
5	$[1, \infty]$

Step (v): we now look at p(k), for $k = 2, 3, 4, 5$ (since m has been 1 only) and p(3) is the minimum. So, set $m = 3$, and repeat steps (iii) and (iv).
Repeat of steps (iii) and (iv) with $m = 3$:
We find the link (3, 4) such that

$$p(3) + c_{3,4} < p(4) \tag{9.137}$$

and so we amend the node 4 label accordingly:

1	[1, 0]
2	[1, 5]
3	[1, 1]
4	[3, 2]
5	[1, ∞]

This step has thus identified the route 1→3→4 as being better than the single link direct route 1→4.

Step (v): we now set $m = 4$
Repeat of steps (iii) and (iv):
We find the link (4, 2) such that

$$p(4) + c_{4,2} < p(2) \tag{9.138}$$

so we amend the label of node 2. This identifies the path 1→3→4→2 as being shorter than the direct path 1→2. We also find the link (4, 5) such that

$$p(4) + c_{4,5} < p(5) \tag{9.139}$$

so we amend the label of node 5:

1	[1, 0]
2	[4, 3]
3	[1, 1]
4	[3, 2]
5	[4, 3]

No further improvements are obtained by setting k equal to 5 and 2. Retracing shows that the best paths from node 1 are:

Node	Path	Length
2	1→3→4→2	3
3	1→3	1
4	1→3→4	2
5	1→3→4→5	3

Inspection of the original network shows the result to be true. These are plotted in Fig. 9.9, which shows that they form a tree. This is a result

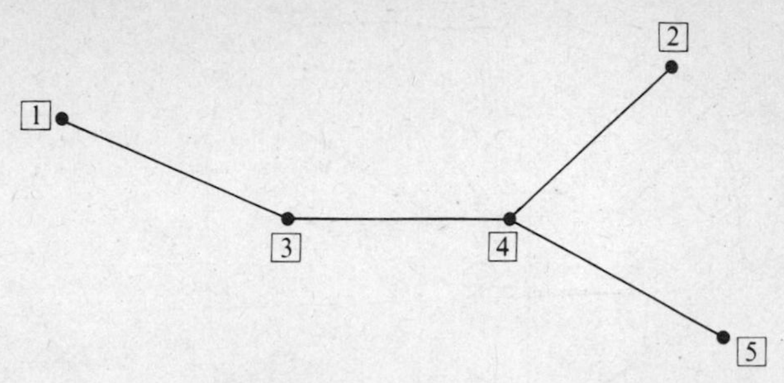

FIG. 9.9. A tree of minimum paths from node 1

which is generally true: the set of best paths from one origin node from a tree. The Moore algorithm is very similar in principle – indeed perhaps it only differs in computational technique. It is fully described in the Bureau of Public Roads (1964) Assignment Manual – indeed, in sufficient detail to make it quite easy to write the computer programme to go with it. In words, either algorithm can be described broadly as follows:

(i) Find the nodes directly connected to the origin node, and associate a label with each such node which is (previous node in path, total cost so far) – the previous node in the path being the origin in this case.

(ii) Take the node nearest the origin, and repeat this process, find the nodes directly connected to it, and create the label. If this process connects to nodes which have been created earlier [e.g. when the link (4, 2) is examined in our example] and a new path is found shorter than that currently recorded in the node label, then the label is updated.

Any model which requires inter-zonal travel costs, such as the c_{ij} s in the model presented in Chapter 3, needs a shortest path algorithm. It is usually assumed that all trips to and from a zone start from one node called the *zone centroid*. Dummy links then connect such nodes to the actual transport network. This is shown schematically in Fig. 9.10. The minimum cost trees can then be found from each centroid, and the travel cost matrix c_{ij} is then built up from this information. Such trees also form the basis of assignment programmes. When origin–destination flows have been calculated, they are then 'loaded' on to the shortest path, and a count is kept of all trips assigned to each link.

9.4.3. Linear programming

A linear programming problem is a special case of the general constrained maximization problem given by equations 9.40 and 9.41 where the objective functions and the constraints are linear. Further, in the most general formulation, the constraints are expressed as inequalities rather

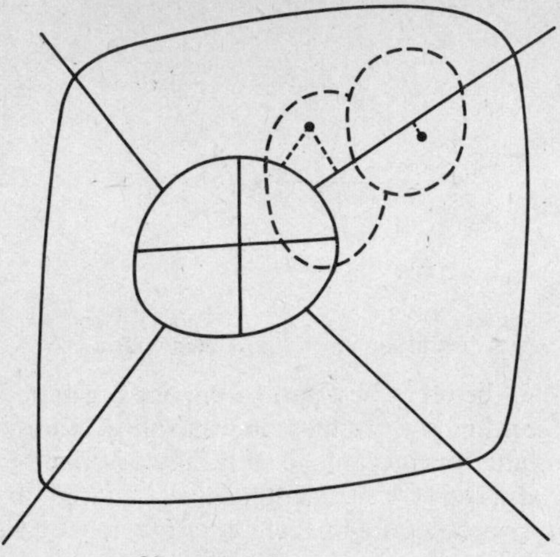

—— Network links
---- Zone boundaries (shown for 2 zones only)
• Zone centroids
----- Dummy links connecting zone centroids to network

FIG. 9.10. Zone centroids and networks

than equations. If we have n variables, $x_1, x_2, \ldots x_n$, and m constraints, the problem may be to maximize z given by

$$z = \sum_{i=1}^{n} \mathrm{a}_i \mathrm{x}_i \tag{9.140}$$

subject to

$$\sum_{j=1}^{n} \mathrm{b}_{ij} \mathrm{x}_j \leqslant \mathrm{c}_i, \; i = 1, \ldots m \tag{9.141}$$

where the a_i s, b_{ij} s and c_i s are all constants. It is easy to see that the Lagrangian method as developed in section 9.2.1 does not help us here since all partial derivatives of z with respect to the x_i s are independent of x_i because of the linearity, and so it is impossible to obtain analytical expressions for the x_i s which maximize z subject to the constraints. For this reason, an algorithm is used instead. We will not describe the algorithm here: rather, we will assume (as we did for the inverse of a square matrix) that if the reader is faced with a problem of the form given by equations 9.140 and 9.141, the coefficients can form the input to a computer programme which will generate the optimum x_i s. We proceed by introducing a range of examples which are of direct interest to geographers and planners, and thus hope to provide the reader with an understanding of the underlying concepts and linear programming outputs in such contexts.

We begin with the example which was introduced to illustrate linear inequalities in section 3.3.6. The problem was to choose investments x and y in two factories which satisfied constraints:

$$x+y \leqslant B \tag{9.142}$$

$$100x + 50y \geqslant E \tag{9.143}$$

and

$$20x + 40y \geqslant S \tag{9.144}$$

We identified a 'feasible region' of such (x, y) s which was exhibited in Figure 3.13. Suppose we now take the example a stage further and assume that the *profit* resulting from investment of x and y is P given by

$$P = 100x + 300y \tag{9.145}$$

and that this is to be maximized subject to the constraints 9.142 to 9.144. Equation 9.145 represents a straight line in the (x, y) plane of gradient $-1/3$. P determines the intercept on the axes. A little thought shows that P will take its greatest value in this case when the line passes through the point A, as exhibited in Figure 9.11(a). Note that if P had been given

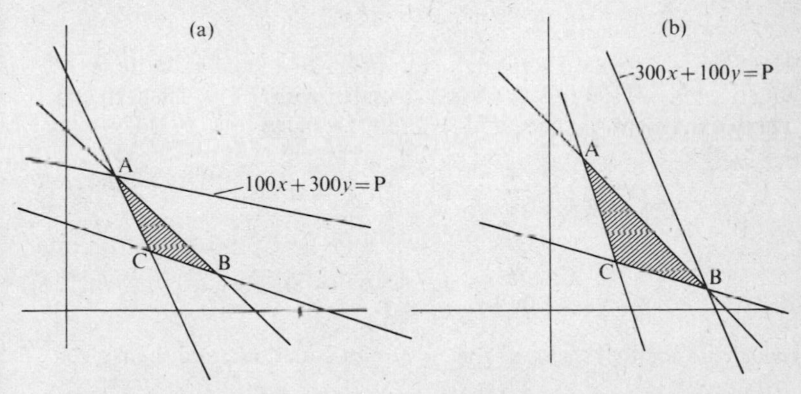

FIG. 9.11. Geometrical representations of a linear-programming problem

not by equation 9.145, but as

$$P = 300x + 100y \tag{9.146}$$

with a gradient of -3, then B would have been the optimum point, as shown in Figure 9.11(b). The computer programme would have produced these points for us, but the geometrical illustration of this simple case does demonstrate a basic feature of the solutions to linear programming problems: that the optimum point is always a 'corner' of the feasible region.

A particularly well known problem is the so-called *transportation problem of linear programming*. We shall state it in a form here in which the constraints are equalities. For some commodity, suppose O_i is produced in region i of a set of regions $i = 1, 2, \ldots n$, and D_j must be supplied to each region j, at *minimum transport cost*. c_{ij} is the unit cost of transport between i and j. Thus, the problem is to find T_{ij} to minimize

$$C = \sum_i \sum_j T_{ij} c_{ij} \tag{9.147}$$

subject to

$$\sum_j T_{ij} = O_i \tag{9.148}$$

and

$$\sum_i T_{ij} = D_j \tag{9.149}$$

Again, a standard computer programme will give these T_{ij} s for us.

In this case, we can gain further insight by proceeding with a Lagrangian formulation. Let

$$L = \sum_i \sum_j T_{ij} c_{ij} + \sum_i \alpha_i \left(O_i - \sum_j T_{ij}\right) + \sum_j \nu_j \left(D_j - \sum_i T_{ij}\right) \tag{9.150}$$

where the α_i s are the Lagrangian multipliers associated with the set of equations 9.148 and the ν_js those associated with 9.149. Then, the optimality conditions can be stated (without proof) as

$$\frac{\partial L}{\partial \alpha_i} = O_i - \sum_j T_{ij} = 0 \tag{9.151}$$

$$\frac{\partial L}{\partial \nu_j} = D_j - \sum_i T_{ij} = 0 \tag{9.152}$$

(which are simply repeats of the constraint equations) and then, given that

$$T_{ij} \geqslant 0 \tag{9.153}$$

$$\frac{\partial L}{\partial T_{ij}} \geqslant 0 \text{ and } T_{ij} \frac{\partial L}{\partial T_{ij}} = 0 \tag{9.154}$$

The pair of conditions 9.154 show that

$$\frac{\partial L}{\partial T_{ij}} = c_{ij} - \alpha_i - \nu_j = 0 \text{ if } T_{ij} > 0 \tag{9.155}$$

and

$$c_{ij} - \alpha_i - \nu_j \geqslant 0 \text{ if } T_{ij} = 0 \tag{9.156}$$

A standard theorem of linear programming tells us that there cannot be more non-zero T_{ij} s than there are constraints – in this case $2N$ if there are N regions. In fact, in this case the number is $2N-1$ since one of the constraints can be taken as dependent on all the others because of the relationship

$$\sum_i \sum_j T_{ij} = \sum_i O_i = \sum_j D_j \tag{9.157}$$

The main use of the equations 9.155 and 9.156 is in relation to the concept of a dual. Any linear programming problem, the so-called *primal* problem also has a dual. There is one dual variable for each constraint; they are in fact, for the transportation problem, the Lagrangian variables α_i and ν_j. The dual of the transportation problem is to maximize

$$C' = \sum_i \alpha_i O_i + \sum_j \nu_j D_i \tag{9.158}$$

subject to

$$c_{ij} - \alpha_i - \nu_j \geqslant 0 \tag{9.159}$$

It can be shown that, at the maximum

$$C' = C \tag{9.160}$$

where C is the minimum cost obtained for the primal problem. We can now use equation 9.155 actually to calculate the dual variables given the non-zero optimal T_{ij} s. Since there are $2N-1$ independent equations in $2N$ unknowns, one of the dual variables can be given an arbitrary value, and then the rest can be calculated in a suitable sequence. These dual variables can often be interpreted as 'rents' or 'comparative advantage' (Dorfman, Samuelson and Solow, 1958).

It turns out that there is a close relationship between the transportation problem of linear programming and the spatial interaction model derived by entropy maximizing methods in equations 9.95 to 9.97 (Evans, 1973, Wilson and Senior, 1973). The linear programming T_{ij} s can be obtained by letting $\beta \to \infty$ in the spatial interaction model equation. Further

$$\lambda_i^{(1)} / \beta \to \alpha_i \text{ as } \beta \to \infty \tag{9.161}$$

and

$$\lambda_j^{(2)} / \beta \to \nu_j \text{ as } \beta \to \infty \tag{9.162}$$

However, a full discussion of this relationship would take us beyond the scope of this book and the interested reader should consult the references cited above.

As a final example of a linear programming model, we outline the residential location model of Herbert and Stevens (1960) which 'operationalizes' the concepts of Alonso on this subject outlined in section 9.2.2 above. There we introduced the concept of the consumer's bid rent for a home. Suppose consumers can be grouped into w-classes (say by income) and that there are P^w households in class w. Let houses be grouped into k-classes (size, condition, and so on). Then let consumer preferences be expressed by the set of bid rents b^{kw} for a type k house by a type w household, and let p_i^{kw} be the price of such a house in zone i *excluding land costs*. Suppose a type k house in i occupies s_i^k acres of land out of a total supply L_i. Then, if T_i^{kw} is the number of w-type households located in type-k houses in zone i when bid rents for land are maximized, this is found by maximizing

$$Z = \sum_i \sum_k \sum_w T_i^{kw}(b^{kw} - p_i^{kw}) \tag{9.163}$$

subject to the constraints

$$\sum_k \sum_w T_i^{kw} s_i^k \leqslant L_i \tag{9.164}$$

and

$$\sum_i \sum_k T_i^{kw} = P^w \tag{9.165}$$

The dual of this problem is to minimize

$$Z' = \sum_i \alpha_i L_i - \sum_w \nu^w p^w \tag{9.166}$$

subject to

$$s_i \alpha^k - \nu^w \geqslant b^{kw} - p_i^{kw} \tag{9.167}$$

and

$$\alpha_i \geqslant 0 \tag{9.168}$$

As usual, the dual variables can be interpreted as rent, and in particular α_i as land rent in zone i. Thus, in this case, the dual problem shows that if the locational pattern is determined by bid rent maximization in the primal problem, the actual rent paid is a minimum! For an extended discussion of this model and developments of it, the reader is referred elsewhere (Senior and Wilson, 1974).

Exercises

Section 9.2

(1) If $y = x_1^{0.2} x_2^{0.4} x_3^{0.4}$, find its maximum value subject to the constraint $x_1 + x_2 + x_3 = 10$.

(2) If $z = x^2 + y^2$, find its maximum value subject to the constraint $2x + y = 3$.

(3) Maximize $S = -\sum_i \sum_j \log T_{ij}!$ subject to $\sum_j T_{ij} = e_i P_i$ and $\sum_i \sum_j T_{ij} c_{ij} =$

C; hence show that $T_{ij} = A_i e_i P_i e^{-\beta c_{ij}}$ in this situation for suitable A_i and β. Show that if c_{ij} is replaced by $c_{ij} - U_j$, and W_j is defined so that $e^{\beta U_j} = W_j^{\alpha}$, the equation for T_{ij} becomes $T_{ij} = A_i e_i P_i W_j^{\alpha} e^{-\beta c_{ij}}$.

Section 9.3

(4) In equation 9.121, take $v(r, \theta) = V$ where V is a constant. What is the differential equation to be satisfied by r and θ? Solve the equation and interpret the result.

Section 9.4

(5) Construct a network using your own data, roughly of the size of that shown in Fig. 9.7 and use Dantzig's algorithm to find all the shortest paths from any one node (of your choice) to all other nodes.

(6) Invent some alternative coefficients for the linear programming problem represented by the objective function in equation 9.145 and the constraints 9.142 to 9.144. Investigate graphically the nature of the feasible region which results, and the optimum solution. Can you envisage situations where the feasible region disappears?

(7) Set up a three zone example of the transportation problem of linear programming with invented data, and see if you can generate the optimum solution by inspection. Calculate dual variables α_i and ν_j and check that the optimality conditions 9.151 to 9.156 are satisfied. If they are not, see if you can improve your solution. If possible, find an appropriate computer programme and run your data through it as a check.

10. Problem solving methods

10.1 Introduction : types of problem solving

In offering the foregoing chapters as 'mathematical methods for geographers and planners' we have, at least implicitly, adopted a problem-solving stance. That is, the utility of such methods stands or falls according to the problem-solving capability generated. Geographers and planners have complementary interests in this respect. The geographer is concerned to achieve the deepest possible understanding of his systems of interest and how they develop. The planner can then utilize this understanding to solve some 'real world' problem of significance to him and society. We are, therefore, concerned with two kinds of problem - which may be called the *analytical* problem and the *planning* problem respectively. They are related by the premise that many planning problems will be difficult to solve without a related analytical capability.

It should be clear, by now, from the examples presented throughout the book that mathematical methods are vital to achieving deeper levels of understanding in an analytical sense. We have paid less direct attention to planning methods as such, though many of the models presented could be used to make forecasts of key variables for planning purposes, such as future populations, or water supply and demand. It is important, however, to keep the two kinds of problem clearly in mind. Otherwise, there is a danger of oversimplified planning – for example, arranging houses and workplaces so as to minimize the journey to work, even though an analysis of existing systems shows that this does not happen because longer work trips are traded for other benefits. The solution of an analytical problem must be sought in two stages: first, identify the associated system of interest and a capability to analyse it, possibly using a model; and second, decide what action can be taken in regard to this system to achieve some desired goals. The second process may also involve mathematical methods in addition to the system models. The reader is referred elsewhere for a more extended discussion of such methods (McLoughlin, 1967, Wilson 1974, Chapters 2, 13 and 14).

10.2 Model building methods : a summary

So far, apart from lengthy accounts of methods of system description in algebraic terms in Chapters 1 and 2, we have concentrated on mathematical techniques, as befits the main purpose of the book. It is worthwhile briefly to examine the broader questions which must be tackled by the model builder facing a *new* problem, however. These questions can be summarized under a number of headings as follows:

(i) *Aggregation, or level of resolution* We saw examples earlier of how systems can be described at different levels of resolution. There are three dimensions to this issue : space, sectors and time. We dealt with space at some length in section 2.1.1 and showed how location could be specified continuously in terms of co-ordinates, or, more usually, in relation to some zone system in which average zone size determines level of resolution. Time, similarly, can be treated continuously or as a sequence of discrete periods. Sector resolution refers to the number of classes or sectors needed to describe some elements of the system – how many population sectors, for example?

When this particular series of questions has been answered, then it should have been possible to complete the description of the system of interest in algebraic terms in the manner of Chapters 1 and 2.

(ii) *Theory* Next, it is important to decide what theory of the structure or operation of the system of interest is to be represented in mathematical terms. Is something being maximized? Is the system in equilibrium? Can the system be treated as deterministic or is its behaviour essentially probabilistic? The answers to such questions may well depend on the level of resolution which has been adopted. System behaviour may be probabilistic at a fine level of resolution, for example, but approximately deterministic at a more coarse scale.

(iii) *Data* It is perhaps a subsidiary qusetion to ask at this stage what data are available, as it could be argued that the best model should be constructed to represent known theory independently of the question of data availability. For some purposes, that is a sound argument. Given a model, suggestions can then be made about data collection for model testing. For other purposes, however, the model builder may be tied to currently existing data for model testing and he may have to 'tailor' his model accordingly.

The answers to these various questions, between them, more or less determine the mathematical techniques which can be applied. It would be good exercise for the reader to back-track through the examples which have been presented earlier in the book, to answer the 'model design questions' in relation to these, and to see how this determines the mathematical techniques to be used. If a system's development through time is being modelled, for example, and if it contains identifiable elements such as people, then a matrix-based accounting framework (section 4.5.2) will be needed. If time is to be treated continuously and explicitly in such a case, then this is likely to generate differential equations (Chapter 6). It should be remembered, of course, that any particular problem is likely to demand a synthesis of the techniques presented here. We hope to have presented a range of techniques in this book which will meet at least elementary needs in these respects.

10.3 Concluding comments

Although we have presented a wide range of examples in this book two things will soon become apparent to the reader. First, there are many more examples to be found in the geography and planning literature, and we hope that the reader now has sufficient grounding (possibly with the help of more advanced mathematics books which he is now equipped to begin reading) to digest and to evaluate these for himself. Secondly, and probably more importantly, there are many analytical problems associated with geographical systems of interest, and many 'real-world' planning problems, which could benefit from a more mathematical treatment. We hope to have communicated enough, by way of example, on model building methods along with mathematical techniques, for the reader to feel equipped to tackle some of these new problems himself.

References

W. Alonso (1964) *Location and land use*, Harvard University Press, Cambridge, Mass.

S. Angel and G. M. Hyman (1970) Urban velocity fields, *Environment and Planning*, 2, pp. 211–224.

S. Angel and G. M. Hyman (1972) Urban spatial interaction, *Environment and Planning*, 4, pp. 99–118.

W. D. Ashton, 1966 *The theory of road traffic flow*, Methuen, London.

N. J. T. Bailey (1957) *The mathematical theory of epidemics*, Griffin, London.

N. J. T. Bailey (1964) *The elements of stochastic processes with applications to the natural sciences*, John Wiley, New York.

G. A. Baker and J. L. Gammel (eds.) (1970) *The Pade approximant in theoretical physics*, Academic Press, New York.

M. Batty and S. Mackie (1972) The calibration of gravity, entropy and related models of spatial interaction, *Environment and Planning*, 4, pp. 205–233.

M. J. Beckmann and T. F. Golob (1971) On the metaphysical foundation of traffic theory: entropy revisited, mimeo, presented at the fifth International Symposium on the Theory of Traffic Flow and Transportation, Berkeley, California.

H. M. Blalock (1960) *Social statistics*, McGraw Hill, New York.

Bureau of Public Roads (1964) *Assignment manual*, Government Printing Office, Washington, D.C.

A. Calver, M. J. Kirkby and D. R. Weyman (1972) Modelling hillslope and channel flows, in R. J. Chorley (ed.) *Spatial analysis in geomorphology*, Methuen, London, pp. 197–218.

H. C. Carey (1858) *Principles of social science*, J. Lippincott, Philadelphia.

G. A. P. Carrothers (1956) An historical review of the gravity and potential concepts of human interaction, *Journal of the American Institute of Planners*, 22, pp. 94–102.

H. S. Carslaw and J. C. Jaeger (1941) *Operational methods in applied mathematics*, Oxford University Press, Oxford.

H. S. Carslaw and J. C. Jaeger (1959) *The conduction of heat in solids*, Oxford University Press, Oxford.

M. A. Carson and M. J. Kirkby (1972) *Hillslope form and process*, Cambridge University Press, Cambridge.

A. Casagrande (1931). Discussion: a new theory of frost heaving. *Proceedings, Highway Research Board*, 11, p. 168–172.

Chicago Area Transportation Study (1960) *Final report*, Chicago.

R. J. Chorley and P. Haggett (1965) Trend-surface mapping in geographical research, *Transactions, Institute of British Geographers*, 37, pp. 47–67.

W. Christaller (1933) *Die zentralen orte in Süddendeutschland*, Jena; English translation by C. W. Baskin, *Central planes in Southern Germany*, Prentice-Hall, Englewood Cliffs, New Jersey.

C. Clark (1951) Urban population densities, *Journal of the Royal Statistical Society, Series A,* **114**, pp. 490–496

D. R. Cox and H. D. Miller (1965) *The theory of stochastic processes*, Methuen, London.

D. R. Cox and W. L. Smith (1961) *Queues*, Methuen, London, and John Wiley, New York.

T. Dalrymple (1960) Flood frequency analysis, *U.S. Geological Survey Water Supply Papers,* 1545-A, U.S. Government Printing Office, Washington, D.C.

G. B. Dantzig (1960) On the shortest route through a network, *Management Science*, **6**, pp. 187–190.

R. Dorfman, P. A. Samuelson and R. Solow (1958) *Linear programming and economic analysis*, McGraw Hill, New York.

E. S. Dunn Jr (1954) *The location of agriculture production*, University of Florida Press, Gainsville.

Suzanne P. Evans (1973) A relationship between the gravity model for trip distribution and the transportation problem of linear programming, *Transportation Research,* **7**, pp. 39–61.

W. Feller (1950) *An introduction to probability theory and its application*, Vol. I. John Wiley, New York.

R. I. Ferguson (1973) Channel pattern and sediment type, *Area,* **5** p 38–41

D. Fine and P. Cowan (1971) Some theoretical aspects of developing networks, in A. G. Wilson (ed.) *Urban and regional planning,* Pion, London, pp. 194–216.

R. C. Folk and W. C. Ward (1957) Braxos river bar: a study in the sifnificance of grain-size parameters. *Journal of Sedimentary Petrology*, **27**, pp. 3–27.

S. Gale (1972) Some formal properties of Hägerstrand's model of spatial interactions, *Journal of Regional Science,* **12**, pp. 199–217.

R. Geiger (1958) 1st Ed, (1965) 2nd Ed. *The climate near the ground*, Harvard University Press, Cambridge, Mass.

P. Gould (1967) On the geographical interpretation of eigenvalues, *Transactions, Institute of British Geographers, No. 42.*

P. Gould (1972) Pedagogic review: Entropy in urban and regional modelling, *Annals, Association of Americal Geographers,* **62**, pp. 689–700.

Greater London Council (1964) *London Traffic Survey,* County Hall, London.

H. A. J. Green (1972) *Consumer theory*, Penguin, London.

S. A. Gustafson and K. O. Kortanek (1972) Analytical properties of some multiple-source diffusion models, *Environment and Planning,* **4**, pp. 31–41.

T. Hägerstrand (1967) *Innovation diffusion as a spatial process,* (A. Pred, translator) Chicago University Press, Chicago.

P. Haggett (1965) *Locational analysis in human geography*, Edward Arnold, London.

P. Haggett (1972) Contagious processes in a planar graph: an epidemiological application, in N. D. McGlashan (ed.) *Medical geography*, Methuen, London, pp. 307–324.

P. Haggett and R. J. Chorley (1969) *Network analysis in geography*, Edward Arnold, London.

B. Hallert (1960) *Photogrammetry*, McGraw-Hill, New York.

J. Herbert and B. H. Stevens (1960) A model for the distribution of residential activity in urban areas, *Journal of Regional Science*, 2, pp. 21–36.

R. E. Horton (1945) Erosional development of streams and their drainage basins: hydrophysical approach to quantitative morphology, *Bulletin of the Geological Society of America*, **56**, pp. 275–370.

D. L. Huff (1964) Defining and estimating a trading area, *Journal of Marketing*, **28**, pp. 34–38.

W. Isard (1956) *Location and space-economy*, M.I.T. Press, Cambridge, Mass.

K. J. Kansky (1963) *Structure of transportation networks*, Research Paper 84, Department of Geography, Chicago University Press, Chicago.

A. V. T. Kirkby (1973) The use of land and water resources in the past and present Valley of Oaxaca, Mexico, *Memoirs of the museum of anthropology*, **1**, University of Michigan, Ann Arbor.

M. J. Kirkby (1969) Infiltration, through flow and overland flow, in R. J. Chorley (ed.) *Water, Earth and Man*, Methuen, London, pp. 215–227.

M. J. Kirkby (1971) Hillslope process-response models based on the continuity equation, in *Slopes: form and process*, compiled by D. Brunsden, IBG Special Publication No. 3, London, pp. 15–30.

T. R. Lakshmanan and W. G. Hansen (1965) A retail market potential model, *Journal of the American Institute of Planners*, **31**, pp. 134–143.

W. Leontief and A. Strout (1963) Multi-regional input-output analysis, in T. Barna (ed.) *Structural interdependence and economic development*, Macmillan, London.

L. B. Leopold and W. B. Langbein (1962) The concept of entropy in landscape evolution, *Geological Survey Professional Paper, 500-A*, U.S. Government Printing Office, Washington, D,C.

P. H. Leslie (1945) On the use of matrices in certain population mathematics, *Biometrika*, **23**, pp. 183–212.

M. J. Lighthill and G. B. Whitham (1955) 'On kinematic waves' II A theory of traffic on long crowded roads, *Proceedings of the Royal Society, Series A*, **229**, 317–45.

A. Lösch (1940) Die Räumliche Ordnung der Wirtschaft, Jena; English translation by W. H. Woglom, *The economics of location*, Yale University Press, New Haven, Connecticut.

R. H. MacArthur and E. O. Wilson (1967) *The theory of island biogeography*. Princeton University Press, Princeton, New Jersey.

H. McConnell and J. M. Horn 1972 Probabilities of surface karst, in R. J. Chorley (ed.) *Spatial analysis in geomorphology'*, Methuen, London, pp. 111–134.

J. B. McLoughlin (1967) *Urban and regional planning*, Faber and Faber, London.

L. March (1971) Urban systems: a generalised distribution function, in A. G. Wilson (ed.) *Urban and regional planning*, Pion, London, pp. 157–170.

J. Maynard Smith (1973) *Models in Ecology*, Cambridge University Press, Cambridge.

Metcalf and Eddy Inc. (1972) *Wastewater engineering: collection, treatment, disposal*, MacGraw-Hill, New York.

M. J. H. Mogridge (1969) Some factors influencing the income distribution of households within a city region, in A. J. Scott (ed.) *Studies in regional science*, Pion, London, pp. 117–144.

E. F. Moore (1959) The shortest path through a maze, *Annals of the Computation Laboratory of Harvard University*, **30**, pp. 285–292.

R. L. Morrill (1968) *Migration and the spread and growth of urban settlement*, Lund Studies in Geography, Series B, 26, Lund.

J. A. Neidercorn and B. V. Bechdolt (1969) An economic derivation of the 'gravity law' of spatial interaction, *Journal of Regional Science*, **9**, pp. 273–282.

J. F. Nye, (1959) The motion of ice sheet and glaciers, *Journal of Glaciology*, 3, p. 493.

J. F. Nye (1960) The response of glaciers and ice sheets to seasonal and climatic changes, *Proceedings of the Royal Society, Series A*, **256**, pp. 559–84.

J. D. Nystuen and M. F. Dacey (1961) A graph theory interpretation of nodal regions, *Papers, Regional Science Association*, **7**, pp. 29–42.

G. Olsson (1965) *Distance and human interaction*, Regional Science Research Institute, Philadelphia.

J. R. Philip (1957) The theory of infiltration' IV Sorbtivity and algebraic infiltration equations, *Soil Science*, **84**, 257–64.

E. G. Phillips (1930) *Analysis*, Cambridge University Press, Cambridge.

E. G. Ravenstein (1885) The laws of immigration, *Journal of the Royal Statistical Society*, **48**, pp. 167–235 and 241–305.

P. H. Rees (1973) private communication.

P. H. Rees and A. G. Wilson (1973) Accounts and models for spatial demographic analysis 1: aggregate populations, *Environment and Planning*, **5**, pp. 61–90.

A. Rogers (1966) Matrix methods of population analysis, *Journal of the American Institute of Planners*, **32**, pp. 177–196.

A. Rogers (1969) Quadrat analysis of urban dispension: 1, Theoretical techniques, *Environment and Planning*, **1**, pp. 47–80.

A. Rogers (1972) *Matrix methods in urban and regional analysis*, Holden Day, San Francisco.

A. J. Scott (1971) *Combinatorial programming, spatial analysis and planning*, Methuen, London.

M. L. Senior and A. G. Wilson (1974) Some explorations and syntheses of linear programming and spatial interaction models of residential location, *Geographical Analysis*, **6**, pp. 209–237.

R. L. Shreve (1966) Statistical law of stream numbers, *Journal of Geology*, **74**, pp. 17–37.

R. L. Shreve (1969) Stream lengths and basin areas in topologically random channel networks, *Journal of Geology*, **77**, pp. 397–414.

S. Siegel (1956) *Non-parametric statistics*, McGraw Hill, New York.

R. Stone (1966) *Mathematics in the social sciences*, Chapman and Hall, London.

S. A. Stouffer (1940) Intervening opportunities: a theory relating mobility and distance, *American Sociology Review*, **5**, pp. 845–867.

S. A. Stouffer (1960) Intervening opportunities and competing migrants, *Journal of Regional Science*, **2**, pp. 1–26.

A. N. Strahler (1952) Hypsometric (area-altitude) analysis of erosional topography, *Bulletin of the Geological Society of America*, **63**, pp. 1117–1142.

A. N. Strahler (1957) Quantitative analysis of watershed geomorphology, *Transactions of the American Geophysical Union*, **38**, pp. 913–20.

A. Stuart (1962) *Basic ideas of scientific sampling*, Griffin, New York.

J. C. Tanner (1961) Factors affecting the amount of travel, *Road Research Laboratory Technical Paper*, **51**, H.M.S.O. London.

M. M. Thompson (1966) *Manual of Photogrammetry* (3rd Edition), American Society of Photogrammetry, Falls Church, Va.

J. H. von Thünen (1826) *Der isolierte Staat in Beziehung auf Landwirschaft und Nationalökönomie*, Hamburg; English translation by C. M. Wartenburg, with an introduction by P. Hall, Oxford University Press, Oxford, 1966.

J. O. Tressider, D. A. Meyers, J. E. Burrell, and T. J. Powell (1968) The London Transportation Study: methods and techniques, *Proceedings of the Institution of Civil Engineers*, **39**, pp. 433–464.

J. P. Waltz (1969) Ground water, in R. J. Chorley (ed.) *Water, earth and man*, Methuen, London, pp. 259–267.

K. E. F. Watt (1968) *Ecology and resources management*, McGraw-Hill, New York.

A. Weber (1909) Über den Standort der Industrien, Tubingen; English translation by C. J. Friedrich, *Theory of the location of industries*, University of Chicago Press, Chicago.

A. Werritty (1972) The topology of stream networks, in R. J. Chorley (ed.) *Spatial analysis in geomorphology*, Methuen, London, pp. 167–196.

A. G. Wilson (1967) A statistical theory of spatial distribution models, *Transportation Research*, **1**, pp. 253–269.

A. G. Wilson (1969) Developments of some elementary residential location models, *Journal of Regional Science*, **9**, pp. 377–385.

A. G. Wilson (1970-A) Inter-regional commodity flows: entropy maximising approaches, *Geographical Analysis*, **2**, pp. 255–282.

A. G. Wilson (1970-B) *Entropy in urban and regional modelling*, Pion, London.

A. G. Wilson (1971) A family of spatial interaction models and associated developments, *Environment and Planning*, **3**, pp. 1–32.

A. G. Wilson (1972) Models of population structure, and some implications for dynamic residential location modelling, in A. G. Wilson (ed.) *Patterns and processes in urban and regional systems*, Pion, London, pp. 217–242.

A. G. Wilson (1973-A) Theoretical geography: some speculations, *Transactions, Institute of British Geographers*, **57**, pp. 31–44.

A. G. Wilson (1973-B) Toward system models for water resource management, *Environmental Management*, **1**, pp. 36–52.

A. G. Wilson (1974) *Urban and regional models in geography and planning*, John Wiley, London and New York.

A. G. Wilson and R. Kirwan (1969) Measures of benefits in the evaluation of urban transport improvements, Working Paper 43, Centre for Environmental Studies, London.

A. G. Wilson and P. H. Rees (1974) Accounts and models for spatial demographic analysis 2: age–sex disaggregated populations, *Environment and Planning*, **6**, pp. 101–116.

A. G. Wilson and M. L. Senior (1974) Some relationships between entropy maximizing models, linear programming models, and their duals, *Journal of Regional Science*, **14**, pp. 207–215.

R. A. Wooding (1965) A hydraulic model for the catchment stream problem. I Kinematic wave theory, *Journal of Hydrology*, **3**, pp. 254–267.

Index